Electric Brain Signals

Foundations and Applications of Biophysical Modeling

Brain activity is commonly studied by measuring the extracellular electric potentials inside or outside the brain. Interpreting such measurements requires knowledge of the physical processes underlying the recorded signals. This book introduces the electromagnetic and biophysical theory required for simulating neural activity and its extracellular potentials. Written by leading experts in the field, it presents results from long-term research into forward modeling of extracellular brain signals and illustrates the link between theory and real recorded signals under various conditions. Practical code examples for modeling real neural systems are included throughout and supported by an online code repository, making this volume a valuable resource for students and scientists who wish to analyze electric brain signals through biophysics-based modeling.

GEIR HALNES is a research scientist with a background in physics and theoretical biology. For the last one-and-a-half decades, he has worked in the Computational Neuroscience Lab at the Norwegian University of Life Sciences. His research focuses on the fundamental physical processes in neurons, glial cells, and brain tissue.

TORBJØRN V. NESS is a research scientist with a background in computational neuroscience who has more than a decade of experience in modeling extracellular potentials, including finite element simulations of neural tissue and measurement equipment. Torbjørn is one of the developers of LFPy, a software for calculating brain signals from simulated neural activity.

SOLVEIG NÆSS is a senior researcher with a background in physics. She received her PhD in computational neuroscience through an international PhD program at the University of Oslo, Simula Research Laboratory, and the University of California, San Diego. Solveig specializes in the biophysical modeling of electric brain signals.

ESPEN HAGEN is a researcher with a background in environmental physics and computational neuroscience. He earned his PhD at the Norwegian University of Life Sciences and has worked on methods and tool development linking the activity of individual neurons and large neuronal networks to electrophysiological recordings of neural activity. Espen is one of the developers of LFPy.

KLAS H. PETTERSEN is CEO of the Norwegian Artificial Intelligence Research Consortium (NORA), spanning eleven Norwegian universities and colleges and four research institutes. He is also the founder of the AI innovation ecosystem NORA.startup. Klas is an active researcher within computational neuroscience and AI and is the director of Norway's national research school on AI.

GAUTE T. EINEVOLL is Professor of Physics at the Norwegian University of Life Sciences and the University of Oslo, working on the modeling of nerve cells, networks of nerve cells, brain tissue, and brain signals as well as the development of neuroinformatics software tools, including LFPy.

Electric Brain Signals

Foundations and Applications of Biophysical Modeling

GEIR HALNES
Norwegian University of Life Sciences

TORBJØRN V. NESS
Norwegian University of Life Sciences

SOLVEIG NÆSS
University of Oslo

ESPEN HAGEN
University of Oslo and Norwegian University of Life Sciences

KLAS H. PETTERSEN
NORA – The Norwegian Artificial Intelligence Research Consortium

GAUTE T. EINEVOLL
Norwegian University of Life Sciences and University of Oslo

CAMBRIDGE
UNIVERSITY PRESS

CAMBRIDGE
UNIVERSITY PRESS

Shaftesbury Road, Cambridge CB2 8EA, United Kingdom

One Liberty Plaza, 20th Floor, New York, NY 10006, USA

477 Williamstown Road, Port Melbourne, VIC 3207, Australia

314–321, 3rd Floor, Plot 3, Splendor Forum, Jasola District Centre, New Delhi – 110025, India

103 Penang Road, #05–06/07, Visioncrest Commercial, Singapore 238467

Cambridge University Press is part of Cambridge University Press & Assessment, a department of the University of Cambridge.

We share the University's mission to contribute to society through the pursuit of education, learning and research at the highest international levels of excellence.

www.cambridge.org
Information on this title: www.cambridge.org/9781316510841

DOI: 10.1017/9781009039826

First published 2024

A catalogue record for this publication is available from the British Library

Library of Congress Cataloging-in-Publication Data
Names: Halnes, Geir, 1976- author. | Ness, Torbjørn V., 1985- author. | Næss, Solveig, 1990- author. | Hagen, Espen, 1979- author. | Pettersen, Klas H., 1976- author. | Einevoll, Gaute, author.
Title: Electric brain signals : foundations and applications of biophysical modeling / Geir Halnes (Norwegian University of Life Sciences, Norway), Torbjørn V. Ness (Norwegian University of Life Sciences, Norway), Solveig Næss (University of Oslo, Norway), Espen Hagen (University of Oslo & Norwegian University of Life Sciences, Norway), Klas H. Pettersen (NORA, The Norwegian Artificial Intelligence Research Consortium, Norway), Gaute T. Einevoll (Norwegian University of Life Sciences & University of Oslo, Norway).
Description: Cambridge, United Kingdom ; New York, NY : Cambridge University Press, 2024. | Includes bibliographical references and index.
Identifiers: LCCN 2024007235 (print) | LCCN 2024007236 (ebook) | ISBN 9781316510841 (hardback) | ISBN 9781009018623 (paperback) | ISBN 9781009039826 (ebook)
Subjects: LCSH: Neural circuitry. | Neural circuitry–Computer simulation.
Classification: LCC QP363.3 .H356 2024 (print) | LCC QP363.3 (ebook) | DDC 612.8/2–dc23/eng/20240314
LC record available at https://lccn.loc.gov/2024007235
LC ebook record available at https://lccn.loc.gov/2024007236

Additional resources for this publication at www.cambridge.org/electricbrainsignals

ISBN 978-1-316-51084-1 Hardback
ISBN 978-1-009-01862-3 Paperback

it was not a closed system

meteorites fell to earth
cracks appeared

through cracks we were born
with our own cracks

we used them
to eat structure and crap chaos

in this way
order could feed in us
and we developed the most wonderful brains

the brains were cracks

trees fell in
and became trees to us
mountains fell in
and became mountains to us

meteors sent electric sprinkles
across the visual cortex
and chills down our spines

Geir Halnes, *Mor Rom* (*Mother Space*). **Transcreated to English by Geir Halnes and Ida Kock.**

Contents

Preface

Neurons in the brain communicate via electric signals that can be recorded with electrodes placed either inside the neurons, in the space between neurons, or outside the brain and head. The foundation for interpreting the signals recorded intracellularly was laid through the pioneering works of Hodgkin, Huxley, Rall, and others in the 1950s. The biophysics-based approach to modeling neurons proposed by these scientists gave predictions that could be compared directly with experimental recordings and represented a "quantitative revolution" in neuroscience.

The relationship between neural activity and what is measured extracellularly is less trivial, partly because an extracellular electrode picks up overlapping signals from many neurons simultaneously and partly because a neuron's contribution to the recorded signal depends on various (typically unknown) factors such as, for example, the neuron's morphology and its position relative to the electrode. As a further complication, extracellular signals may contain "noisy" contributions that are difficult to control for in an experimental setting – for example, from ongoing muscle activity or recording equipment.

The extracellular signal that is the easiest to interpret is the "spike," which is the extracellular signature of a neuronal action potential. For this reason, the analysis of extracellular potentials has largely been limited to inferring spike times of a relatively small number of neurons located in the vicinity of recording contacts. Although this is well and good, the extracellularly recorded potentials contain information beyond spike times that is not taken advantage of in these kinds of analyses.

Given the proper theoretical tools, extracellular recordings could putatively be used to answer questions regarding, for example, the positioning of contributing neurons, the number of contributing neurons, the types of neurons that contribute the most, the distributions of synaptic input onto neurons, etc. Such questions have, however, lacked quantitative answers grounded in biophysics. In the absence of precise biophysics-based tools for interpreting extracellular potentials, "rules of thumb" for interpretation have been spread in our neuroscience community. These are not always well-founded and have often been based on intuition and "folk physics" rather than actual biophysics.

Over the last 20 years, our research group at the Norwegian University of Life Sciences and the University of Oslo has worked on forward-modeling of brain signals from biophysically detailed (multicompartment) models of neurons and networks of neurons. In the process, we have developed the open-source simulation tool LFPy (https://LFPy.readthedocs.io) and related tools to facilitate the biophysics-based modeling of extracellular signals. Most of our work has focused on electric signals recorded with electrodes placed inside brain tissue to detect spikes, which are revealed in the high-frequency part of the signal, or the local field potential (LFP), which is the term for the low-frequency part of the signal. However, we have in recent years extended our research interests to include also electric signals recorded on the cortical surface (ECoG) or on the scalp (EEG), as well as magnetic signals (MEG). As all these types of signals essentially have the same neural origin, they can be modeled using conceptually and methodologically similar computational schemes. By establishing a link between neural activity and extracellularly measured signals, these biophysically founded schemes can make the analysis of extracellular signals more quantitative and informative. These modeling schemes and the insights drawn from them are the main topics of this book.

The amount of data being routinely gathered in modern-day neuroscience experiments is enormous. For example, the number of micro-electrode contacts used to record spikes and LFPs within a single in vivo experiment is often in the hundreds, and is steadily increasing with the ongoing technological development. We anticipate that the biophysics-based modeling that we advocate will become an increasingly important tool for interpreting these ever-expanding data sets. However, getting to grips with this ever-growing amount of incoming data requires a team effort. This book has been written with the hope that students and scientists wish to join in the ambitious endeavor of trying to understand the brain through biophysics-based modeling of neurons, neural networks, and the extracellular signatures of their activity. For this reason, the computer code used in many of the simulations and corresponding figures presented in this book has been made publicly available in the online code repository (https://github.com/LFPy/ElectricBrainSignals), also accessible via www.cambridge.org/electricbrainsignals, so that interested readers can download and modify the simulation codes. Unless otherwise noted, figures in this book are provided by the authors (CC-BY 4.0 International license).

Our work on extracellular brain signals over the last 20 years has been based on collaborations with numerous scientists, including Evrim Acar, Jose-Manuel Alonso, Sebastian Amundsen, Costas Anastassiou, Ole A. Andreassen, Anton Arkhipov, Pooja N. Babu, Aslak Wigdahl Bergersen, Kosio Beshkov, Yassan Billeh, Erin Bjørkeli, Daniel M. Bjørnstad, Patrick Blomquist, Christine Brinchmann, Alessio Buccino, Miguel Cavero, Chaitanya Chintaluri, Hermann Cuntz, David Dahmen, Anders M. Dale, Alain Destexhe, Anna Devor, Markus Diesmann, Andrew G. Edwards, Ada Ellingsrud, Torbjørn Elvsåshagen, Rune Enger, Janne C. Fossum, Felix Franke, Marianne Fyhn, Justen Geddes, John Gigg, Torbjørn L. Gjerberg, Helena Głąbska, Sergey Gratiy, Sonja Grün, Torkel Hafting, Don Hagler, Eric Halgren, Ken Harris, Eivind Hennestad, Jørgen Hoel, Ulf Indahl, Karoline Jæger, Daniel Keller, Amir Khosrowshahi, Jan Fredrik Kismul, Christof Koch, Anders Lansner, Szymon Łeski, Henrik Lindén, Glenn T. Lines, Charl Linssen, Steinn H. Magnusson, Tuomo Mäki-Marttunen, Anders Malthe-Sørensen, Pablo Martínez-Cañada, Alberto Mazzoni, Nicolo Meneghetti, Stéphanie Miceli, Abigail Morrison, Jonas Nilsen, Eivind S. Norheim, Stig Omholt, Stefano Panzeri, Łukasz Paszkowski, Bijan Pesaran, Hans Ekkehard Plesser, Jan Potworowski,

Christophe Pouzat, Michiel Remme, Atle Rimehaug, Marie E. Rognes, Stefan Rotter, Dirk Schubert, Johanna Senk, Jan-Eirik Welle Skaar, Andreas Solbrå, Alexander Stasik, Maria L. Stavrinou, Karianne Strand, Kamilla Sulebakk, Joakim Sundnes, Harvey Swadlow, Marte-Julie Sætra, Øystein Sørensen, Tom Tetzlaff, Daniel Torres, Beth Tunstall, Aslak Tveito, Kristin Tøndel, Istvan Ulbert, Sacha van Albada, Jonas van den Brink, Jonas Verhellen, Sigrid Videm, Daniel Wójcik, John Wyller, Ivar Østby, and Leiv Øyehaug. We thank them all.

We direct special thanks to friends and colleagues who have provided critical and encouraging feedback during the making of this book. These include Alessio Buccino, Mike X. Cohen, David Dahmen, Bruce Graham, Risto Ilmoniemi, Mathijs Janssen, Ørjan Martinsen, Hamish Meffin, Charles Nicholson, Supratim Ray, Marte-Julie Sætra, Tom Tetzlaff, Arnt-Inge Vistnes, and Daniel Wójcik.

Reserved Physical Symbols and Quantities

Symbol	Unit	Description
t	s	time
f	Hz	frequency
q	C	electric charge
T	K	absolute temperature
$\mathbf{r}$	m	position vector
$u \in \{x, y, z\}$	m	location on u-axis
$E, \mathbf{E}$	V/m	electric field (scalar, directed)
$B, \mathbf{B}$	T	magnetic field (scalar, directed)
$F_k, \mathbf{F}_k$	N	force (scalar, directed) on particle k
V, V_i, V_e, V_m	V	electric potential (general, intracellular, extracellular, membrane)
$[k]$	mol/m^3	ion concentration of species k
D_k	m^2/s	diffusion constant of ion species k
f_k	mol/(m^{3}s)	source density of ion species k
$J_k, \mathbf{J}_k$	mol/s	flux of ion species k (scalar, directed)
$j_k, \mathbf{j}_k$	mol/(m^{2}s)	flux density of k (scalar, directed)
$I, \mathbf{I}$	A	electric current (scalar, directed)
$\mathcal{I}$	A/m	electric current per unit length (scalar)
$i, \mathbf{i}$	A/m^2	electric current density (scalar, directed)
$P, \mathbf{P}$	Am	current-dipole moment (scalar, directed)
C	A/m^3	current-source density
σ, σ_t	S/m	conductivity (general, for tissue) for volume currents
r_m	Ωm^2	specific membrane resistivity
r_a	Ω/m	specific axial resistivity
c_m	F/m^2	specific membrane capacitance
g_k	S/m^2	specific membrane conductance for ion species k
E_k	V	reversal potential for ion species k
l	m	length of dendritic stick
τ, τ_m	s	time constant (general, membrane)
$h(\tau)$	arbitrary	temporal filter/impulse response (1D)
$H(\mathbf{r}, \tau)$	arbitrary	spatiotemporal filter/impulse response

F	96 485.332 12 C/mol	Faraday constant
R	8.314 462 618 153 24 J/(Kmol)	gas constant
N_A	6.022 140 76 $\times$ 10^{23} 1/mol	Avogadro constant
c	299 792 458 m/s	speed of light
e	1.602 176 62 $\times$ 10^{-19} C	elementary charge
ϵ_0	8.854 187 812 8 $\times$ 10^{-12} F/m	vacuum permittivity
μ_0	1.256 637 062 12 $\times$ 10^{-6} N/A^2	vacuum permeability

Abbreviations

AP	action potential
CSD	current-source density
CSF	cerebrospinal fluid
ECoG	electrocorticography
EEG	electroencephalography
FEM	finite element method
LFP	local field potential
MC	multicompartment (model of neuron)
MC+VC	multicompartment (neurons) + volume-conductor (theory)
MEA	micro-electrode array
MEG	magnetoencephalograpy
MoI	method of images
MUA	multi-unit activity
PSD	power-spectral density
VC	volume conductor

1   Introduction

Understanding the principles behind how the brain works is a challenging scientific quest, driven in part by curiosity and in part by the urge to develop drugs and other treatment strategies for brain disorders.

We know today that the basic processing units of the brain are neurons, or nerve cells, that communicate via what is known as *action potentials* (APs). An AP is a brief, high-amplitude fluctuation in the neural membrane potential, and it can be recorded with an intracellularly inserted electrode. An AP also causes a brief fluctuation, or *spike*, in the electric potential in the extracellular space surrounding the active neuron. Extracellular spikes are typically obtained by high-pass filtering the extracellular potentials recorded inside the brain tissue. The signal used to detect spikes is often called the *multi-unit activity* (MUA) and is usually believed to reflect the aggregate AP firing of a relatively small number of nearby neurons.

Depending on what electrode one uses, where one positions it, and how one filters the recorded signals, one can – instead of recording spikes – choose to record more mesoscopic or macroscopic extracellular signals, reflecting the activity of larger populations of neurons. In addition to the spikes, commonly recorded extracellular signals include the *local field potential* (LFP), which is the low-frequency part of extracellular potentials recorded inside brain tissue;[1] the *electrocorticographic* (ECoG) signals, which are electric potentials recorded by electrodes placed on the cortical surface; the *electroencephalographic* (EEG) signals, which are electric potentials recorded by electrodes placed on the scalp; and the *magnetoencephalographic* (MEG) signals, which are magnetic fields recorded immediately outside the head. These different electric and magnetic measurement modalities are illustrated in Figure 1.1. This figure does not show real experimental data, but instead shows the results from a computer simulation based on the methodology and computational schemes that will be presented throughout this book. The simulation illustrates how a 200 ms period of activity in a single biophysically detailed model neuron contributes to the various brain signals.

In addition to the electric and magnetic measurement modalities illustrated in Figure 1.1, brain activity can be probed with optical methods such as *voltage-sensitive dye imaging* (VSDI) measuring membrane potentials, *intrinsic optical signal imaging*

---

[1] While others have defined the LFP to simply mean the extracellular potential recorded outside neurons, this book defines the LFP to be the low-frequency part of the potential. With this definition, LFP could thus more aptly be thought of as an acronym for "low-frequency potential" (Głąbska et al. 2017).

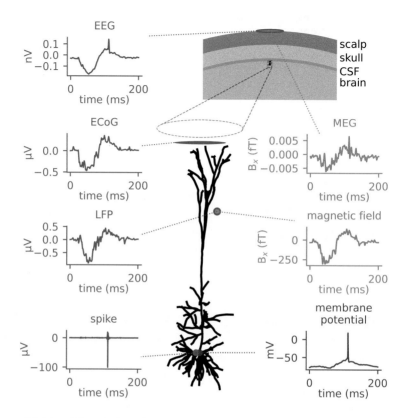

Figure 1.1 **Multimodal modeling of a neuron receiving synaptic inputs and firing an action potential.** Simulations were done on a multicompartment model of a layer 5 pyramidal cell developed by Hay et al. (2011). Extracellular signals (produced by this single neuron) were computed using methods that will be presented throughout this book. Code available via www .cambridge.org/electricbrainsignals.

(IOSI) measuring hemodynamics, and *two-photon* Ca^{2+} *imaging* (2PCI) measuring changes in intracellular Ca^{2+} concentrations. Further, *functional magnetic resonance imaging* (fMRI) measures blood flow and blood oxygenation in brain tissue, while *positron emission tomography* (PET) measures metabolic activity.

Taken together, the techniques listed above probe brain activity on a wide spectrum of spatial and temporal scales (Figure 1.2A). The focus of this book will be on the electric and magnetic measurement modalities summarized in Figure 1.1. Other measurement modalities will not be covered further, but for an overview of the physical principles governing other techniques, see Brette & Destexhe (2012).

The last decades have seen the development of better electrode systems, allowing experimentalists to perform electric measurements with steadily increasing numbers of electrode contacts and improved spatial resolution. Cheaper hard drives have also facilitated en masse data storage at higher temporal resolutions. In parallel with the improvements in measurement techniques comes the need for better theoretical models

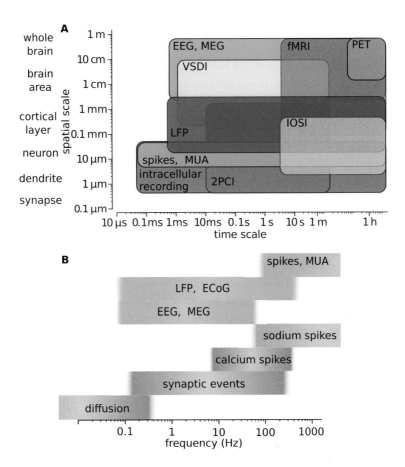

Figure 1.2 Overview of measurements of neural activity A: Spatiotemporal scale covered by various measurements. **B**: Main frequency bands of various extracellular measurements (blue boxes) and various events/processes (grey boxes).

and computational tools that can help us interpret what these measurements tell us. Spike recordings are in principle straightforward to interpret in terms of underlying neural activity: a spike means there is an AP in a nearby neuron. The meso- and macroscopic signals are likely to contain more information about what the brain is doing on the systems level. However, interpreting these systems-level measurements in terms of what they tell us about the underlying network activity is challenging and has largely been done in a qualitative manner only.

We note that the terms "spike" and "AP" are often used interchangeably in the literature. In this book, we will mostly preserve the term "spike" for the extracellular signature of the intracellular AP. However, we do make some exceptions from this "rule" to accommodate established terminology – for example, when we refer to "spiking neuron models," which is the common term for models that generate APs or at least predict the times of their occurrence, or in reference to a "spike train," which refers to a sequence of APs.

1.1   Forward Modeling of Extracellular Signals

Over the last decades, the field of computational neuroscience has grown both in size and scope, and the use of mathematics is now ubiquitous in neuroscience. The types of modeling pursued typically belong to one of three categories (Dayan & Abbott 2001):

- In *mechanistic* modeling, the goal is to explain a phenomenon in terms of the physical processes that give rise to it. This is the traditional physics approach. Its most prominent application in computational neuroscience is the Hodgkin-Huxley model (Hodgkin & Huxley 1952*a*), where the AP is modeled based on the electrical properties of the cell membrane and ion channels.
- In *descriptive* or *statistical* modeling, the goal is rather to summarize experimental data compactly. While such models may be motivated by neurobiological insights, the goal is to account mathematically for a phenomenon, not to explain it. An example of such a model in neuroscience is the Gabor-filter description of receptive fields of so-called simple cells in the primary visual cortex in mammals (Dayan & Abbott 2001, chapter 2).
- In *normative* or *interpretive* modeling, the goal is to explain a phenomenon in terms of its function. Normative modeling is unique to biological systems and is not pursued in physics. A question of *why* a stone falls to the ground is not fruitful, as the law of gravity is not set up to perform a certain task. Biological systems, however, have developed under evolutionary pressure, and the "why" question is sensible. An example of normative modeling in neuroscience is the use of information theory to explain why salient features of receptive fields in the early visual system are well suited to convey information about the world (Dayan & Abbott 2001, chapter 4). Here the basic idea is that these receptive fields are optimized to convey the maximum amount of information about the natural world.

In this book, we will mostly deal with the mechanistic modeling of extracellular electric and magnetic brain signals based on biophysics-based models of neurons and networks. Although simplistic, physics-based neuron models existed earlier (Lapicque 1907), mechanistic modeling in neuroscience arguably started in full with the pioneering works of Alan Hodgkin and Andrew Huxley on axonal action-potential generation and propagation (Hodgkin & Huxley 1952*a*). Together with the work of Wilfred Rall on how neuronal dendrites integrate synaptic inputs (Rall 1959, Segev et al. 1995), the work by Hodgkin and Huxley established what still today remains the standard formalism for the biophysical modeling of neurons. These works typically form a core in courses in computational neuroscience lectured at universities and are covered in many excellent textbooks (see e.g. Koch (1999), Dayan & Abbott (2001), Gerstner et al. (2014), and Sterratt et al. (2023)).

Despite groundbreaking works by Rall (1962) and Holt & Koch (1999) providing the physical foundation, the mechanistic modeling of the link between neural activity and what is measured extracellularly in different types of experiments has received relatively little attention in the computational neuroscience community. As a consequence, it has been difficult to compare predictions from network modeling with experimental data other than spikes (the extracellular signatures of neuronal APs).

The type of mechanistic modeling of extracellular signals pursued in this book is often called *forward modeling*, since its purpose is to take us "forward" from a mechanistic model of neural activity to the corresponding extracellular measurement signal. In contrast, *inverse modeling* aims to estimate the underlying activity from a set of measurements. The signals that we focus on, and the typical signal-frequency ranges that they cover, are summarized in Figure 1.2B (blue bands). The figure also indicates which kinds of neural events (grey bands) are expected to contribute within the various frequency ranges.

A key reason for focusing on electric and magnetic extracellular signals is that a (forward) link from neural activity to these signals is well established by the so-called *volume-conductor* theory. Another reason is that the high time resolution of these signals, typically on the order of a millisecond or so, makes their quantitative modeling particularly important since much of the relevant neural activity in the brain takes place on this millisecond time scale. A third reason is the rapid development of new electrode systems and experimental techniques for measuring electric and magnetic brain signals.

We believe that forward modeling of extracellular signals has many uses, some of which are listed below:

- Forward modeling allows us to run controlled neural simulations to systematically explore how various aspects of neuronal activity are reflected in extracellular signals. Insights from forward modeling can be used (inversely) to interpret experimentally recorded signals in terms of what they tell us about the underlying neural activity, at least on a qualitative level. Many neuroscientists have the notion, for example, that LFPs recorded in a particular layer of cortex necessarily stem from neurons with somas in the same layer, while biophysics-based modeling shows that this is not correct.
- Forward modeling allows for a quantitative comparison between network model predictions and experimentally recorded LFP, ECoG, EEG, and MEG signals. Such comparisons will in turn allow the data to be used (inversely) to estimate model parameters, facilitating the development of models that bridge the scales between the single-neuron level and the systems level in the brain (Einevoll et al. 2019).
- Forward models can be used as starting points when developing new (inverse) methods for data analysis – that is, the estimation of neural activity from experimental data. An example is the development of methods for estimating the current-source density (CSD) from multielectrode LFP recordings based on a corresponding forward model (Pettersen et al. 2006, Potworowski et al. 2012, Cserpan et al. 2017).
- Forward models allow us to produce simulation-based benchmarking data for the validation of data analysis methods, such as spike-sorting methods (Einevoll et al. 2012, Hagen et al. 2015, Buccino et al. 2018) or methods for estimating

current-source densities (CSD) from multielectrode LFPs recordings (Pettersen et al. 2008, Hagen et al. 2017). Another example is test-data generation for the validation of laminar population analysis (LPA) (Głąbska et al. 2016).

1.2   Overview of the Contents of This Book

In brief, this book presents (i) a biophysical theory for computing electric and magnetic brain signals, (ii) computations of such signals for various scenarios (mostly using computer simulations), and (iii) extractions of key insights from these computations.

Throughout the book, what we will refer to as the "MC+VC scheme" will hold a central position. This is the standard scheme for biophysics-based computation of electric and magnetic brain signals. The scheme will be defined properly later on, but in brief, it entails the computation of neural electric activity using multicompartment (MC) neuron models as well as the computation of extracellular signals using volume-conductor (VC) theory. Our main focus will be on electric signals, whereas magnetic signals will be covered more briefly. The reason for this is partly that we, the authors, have our main expertise in electric signals and partly that magnetic brain signals are well covered in the book by Ilmoniemi & Sarvas (2019).

Our focus will be on general principles of the generation of extracellular signals. We will, however, have a slight bias towards extracellular potentials generated by cortical neurons. For spikes (Chapter 7), the principled insights from considering cortical neurons should also directly apply to spikes from neurons in other brain areas, as there is nothing principally different in the action potential and spike generation of cortical neurons compared to other neurons. However, the LFP (Chapter 8) generated by cortical neurons and populations of such can be uniquely large due to the so-called open-field dendritic structure and layered organization of the dominant pyramidal neurons. Example results for cortical LFPs will thus be less representative for the various LFPs that can be recorded around the brain. The EEG (Chapter 9), ECoG (Chapter 10), and MEG signals (Chapter 11) are in any case expected to be dominated by contributions from cortical neurons due to their proximity to the recording devices so, for these signals, our choice of using cortical neurons as examples is warranted.

The book is outlined as follows. In the next five chapters (Chapters 2–6), we establish the theoretical foundation for computing brain signals, focusing mainly on extracellular potentials. In the first of these theory chapters (Chapter 2), we walk the reader through fundamental physical laws and concepts as well as define the MC+VC scheme (Section 2.6.2). The main theory for computing neurodynamics using multicompartment neuron models (the MC part) is introduced in Chapter 3. The volume-conductor theory (the VC part) that takes us (forward) from neurodynamics to predictions of extracellular signals is introduced in Chapter 4.

An important component in VC theory is the conductivity of brain tissue. In Chapter 4, we assume it to be constant since this makes the theory much simpler. In Chapter 5, we discuss this assumption and introduce ways to extend the VC theory to more general cases where the conductivity depends on the position, direction, or signal frequency. The last theory chapter, Chapter 6, describes ways to implement the MC+VC framework for running numerical simulations on computers, as well as ways to reduce model complexity in order to obtain more efficient simulations.

Following the theory chapters is a series of chapters where we apply the defined theory to simulate the kinds of extracellular signals summarized in Figure 1.2B. In Chapter 7, we consider spikes, the extracellular signatures of neuronal APs. As they can normally just be picked up by electrodes in close proximity to the AP-firing neuron, spikes are the most local extracellular signals that we consider. In Chapter 8, we follow up with another (albeit less) local measure, namely the local field potential (LFP). The LFP is the low-frequency part of electric potential recorded in the grey matter of the brain. In Chapters 9 and 10, we follow up with simulations of extracellular potentials on the largest spatial scale, which are those recorded on the scalp (EEG) and cortical surface (ECoG) respectively. In all these chapters, the main aim is to provide insight into how various features of neuronal structure and activity, such as synaptic positions, neural morphology, firing frequency, and degree of synchrony within a network, are related to various features in the extracellular potential.

The prediction of magnetic brain signals, and the physical theory needed to do so, is covered separately in the rather brief Chapter 11. The focus is on the MEG signal (the magnetic signal recorded immediately outside the head), but we also briefly outline methods for computing magnetic fields inside the brain. The origin of the magnetic signals are the same as for the electric signals, and both can be simulated using the MC+VC scheme.

As implied earlier, we are mainly concerned with the forward computation of extracellular signals resulting from a given neural activity. Computing how extracellular signals affect neural activity is in principle described with the same physics. The governing theory for electric and magnetic stimulation is briefly described in Section 4.6.

Although it is fairly well established that the main source of extracellular brain signals are currents produced by neuronal activity, the signals may in principle contain additional contributions from glial activity or the diffusion of ions along extracellular concentration gradients. These additional sources are, under most circumstances, believed to be relatively small and to predominantly affect the very-slow frequency components of extracellular potentials. As there is no principal difference between neuronal and glial membrane currents, both can be computed with the same MC+VC framework, and when we make general references to neural current sources, they can include all cellular sources. Extracellular diffusive currents do, however, belong to another category and are not accounted for in the MC+VC framework. Theory for computing diffusion potentials is presented in Chapter 12, where we also give some estimates of their relative contributions to extracellular potentials.

In the final Chapter 13, we first describe some common misconceptions about extracellular potentials. Next, we discuss the general reliability of MC+VC-based modeling of electric and magnetic extracellular signals before we round off the book by listing some future applications of forward modeling.

1.3   Guide to Reading This Book

Different readers may have different motivations for consulting this book. It is largely a physics book, focusing on the physical and biophysical processes governing extracellular brain signals. However, we have made some effort to present the material in a way that should make it accessible not only to neuroscientists with strong schooling in physics and mathematics, but also to other neuroscientists trying to make sense of extracellularly recorded brain signals.

All readers are advised to read the basic theory chapter (Chapter 2), as key concepts and variables that will be used throughout the book are introduced there. Parts of this theory chapter will be challenging for readers without a physics background, but understanding all of it will not be necessary. The most important takeaway from Chapter 2 is the description of neural activity in terms of membrane current sources in Section 2.6.1 and especially the definition of the MC+VC scheme in Section 2.6.2. This standard scheme for computing extracellular brain signals will be used throughout most parts of the book.

The MC part, which is about how one uses multicompartment neural models to compute neural membrane currents, is described in further detail in Chapter 3. The first two sections (Sections 3.1–3.2) should cover the material that is most important. The VC part, about how one uses volume-conductor theory to compute the extracellular potentials resulting from the MC simulations, is described in further detail in Chapter 4. The main idea is established already in Section 4.1 of the chapter. Although the theory can be extended to more general cases, readers who understand the sections just suggested should have a good foundation for following the later parts of this book.

After following the instructions in the previous paragraphs, readers that consult this book for help in interpreting various kinds of extracellular recordings may choose to go directly to the relevant chapter. The book contains chapters devoted to computing and examining spikes (Chapter 7), LFPs (Chapter 8), EEGs (Chapter 9), ECoGs (Chapter 10), MEGs (Chapter 11), and diffusion potentials (Chapter 12). In many cases, we will be analyzing the extracellular signals by examining their frequency content, presented in terms of amplitude or power spectra. Readers that are not familiar with these concepts can get a brief introduction in Appendix F.

A more biophysically oriented reader will probably choose to delve deeper into the theoretical material in Chapters 2–5. When it comes to the biophysical modeling of neurons, the introduction in Chapter 3 is kept rather brief as this topic is covered extensively in several other books, such as those by Koch (1999), De Schutter (2009), or Sterratt et al. (2023). The focus of this book is rather on extracellular signals and on the tissue (medium) they propagate through. The later parts of Chapter 2 deal with the problems and approximations involved when applying electromagnetic theory to a complex medium such as brain tissue, while the equations for signal propagation in brain tissue are derived in Chapters 4 and 5. In Chapter 5, we review both the experimental and theoretical efforts to understand the properties of brain tissue and its electric conductivity.

Readers who consult the book to learn how to develop computational models of extracellular signals should get a solid theoretical platform through Chapters 2–4. Technical details useful for numerical implementation, as well as methods for computing extracellular signals using more simplified and computationally efficient models of networks of neurons, are found in Chapter 6.

1.4 Guide to Simulations, Codes, and Figures

Much of the insight into the nature of extracellular electric signals of neural origin presented throughout this book has been gained through computer-based simulations, such as those shown in Figure 1.1. This figure, as well as many others in this book, are based on simulations using the MC+VC scheme and using our own software tool LFPy (Hagen et al. 2018). In the following chapters, we use example simulations of this kind both to illustrate the methodology for the forward modeling of extracellular signals as well as to gain insights into how various features of neuronal activity affect various features of spikes and LFP, ECoG, EEG, and MEG signals. As a supplement to this book, the simulation codes used to produce the figures are freely available and hosted in an external code repository at https://github.com/LFPy/ElectricBrainSignals (Hagen & Ness 2023), also accessible via www.cambridge.org/electricbrainsignals. The codes are released under open, free-to-use licenses.

As we shall see later on, a neuron's contribution to the extracellular signal depends on its morphology, as well as on how synapses and ion channels are distributed onto this morphology. There are thus certain salient features in extracellular signals that can only be simulated with rather complex neuron models that account for morphological and biophysical detail. The simulation in Figure 1.1 was done using a biophysically detailed model of a rat cortical layer 5 pyramidal cell developed by Hay et al. (2011). This model, which we will simply refer to as the *Hay model*, is presented in further detail in Section 3.2.5. Throughout this book, the Hay model will serve as our default go-to model when we need a biophysically detailed neuron model. We will, however, also present simulations on simpler, generic neuron models, as well as on biophysically detailed models representing other types of neurons. These other models will be described briefly in the sections where they are first used.

Readers of this book are cordially encouraged to download the simulation codes used throughout this book, re-run them, and modify them in an exploratory manner. Playing around with the codes interactively while consulting this textbook for interpretations and explanations might be a good learning strategy. Demonstrations in this book include the effect of neuronal morphology, the effects of synapse distributions, and the effects of the distribution of different ion channels on extracellular signals. However, there are several uncharted paths for such ventures, and readers that want to pursue particular research questions may use the codes as a starting point for setting up their own simulations. Thus, our hope is that this book and these codes may also serve as a basis for future research – for instance, aiding the interpretation of experimental recordings of neural activity.

In case the provided codes result in new publications, kindly cite this book, the relevant original publication, and Hagen & Ness (2023).

The main figure-generating files are set up as Jupyter[2]) notebooks and rely on the Python programming language,[3] with a few other package dependencies such as the aforementioned LFPy and neural simulation software such as NEURON (Hines & Carnevale 1997, Hines et al. 2009). The notebook codes may readily be employed in interactive cloud computing services such as the EBRAINS Collaboratory (www.ebrains .eu/tools/collaboratory).

[2] https://jupyter.org
[3] https://python.org

2 Charges, Currents, Fields, and Potentials in the Brain

Neurons communicate via electric signals, which they generate by letting ionic currents pass through their membranes in a selective manner. For example, the main signaling unit in the brain, the *action potential* (AP), is essentially a brief depolarization of the membrane caused by a flux of Na^+ ions into the neuron, followed by a repolarization largely caused by a flux of K^+ out from the neuron. Whereas the main effect of membrane currents[1] is to evoke changes in the membrane potential of neurons, they also have the secondary effect that they cause the electric potential to vary in the extracellular space of the brain.

A neuron's membrane potential can be recorded experimentally with an intracellular electrode, normally inserted into the soma. Recording from dendrites or axons is also possible but generally more challenging, since these structures tend to be very thin. Intracellular recordings give us good insight into the dynamics of the neuron. However, it is difficult to record intracellularly from more than a few neurons at the same time, and if we want to study what the brain is doing on a more systemic level, the doings of a couple of individual neurons might not tell us that much. Many experimental studies are therefore instead based on recording the extracellular potential. An electrode is then, for example, inserted into the tissue of some brain region that the experimentalist wants to study so that the potential it records is an indirect testimony of the activity of a larger population of surrounding neurons.

Extracellular signals are more difficult to interpret than intracellular signals, but forward modeling can help us in this regard. The main focus of this book is on extracellular electric potentials, and in the following chapters, we will first present the theory needed to model them and later present simulations set up to explore the relationship between the extracellular potentials and the neural activity they originate from. We will also touch briefly upon the theory behind magnetic signals in the brain.

We will, in this book, be dealing with so-called forward modeling of extracellular signals. For most parts of the book, we will perform such modeling using a scheme where the neurons and the extracellular signals are modeled in two separate steps:

- Step 1: The complex spatiotemporal aspects of the membrane potential dynamics in individual neurons are modeled using so-called *multicompartment* (MC) models of neurons. MC models can account for the special properties of neural membranes and can, for example, describe how the AP propagates (like a wave)

---

[1] What we refer to as "membrane current" is also commonly referred to as "transmembrane current."

along the axon of a neuron. Since physics-based theory for modeling neurons has been covered in many textbooks, MC modeling will be covered rather briefly in the current book, in Chapter 3.

- Step 2: Once the neurons have been modeled, the extracellular electric and magnetic fields are computed by treating the tissue medium that the neurons are embedded in as a continuous *volume conductor* (VC). VC theory is further introduced in Chapter 4.

We will refer to the two-step scheme for modeling extracellular signals as the *MC+VC scheme*. This scheme has become the standard scheme for biophysical modeling of extracellular potentials, and it will be defined more properly in Section 2.6.2, after we have introduced the physical concepts that it is based upon. The reason for mentioning it here is that we hope to avoid confusion by emphasizing that the MC and VC parts of the modeling scheme are not based on the same set of physical equations and assumptions. Not all the assumptions that are made in the MC framework apply to the VC framework, and vice versa. However, the physical processes involved both in the MC and in the VC frameworks, and arguably the physical basis of most brain functions, appear to be well described with classical physics (Koch & Hepp 2006).

Before addressing the brain specifically, we will in the current chapter give a general introduction to electromagnetic theory, which is the foundation for understanding brain signals. The aim of this rather incomplete introduction is to establish an elementary understanding of the main concepts that we will use throughout this book.

Electromagnetic theory is summarized by Maxwell's equations and the Lorentz force law, and we might have kick-started this chapter by listing up those. However, Maxwell's equations may appear a bit challenging to the untrained eye, and both Maxwell's equations and the Lorentz force law contain terms that are not so relevant for the kind of brain dynamics we will be focusing on. We will therefore be taking a softer path, which we deem sufficient for establishing the main concepts that we will be working with. For good taste and later reference, we nevertheless go through the theoretical pillars of electromagnetism in the final part of the chapter (Section 2.7), where we also discuss some of the most important assumptions that we make when applying the theory to study signaling in the brain.

Readers without prior schooling in physics might not be able to follow all parts of this introduction, but it will hopefully still give them a basic understanding of the concepts of *electric charge*, *electric currents*, *electric fields*, and *electric potentials* and some insight into how they are related to one another. Readers that do have a physics background might see the first part of this introduction as a useful refresher and may appreciate the later parts, which deal with the applications of electromagnetic theory to a complex medium such as brain tissue.

2.1   Electric Charge

Let us begin with the fundamental quantity for electricity – the electric charge. Nature's fundamental charges are those carried by the protons and electrons that build up the atoms that build up the material world. The proton carries the charge e, while the electron carries the charge $-e$, where $e = 1.602 \times 10^{-19}$ coulomb (C) is called the *elementary charge*. A fundamental law of nature is *charge conservation*, meaning that it seems fundamentally impossible to create or destroy a net amount of positive or negative charge. The universe as we know it is exactly electrically neutral, and atoms in their neutral state contain an equal number of electrons and protons.

Although most matter is electrically neutral, some mediums are nevertheless *conductive*, which means that they contain so-called *free charge carriers* that may move through the medium. In a copper wire, such as that carrying current to a desktop lamp, the free charge carriers are electrons that are not bound to specific locations (copper atoms), but can move freely through the copper material. In an electrolyte solution – for example, salt water (saline) – the free charge carriers are instead ions, that is, atoms or molecules that have gained or donated one or several electrons, and therefore have become electrically charged. For example, table salt (sodium chloride ($NaCl$)) dissolves in water into sodium ions (Na^+) with charge $1e$ and chloride ions (Cl^-) with charge $-1e$. The saline solutions that fill the intracellular and extracellular space of the brain are such electrolytic solutions, where sodium and potassium ions (Na^+ and K^+ with charge $1e$) and chloride ions (Cl^- with charge $-1e$) are the most important charge carriers. Because of these saline solutions, brain tissue as a whole can largely be seen as a conductive medium.

At a fundamental level, the movement of electrically charged particles is affected both by electric and by magnetic forces. When it comes to charges moving in the brain, magnetic forces can be neglected (see Section 2.7.3), so we here introduce only the electric force. A pair of charges, q_1 and q_2, in free space, will act on each other with an electric force F (with unit newton (N)) given by Coulomb's law:

$$F = \frac{q_1 q_2}{4\pi\epsilon_0} \frac{1}{r^2}, \qquad (2.1)$$

where the constant ϵ_0 is the permittivity of free space, and r is the distance between the two charges. The Coulomb force is inversely proportional to the square of the distance between the two charges. Equation (2.1) is a so-called scalar expression, meaning that the direction of the force is not explicitly expressed in the equation. However, we know that the force will be directed along the line between the two charges, and it will be repelling (positive) if the charges have the same sign and attractive (negative) if they have the opposite sign.

We can express the direction of the force (acting on q_1) mathematically by use of vector notation (Figure 2.1):

$$\mathbf{F}_1 = \frac{q_1 q_2}{4\pi\epsilon_0} \frac{1}{\|\mathbf{r}_1 - \mathbf{r}_2\|^2} \mathbf{e}_{1,2}, \qquad (2.2)$$

where we have let a boldface notation indicate that an entity is a vector and where the positions of the charges q_1 and q_2 have been denoted by $\mathbf{r}_1$ and $\mathbf{r}_2$, respectively. In equation (2.2), the unit vector $\mathbf{e}_{1,2}$ (directed from 2 to 1) defines the direction of the force but says nothing about its value, while the fraction in front of it is a scalar (a number

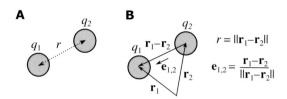

Figure 2.1 Electric interaction between charges. A: The force between two charges depends on the distance r between them. The direction of the force is along the imaginary line connecting the two charges. To express this direction mathematically, we need a vector notation. **B:** The position vectors $\mathbf{r}_1$ and $\mathbf{r}_2$ can be visualized as arrows from some reference point $r = 0$ to the positions of the charges q_1 and q_2, respectively. Likewise, $\mathbf{r}_1 - \mathbf{r}_2$ can be visualized as an arrow from the position of q_2 to the position of q_1, defining both the distance and direction of the (imagined) line connecting them. The direction alone (without distance) is defined by the *unit vector*, $\mathbf{e}_{1,2}$ (read "from 2 to 1"), computed by dividing a vector $\mathbf{r}_1 - \mathbf{r}_2$ by its own length $\|\mathbf{r}_1 - \mathbf{r}_2\|$, so that its length is one (unity).

with no direction) that defines the value of the force but says nothing about its direction. It is thus easy to verify that the value of the force is the same as in equation (2.1).

Whereas the scalar expression in equation (2.1) works well for describing pairwise particle interactions, a vector notation is more convenient when we need to describe interactions involving a larger number of charges. When many charges are present, the contributions from each charge to the total force sum up linearly. Hence, if we have N charges, the force acting on one of them (q_1 at position $\mathbf{r}_1$) from the $N - 1$ other charges ($q_2, q_3, q_4 ... q_N$ at positions $\mathbf{r}_2, \mathbf{r}_3, \mathbf{r}_4 ... \mathbf{r}_N$) will be

$$\mathbf{F}_1 = \sum_{n=2}^{N} \frac{q_1 q_n}{4\pi\epsilon_0} \frac{\mathbf{e}_{1,n}}{\|\mathbf{r}_1 - \mathbf{r}_n\|^2} . \tag{2.3}$$

If the number of charges (N) is small, we could use N instances of equation (2.3) to compute the force acting on all the N individual charges. However, when dealing with macroscopic systems, keeping track of individual charges and the forces acting on them is not feasible. On larger spatial scales, it is therefore more useful to describe electric phenomena through concepts like electric currents, fields, and potentials.

2.2 Electric Fields and Potentials

The electric field $\mathbf{E}$ (with unit volts per meter (V/m)) at a given location can be defined as the electric force that *would* act on a reference charge q if it were present there:

$$\mathbf{E}(\mathbf{r}) = \mathbf{F}_q(\mathbf{r})/q . \tag{2.4}$$

When the electric force is due to the Coulomb force, the electric field generated by N charges is obtained by inserting equation (2.3) into equation (2.4):

$$E(r) = \sum_{n=1}^{N} \frac{q_n}{4\pi\epsilon_0} \frac{e_n}{\|r - r_n\|^2} .$$ (2.5)

Here we have let e_n denote the unit vector from charge n to the position r where the field is being evaluated. Equation (2.5) indicates that the existence of an electric field ultimately reflects an underlying distribution of charges.

We note that, in general, electric fields can be also affected by the presence of magnetic fields. However, when it comes to endogenous fields in the brain, such field interactions are negligibly small (see Section 2.7). When this is the case, the electric field is said to be conservative. Mathematically, this means that it can be written as the gradient (spatial derivative) of an electric potential V (with unit volts (V)):

$$E(x, y, z) = -\nabla V(x, y, z) = -\left(\frac{dV}{dx}e_x + \frac{dV}{dy}e_y + \frac{dV}{dz}e_z\right).$$ (2.6)

The operator ∇ computes a scalar field's spatial rate of change in the various spatial directions, and e_x, e_y, and e_z are the unit vectors in the three spatial directions x, y, and z, respectively.

It is the potential V that we normally measure experimentally with electrodes. While E is a vector field, V is a scalar field, and as such V tends to be easier to deal with because measuring the value of a vector field requires measurement in all three spatial dimensions. Unlike for E, however, V is always measured relative to some arbitrarily selected reference point (often called *ground*), as we explain in the next subsection.

For a spherically symmetric potential, equation (2.6) reduces to

$$\nabla V(r) = \frac{dV(r)}{dr}e_r ,$$ (2.7)

where e_r is the unit vector in the radial direction. Since the field contribution from each point charge is spherically symmetric around its position r_n, the electric field in equation (2.5) can be written as the (negative) gradient of the potential

$$V(r) = \sum_{n=1}^{N} \frac{q_n}{4\pi\epsilon_0} \frac{e_{1,n}}{\|r - r_n\|} .$$ (2.8)

In practical physics problems, the positions of all present charges q_n are rarely known, and the computation of the electric field from equation (2.5) or the potential from equation (2.8) can therefore only be done in idealized cases. Fields and potentials are nevertheless useful quantities at a systems level. They can be measured experimentally, and as we will come back to later, they can be computed from other physical principles than interactions between individual charges.

2.2.1   Reference Point for Potentials

To get an intuitive understanding of the relationship between E and V, it helps to consider an idealized one-dimensional (1D) scenario with a constant field in the x-direction. Then, equation (2.6) simplifies to

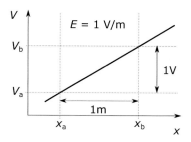

Figure 2.2 Relationship between the electric field and electric potential. With a constant electric field $E = 1$ V, the electric potential V increases linearly with distance x. With two locations x_a and x_b 1 m apart, we know that $V(x_b) = V(x_a) + 1$ V. We are free to define an arbitrary reference point (ground) for V, and if we take $V(x_a) = 0$, it follows that $V(x) = (x - x_a)E$. In x_b, we then get $V(x_b) = (1\text{ m}) \times (1\text{ V/m}) = 1$ V. Equivalently, we might define x_b as our reference point, which would mean that $V(x_b) = 0$ and $V(x_a) = -1$ V. While V depends on the distance between the measurement point and the reference point, E is independent of this choice.

$$E = -\frac{dV(x)}{dx} = -\frac{V(x_b) - V(x_a)}{x_b - x_a}, \tag{2.9}$$

where the first equality follows from the 1D assumption, and the second follows from the assumption that E is constant, with x_a and x_b representing any two arbitrary points in space. For example, if $E = 1$ V/m, and the distance $x_b - x_a$ between our points is 1 m, equation (2.9) tells us that V will be 1 V lower in x_b compared to x_a (Figure 2.2).

In the example, the field $E = 1$ V/m would be consistent with any pair of potentials V_a and V_b as long as $V_b - V_a = 1$ V. Hence, the field **E** generally determines the potential only up to a constant. We can therefore not speak of the potential in a certain point as an absolute entity, but only the potential *difference* between two points. When we measure the potential in a given location, it is always measured relative to some reference electrode that is placed in another location where we define $V = 0$.

For a non-constant field $\mathbf{E}(\mathbf{r})$, the potential relative to a reference position $\mathbf{r}_{ref}$ can be found by solving the integral

$$V(\mathbf{r}) = -\int_{\mathbf{r}_{ref}}^{\mathbf{r}} \mathbf{E}(\mathbf{r}') \cdot d\mathbf{r}'. \tag{2.10}$$

By definition, the electric potential is the energy needed to move a unit of electric charge q from the reference position to a specific location in the electric field. The unit (V) of the electric potential is thus equivalent to energy per charge, or joule per coulomb (1 V=1 J/C). As such, the concept of an electric potential is closely related to the concept of a potential energy. The potential energy (U_E) of a charge q in the presence of an electric potential is

$$U_E(\mathbf{r}) = qV(\mathbf{r}). \tag{2.11}$$

Like the electric potential, the potential energy must also be defined relative to a reference point.

When we record extracellular potentials in the brain, the positioning of the reference electrode can be chosen to be either close to or far away from the recording electrode, depending on the kind of signal one wishes to measure (see Section 4.5.4).

2.2.2   Electroneutrality and Debye Shielding

In Section 2.2, we saw that the electric potential resulting from a certain known arrangement of charges is given by equation (2.8). This formula works very well for textbook physics examples considering a few test charges in vacuum or a dielectric material where the surrounding charges are bound to specific locations. However, it is impractical for describing large-scale processes in a conducting material such as brain tissue. The reason is that a conducting material is densely packed with free charges that are constantly interacting and in constant motion. The Coulomb force (equation (2.1)) will make each of these charges attract charges of the opposite sign and repel charges of the same sign. An individual positive charge in a conductive medium will therefore be immediately surrounded by negative charges, and vice versa, which very effectively shields the potential from the charge. This phenomenon is known as Debye screening or Debye shielding.

We know from *Debye–Hückel theory*, which we do not present in any detail here, that the shielding effects in a conducting medium will cause the potential contribution from a charge q to decay *not* as

$$V(r) = \frac{q}{4\pi\epsilon_0 r} \, , \tag{2.12}$$

which is the single charge version of equation (2.8), but instead as (Robinson & Stokes 2002)

$$V(r) = \frac{q}{4\pi\epsilon_0\epsilon_r r} e^{-r/r_\mathrm{D}} \, . \tag{2.13}$$

Here, r_D is called the *Debye shielding distance*, and ϵ_r is the *relative permittivity* of the medium, which has a value of about 80 in water and dilute saline solutions (Hasted et al. 1948). In the saline solution in the brain, r_D is typically just below 1 nm (Hille 2001), which implies that any given reference charge will give a negligible contribution to the potential at distances larger than a nanometer away.

Closely related to the shielding effects is the concept of electroneutrality. When arranged by Coulomb interactions in a conductive medium, the numbers of positive and negative charges within a finite and not-too-small volume of space will be kept in close balance. If this were not the case, the very strong Coulomb force associated with a finite charge density ρ would cause ρ to decay towards zero at a rate proportional to the so-called *charge-relaxation time*. In dilute saline solutions, such as that filling the extracellular space of the brain, the charge-relaxation time is on the order of a nanosecond (see Section 5.4.2.1). On spatial scales larger than nanometers and temporal scales larger than nanoseconds, the extracellular saline solution of the brain is therefore – to a good approximation – electroneutral (Grodzinsky 2011). The same goes for the saline solution in the intracellular space of neurons.

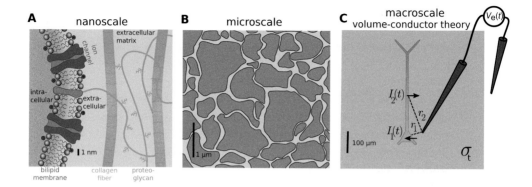

Figure 2.3 Brain tissue on different spatial scales. A: Nanometer scale: single charge interactions. The capacitive membrane can separate an amount of negative charges (blue) on one side from the same amount of positive charges (red) on the other side. **B**: Micrometer scale: continuous variables such as concentrations and currents. **C**: Macroscopic scale: brain tissue considered as a continuous porous medium defined by its conductivity σ_t. The electric potential (V_e) in this medium depends on the distances from membrane current sources.

While the intra- and extracellular mediums are conducting, the membrane that separates them is largely dielectric (insulating). A patch of membrane therefore acts as a capacitor (see Section 2.4.2) and can separate a net charge $-q$ on the interior side from a charge q on the exterior side. It is this breach of electroneutrality, indicated in Figure 2.3A, that gives rise to the membrane potential, and it must thus be accounted for when modeling neural dynamics. The time constant (or charge-relaxation time) for the membrane dynamics is on the order of milliseconds (see Section 5.4.2.2).

The membrane is, however, just a few nanometers thick, and the breach in electroneutrality associated with the membrane charges thus occurs on a nanometer scale. When we describe the tissue medium on a larger spatial scale (see Section 2.3.2), the inside and outside membrane charges effectively cancel to a charge density of zero. On this coarser spatial scale, the conductive properties of the saline solutions in the intra- and extracellular spaces dominate, and brain tissue can – to a good approximation – be modeled as an electroneutral, conductive medium (see Section 5.4.2.3). The practical implication of electroneutrality and shielding effects is that we do not need to compute extracellular electric potentials or fields from a precisely known distribution of charges, but we can instead compute them from the constraint of *current conservation* or, equivalently, the constraint that there should be no charge accumulation anywhere in the tissue.

We return to this topic in Section 2.4, but we must first say some words about the jump in spatial scale when going from single particles (atoms and charges) to a continuum description (concentrations and currents).

2.3   Coarse-Graining

A law of physics is typically restricted to a limited range of spatial and temporal scales. For example, equation (2.1) for the Coulomb force between two individual charged particles only applies for inter-charge distances r greater than the extension of the individual charge. For the problems that we are tackling in this book, this is not a serious limitation. As mentioned previously, the limitation with equation (2.1) is rather that it is impractical to use for describing macroscopic processes.

In the saline solutions of the brain, the individual ions are bumping around due to Brownian motion, with thermal velocities $u \approx (k_B T/m)^{1/2} \approx 100\text{--}1000\,\text{m/s}$, constantly colliding with water molecules or each other. The mean free path length between two collisions is on the order of a nanometer and is traveled in much less than a nanosecond. As we argued in the previous section, we do not (and can not) study a macroscopic system by keeping track of individual particle interactions. To study brain dynamics, we therefore use so-called *coarse-grained* models.

2.3.1   Coarse-Graining 1: From Individual Atoms and Charges to Concentrations and Currents

On a larger spatiotemporal scale of, say, micrometers and microseconds, details of interactions between individual particles will be lost. At this scale, the system dynamics can be described with continuous variables, such as the charge density ρ (with unit C/m^3), the concentration $[k]$ of a particle species k (with unit millimolar ($1\,\text{mM} = 1\,\text{mol/m}^3$ $= 6.02 \times 10^{23}$ particles per m^3)), the electric current density i (with unit ampere per square area ($\text{A/m}^2 = \text{C/m}^2\text{s}$)), and particle flux density j_k (with unit $\text{mol/m}^2\text{s}$). These continuous variables basically represent spatiotemporal averages of very large numbers of individual particle interactions.

How coarse the spatial resolution must be for the continuum description to be warranted depends on how densely packed the medium is with the particle species that we are interested in. Among the concentrations of the main charge carriers in the brain (Na^+, K^+, and Cl^+), the extracellular K^+ concentration is the lowest, so let us use that as an example. It can have a value of around 3 mM (Somjen 2004), which means that a $1\,\mu\text{m}^3$ volume still contains $3 \times 6.02 \times 10^{23} \times 10^{-18} \approx 2 \times 10^6$ K^+ ions. Since this is a fairly high number, it should not be subject to significant stochastic fluctuations due to the (Brownian) randomness on the single ion level. When considering processes taking place on the microscale and above, we therefore do not need to be concerned with details of single charge interactions taking place on the nanoscale and below, and a continuum description should therefore be well justified. In contrast, a billion times smaller volume of $1\,\text{nm}^3$ of extracellular space should (on average) contain only ≈ 0.002 K^+ ions. On such a fine spatial scale, the continuum description is likely to be inadequate, at least for a detailed description of the physical system at hand.

The continuous variables introduced above are used both in the multicompartment (MC) modeling of neurons and in the description of the surrounding tissue as volume conductor (VC). However, the VC model also requires that we make a second coarse-graining.

2.3.2 Coarse-Graining 2: From a Mismmash of Entangled Neurites to Smooth Neural Tissue

Brain tissue is a dense mishmash of neurites and other cellular structures. The composition varies with brain state and brain region, but common estimates are that about 80 percent of the volume is occupied by cells, while the remaining 20 percent is the highly tortuous extracellular space (Figure 2.3B). The diameter of a typical neurite is on the order of $1\,\mu m$, and estimates of the typical extracellular distance between cellular membranes range between 10 and $80\,nm$ (Syková & Nicholson 2008, Kinney et al. 2013). Electric currents passing through brain tissue must somehow make their way through this mishmash.

Locally, electric currents in the narrow confines of extracellular space may have a free passage through the extracellular saline solution only in certain spatial directions, while passage in other directions may be blocked by cellular membranes with a much higher resistance than saline. Hence, a micrometer-precise insight into the pathways taken by extracellular currents requires knowledge of the highly inhomogeneous microstructure of the tissue. Although small volumes of tissue can be partially reconstructed using *ex vivo* techniques like scanning electron microscopy (see for example Holter et al. (2017)), the overall microstructure is almost always unknown. Even if it were known, including it in modeling or analysis of extracellular potentials would require prohibitively large computational resources. Keeping track of the exact microstructure of neural tissue is therefore practically impossible. Luckily, it is also for most purposes unnecessary.

Earlier, we explained how we, by jumping from a nanometer-scale to a micrometer-scale system description, can stop worrying about single ions and charges and instead work with continuous variables such as ion concentrations and electric currents. When we want to consider extracellular currents through brain tissue, we need to make a second jump in scale, this time from cellular structures to smooth brain tissue.

While tissue is highly inhomogeneous on the micrometer scale, we may expect the microscale inhomogeneities to average out on a coarser spatial scale of, say, some tens of micrometers (Nicholson & Freeman 1975, Okada et al. 1994, Logothetis et al. 2007, Goto et al. 2010, Ness et al. 2015). By that we mean that the average conductivity of a cubic tissue sample with sides of some tens of micrometers will be the same no matter where the sample is taken from (at least if we stay within a given brain region). At the spatial scale of some tens of micrometers, we can then assume that an extracellular current experiences tissue as a continuous volume conductor with a conductivity σ_t that is the same everywhere (Figure 2.3C).

The tissue currents and the tissue conductivity will be defined properly in the next section, and the treatment of tissue as a continuous medium is discussed in greater detail in Chapter 5.

2.4   Electric Currents and Current Conservation

An important constraint that we will use many times in this book is the requirement that the sum of currents into a finite volume of space must be zero. We will refer to this constraint as *current conservation*. Current conservation is a slightly less fundamental version of the charge conservation discussed in Section 2.1. When charge conservation is combined with electroneutrality (Section 2.2.2), current conservation automatically follows: the net amount of charge entering a finite volume must be zero because otherwise charge would pile up there and violate electroneutrality.

As we explained in Section 2.2.2, electroneutrality is not always strictly fulfilled, but it is violated locally on the neural membrane, which can separate a net positive charge on one side from a net negative charge on the other side. However, since charge is conserved, the accumulated charge can be released as a current at a later time. As we shall see, this effect is equivalent to what occurs in circuit elements known as capacitors, where charge piles up locally on both sides of an insulating layer. Furthermore, it is customary to describe the charge accumulation process in terms of a *capacitive current*, which allows the concept of current conservation to be applied also to the charging of a capacitor.

To understand how the principle of current conservation helps us to compute extracellular potentials, consider the illustration in Figure 2.3C. For simplicity, there are in this illustration only two membrane currents: an inward current I_1 into the soma and an outward current I_2 from the dendrite. In the region of extracellular space outside the soma, current conservation requires that the membrane current I_1 (leaving this region) must be compensated with an equal amount of positive current (entering this region) from elsewhere in the extracellular space. If it weren't so, the region would become negatively charged (not electroneutral). Moreover, since electric currents always go in the direction towards a lower electric potential, the positive extracellular current towards the soma region tells us that V_e must be lower there than in the surroundings.

A real neuron will, of course, have membrane currents distributed over its whole morphology, but the same principle can still be applied, and as it turns out, the distribution of membrane currents determines what the value of V_e must be in all positions in order for current conservation not to be violated. Equations for the relationship between membrane sources and V_e will be given later.

By Figure 2.3C and the earlier description, we may unintentionally have formed a somewhat static mental picture of neural dynamics. We therefore rush to note that the illustration only represents a snapshot in time and that the membrane currents can be used to determine V_e at any such snapshot. This is true since, as we noted earlier, the charge-relaxation time in the extracellular saline solution is on the order of a nanosecond. This gives them a good margin for keeping up with variations in neural membrane currents, which normally occur at the timescale of milliseconds (we will see this in Chapter 3). For practical purposes, the relationship between membrane currents and the extracellular potential can therefore be considered to be instantaneous.

Before we start applying the principles of current conservation, we should define the kind of various currents that may be involved. There exist several kinds of electric currents, differing in terms of what drives them, and the content of the previous paragraphs will hopefully become more clear when we below introduce the three kinds of currents that are most important for understanding large-scale brain dynamics. These are conductive currents (Section 2.4.1), capacitive currents (Section 2.4.2), and diffusive currents (Section 2.4.3). We also briefly comment on inductive and advective currents (Section 2.4.4), although these are generally assumed to be of negligible importance for the kind of brain dynamics that we focus on in this book.

2.4.1   Conductive Currents

A conductive material is one that free charges can move through when exposed to an electric field. The field-driven motion of charges can be quantified in terms of a conductive current, defined as the amount of charge that moves through some reference cross-section area of the material per time unit.

In electric circuits composed of metallic wires and various circuit elements, currents are practically running exclusively in the direction defined by the circuitry – that is, longitudinally along wires and through circuit elements, having no radial components. The reference cross-section can then be taken to be that of "the whole wire" or "the whole circuit element," which in practice means that we do not need to define it further since our variable of interest is the total current at each point along the circuit.

A simple example is shown in Figure 2.4A, where a current passes through a wire with a resistor. The current is then given by Ohm's law:

$$I = -\frac{\Delta V}{R} \, . \tag{2.14}$$

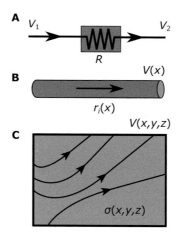

Figure 2.4 Ohm's law for currents and current densities. A: Current through wire and resistor with resistance R (Ω): $I = -(V_2 - V_1)/R$. **B**: Current in continuous resistor with internal resistance per unit length r_i ($\Omega\,\mathrm{m}^{-1}$). $I = -(1/r_i)dV/dx$. **C**: Current density (A/m^2) for current passing through a volume conductor with conductivity σ (S/m): $\mathbf{i} = \sigma\mathbf{E} = -\sigma\nabla V$. Current paths will depend on $\sigma(x,y,z)$, which generally can vary with position.

Here $\Delta V = V_2 - V_1$ is the voltage difference across the resistor, and R (with unit ohm (Ω)) is its resistance. The negative sign on the right-hand side of the equation implies that the current goes in the direction towards lower voltage. In this example, the wire itself is assumed to have a zero resistance so that the entire resistance in the system, and the entire voltage drop, occurs over the resistor. To describe this system, we only need to consider the potential at two locations, 1 and 2, representing the two sides of the resistor.

For most standard electrical circuits, it is an excellent approximation to assume that the resistance of the wire is zero. However, in reality, this is not strictly true, and if the wire is very long and thin, or if it is made of a less-conductive material, it may be necessary to treat it as a continuous resistor. If we define the internal resistance per unit length as r_i (with unit $\Omega\,\mathrm{m}^{-1}$), Ohm's law becomes

$$I = -\frac{1}{r_i}\frac{dV}{dx}\,,\tag{2.15}$$

where $r_i = r_i(x)$ in the general case can vary along the wire (Figure 2.4B). With equation (2.15), the potential $V(x)$ will vary continuously along the wire. If r_i is constant along the wire and the wire has length L, r_i is related to the resistance of the total wire through $r_i = R/L$. As we shall see in Section 3.2, conductive currents running axially inside neural dendrites and axons are typically modeled using equation (2.15).

When treated as a continuous medium, brain tissue can largely be seen as a volume conductor. Unlike in the earlier examples – where the current was restricted to run exclusively in the direction defined by the wire (1D) – currents in a 3D volume, such as brain tissue, may run in all spatial directions (Figure 2.4C). It is then convenient to describe these currents in terms of a current density, $\mathbf{i}$ (with unit $\mathrm{A/m}^2$), a vector that defines the current per unit area and its spatial direction. The current density can be defined in all points in 3D space. The conductive properties of a material can be specified either through its resistivity (with unit $\Omega\,\mathrm{m}$) or its inverse, the conductivity σ (with unit $\mathrm{S/m}$), both being material properties. We will here use the latter convention, and with that, Ohm's law takes the form

$$\mathbf{i} = -\sigma\nabla V\,,\tag{2.16}$$

or, equivalently, the form

$$\mathbf{i} = \sigma\mathbf{E}\,,\tag{2.17}$$

since the field and potentials are related through equation (2.6). Note that the conductivity can generally be inhomogeneous (position-dependent), anisotropic (direction-dependent), and depend on the frequency of the electric field. We discuss this further in Chapter 5.

2.4.1.1   A Note on Ohm's Law

Note that Ohm's law (equation (2.17)) represents the average movement of charge on larger spatial scales and does not apply to the movement of individual charges on the nanometer scale. If we compare it with the physical foundations for electric interactions, it is easy to get confused, so let us dive into that confusion and try to clear it up.

According to equation (2.4), an electric field acts on a reference charge (q) with a constant force ($\mathbf{F}_q$), which according to Newton's law ($\mathbf{F}_q = m_q \mathbf{a}_q$) should give it a constant acceleration in the field-direction. Conversely, equation (2.17) states that $\mathbf{E}$ gives rise not to a constant acceleration of charges but rather to a constant current, that is, a constant average *velocity* of charges.

The reason for the discrepancy between the single charge (constant acceleration) and many-charge (constant average velocity) scales is that the constant acceleration of our single protagonist charge q will go on for only a tiny time period, much smaller than the charge-relaxation time, before it will collide with some other particle. The collision will scatter the protagonist charge out in some random direction so that the work done by the electric force in giving it a preferred motion along the field direction will be lost. This acceleration/collision cycle will repeat itself: the acceleration will start anew and go on until the next collision takes place, and so forth. Whereas the scattering events tend to make the motion of q a random walk (zero average velocity in any preferred direction), the small periods of acceleration between collisions will tend to give it a nonzero average velocity in the field direction. This so-called *drift velocity* is roughly determined by the average acceleration time that the charge gets between two consecutive collisions. As the same will happen for all other charges present, there will be a net drift velocity of charge in the field direction. The current density given by equation (2.17) is therefore often referred to as the *drift current density*.

Admittedly, the explanation that we have proposed here was somewhat hand-wavy, and the fact that we get the linear relationship between the current and the electric field in equation (2.17) is constitutive, meaning that it is observed experimentally rather than rigorously derived from first physical principles. However, it is found to be a good approximation for many mediums under many conditions and an excellent approximation for brain tissue (Nicholson & Freeman 1975, Nunez & Srinivasan 2006, Logothetis et al. 2007, Miceli et al. 2017).

The maximum drift velocity of electrons in copper wires carrying electricity in a household is on the order of millimeters per second, while the drift velocity of ions for large-scale currents in brain tissue is on the order of some nanometers per second. Importantly, this perhaps counterintuitively slow drift velocity should not be confused with the "speed of electricity." When turning on a light switch, an electric field will spread throughout the wire with a speed close to the speed of light, meaning that the electric current will be established (electrons will obtain their drift velocity) throughout the entire cable almost instantaneously.

2.4.2   Capacitive Currents

A common element in electrical circuits is the capacitor. As previously mentioned, capacitors are devices that can accumulate electric charge. The simplest form of a capacitor, a parallel plate capacitor, consists of two conducting metallic plates or surfaces separated by a vacuum or an insulating (dielectric) medium. By definition, a dielectric medium is one where charges are bound to stay in confined regions of space. An electric

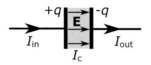

Figure 2.5 Capacitor. In the simplest form, a capacitor consists of two metallic plates or surfaces separated by an insulating (dielectric) medium. Charges q and $-q$ on the metal plate generate an electric field E in the dielectric medium, or a potential difference, $V = Ed$ where d is the distance between the metal plates.

field will only slightly shift their average equilibrium positions, causing a polarization of the material.

A parallel plate capacitor can separate a charge q on one plate from a charge $-q$ on the other plate to obtain a voltage difference

$$V = \frac{q}{C}, \tag{2.18}$$

where C (with unit farad (F $= $ C/V)) is the capacitance.

To understand how a capacitor works, consider the illustration in Figure 2.5, where a conductive current I_{in} enters the capacitor from the left and a conductive current I_{out} leaves to the right. Whereas I_{in} and I_{out} are mediated by charges moving freely through cables, no charges actually move across the capacitor itself. Instead, I_{in} leads to an accumulation of charge (dq/dt) on the left metal plate of the capacitor and to a corresponding electric field across its dielectric core, which drives positive charges away from the rightmost metal plate, charging it negatively $(-dq/dt)$. The temporal rate of change of the electric field across the dielectric medium can be described in terms of a *capacitive current* defined by

$$I_c = \frac{dq}{dt} = C\frac{dV}{dt}. \tag{2.19}$$

Although I_c differs from conductive currents in the sense that it does not involve the free transfer of charge through a medium, it has the same unit (A) and ensures current conservation so that $I_{in} = I_c = I_{out}$.

The capacitive current was introduced here because neural membranes have capacitive properties, which allows the neural membrane potential to change when currents pass through the membrane. It is customary to express the capacitive membrane current in terms of a specific current density (current per membrane area)

$$i_c = c_m\frac{dV_m}{dt}, \tag{2.20}$$

where c_m (with unit F/m^2) is the specific capacitance, defined as the capacitance per unit membrane area. In this case, the current goes only in the direction normal to the membrane.

2.4.3   Diffusive Currents

The main charge carriers in the brain are ions floating around in the saline solutions that fill up the intra- and extracellular spaces. Ions in salt water will, like electrons in a metal conductor, be accelerated by the presence of an electric field and will thus carry conductive currents in accordance to Ohm's law (equation (2.17)). In addition to movement propelled by the electric field, ions in water may also move due to diffusion. If the concentration ($[k]$) of an ion species k varies with position, diffusion will drive ions towards regions with low concentration, and since ions carry electric charge, this movement amounts to an electric current that comes in addition to the conductive current.

The electrodiffusive nature of ionic motion is described by the Nernst-Planck equation for the flux density $\mathbf{j}_k$ (with unit mol/(m^{2}s)) of an ion species k:

$$\mathbf{j}_k = -D_k \nabla[k] - \frac{F}{RT} D_k z_k [k] \nabla V . \tag{2.21}$$

Here, $R = 8.314\,\mathrm{J/(molK)}$ is the gas constant, $F = 96\,458.3\,\mathrm{C/mol}$ is Faraday's constant, and T (with unit K) is the temperature. The valency z_k is a unit-less number that denotes the number of elementary charges associated with a single ion of species k.

The first term on the right-hand side of equation (2.21) is Fick's law for diffusion, and it states that the diffusive flux of ion species k is proportional to the diffusion constant D_k (with unit m^2/s) times the concentration gradient $\nabla[k]$. D_k depends on both the ion species and the medium it exists in, and it determines how "easy" it is for the ion k to move through the medium.

The second term on the right-hand side of equation (2.21) accounts for ionic drift due to the electric field. The derivation of this term builds on two assumptions. The first is that the electric drift velocity $\mathbf{u}_k$ of ion k when exposed to an electric field $\mathbf{E}$ is given by

$$\mathbf{u}_k = \mu_k \mathbf{E} , \tag{2.22}$$

where μ_k is the electric mobility of ion k. The second assumption is that the mobility and diffusion constant are related via the *Einstein relation*:

$$D_k = \frac{RT}{F|z_k|} \mu_k . \tag{2.23}$$

The Einstein relation does not apply generally, but it is a good approximation for many mediums, including dilute saline solutions such as the saline solution in the brain (Grodzinsky 2011).

Faraday's constant F is defined as the charge per mole (6.02×10^{23}) of elementary charges. The flux density $\mathbf{j}_k$ of an ion species $[k]$ can thus be converted to a current density by multiplying it with F and the ion's valency z_k. A salt-water solution contains several ion species, and to obtain the total electric current density, we must sum over the contributions from all of them to obtain

$$\mathbf{i} = \sum_k z_k F \mathbf{j}_k = -F \sum_k z_k D_k \nabla[k] - \frac{F^2}{RT} \sum_k D_k z_k^2 [k] \nabla V . \tag{2.24}$$

The last term on the right-hand side is the drift current density,

$$\mathbf{i}_{\text{drift}} = -\frac{F^2}{RT} \sum_k D_k z_k^2 [k] \nabla V = -\sigma \nabla V \, . \tag{2.25}$$

As we indicated with the last equality, the drift current density is the same as the ohmic current density that we defined in equation (2.16), which means that we can identify the conductivity σ of a saline solution as

$$\sigma = \frac{F^2}{RT} \sum_k D_k z_k^2 [k] \, . \tag{2.26}$$

The first term on the right-hand side of equation (2.24) is the diffusive current density

$$\mathbf{i}_{\text{diff}} = -F \sum_k z_k D_k \nabla [k] \, . \tag{2.27}$$

Hence, in a conductive medium that contains concentration gradients, the total current is not purely ohmic but can contain an additional contribution from ionic diffusion.

In the brain, the most dramatic concentration gradients are those over the cellular membranes. Diffusive currents carry a large fraction of the membrane currents and are accounted for when neurons are modeled (see Chapter 3). The concentration gradients in the extracellular space are much smaller, and extracellular currents are normally assumed to be dominated by drift currents. In models, one therefore often assumes that extracellular diffusive currents give negligible contributions to the extracellular potential V_e. This assumption is probably acceptable for most cases, although there exist scenarios where it may be violated (see Chapter 12).

2.4.4   Other Currents

Generally, currents additional to those defined earlier may arise due to advection or due to magnetic induction. An advective current

$$\mathbf{i}_{\text{adv}} = \rho \mathbf{u} \, , \tag{2.28}$$

arises in a bulk solution if the solution has a bulk flow with velocity $\mathbf{u}$ and drags along a net charge density ρ. However, as we argued in Section 2.2.2, the saline solution in the brain is practically electroneutral ($\rho \simeq 0$), and the same holds for blood in blood vessels. Hence, the fluids in the brain do not have much of a net charge density to drag along with them. Convincing evidence that advective currents around cells are negligible (even compared to diffusion) was presented by Holter et al. (2017).

As we briefly touched upon earlier, electric currents may generally also be affected or evoked by magnetic fields. However, magnetic fields of physiological origin can generally be assumed to have negligible effects on the electrodynamics in the brain, and inductive currents can therefore be neglected. Arguments for this are presented in Section 2.7.

2.5   Electric Currents in the Brain

The main focus in this book is on modeling and interpreting the extracellular potential (V_e) in the brain. As we indicated earlier, an expression for the extracellular potentials originating from neural activity can be derived from the principle of current conservation. Let us therefore start this endeavor by defining the types of currents that are relevant in different components of brain tissue.

Having introduced different kinds of currents in Section 2.4, we can now summarize the role that they play within a brain-specific context. It is useful to make a distinction between currents existing in one of three different "domains":

- intracellular currents
- extracellular currents
- membrane currents

These are schematically illustrated in Figure 2.6A.

Among these three currents, the membrane currents are the odd ones out. A bit cartoonishly, we may think of the cellular membrane as an insulator with small holes. The insulating properties give the membrane the ability to separate charges in the conductive solution on the cellular outside from charges in the conductive solution on the cellular inside. Nonzero charge densities are stored on membrane surfaces and in so-called *Debye layers* on the inside and outside of the membrane, equivalent to how charges are stored on the metal plates in a parallel plate capacitor (Figure 2.5). Variations in the amount of stored charge cause the membrane potential to vary with time and can be described in terms of a capacitive current, as defined in equation (2.20).

The holes represent various kinds of ion channels, which are conductive pores in the membrane allowing ions to pass through. Ionic currents through ion channels come in addition to the capacitive current. Both the capacitive and ionic membrane currents are essentially one-dimensional and perpendicular to the membrane. The ionic currents depend both on voltage- and ion-concentration differences between the inside and outside of the membrane and are thus electrodiffusive in their nature (see equation (2.24)) but in one dimension, so that $\nabla \rightarrow d/dx$.

Concentration gradients within the intra- and extracellular spaces are typically much smaller than those across membranes. The currents in the intra- and extracellular saline solutions are therefore normally assumed to be purely conductive and thus described by equation (2.16).

As indicated in Figure 2.6A, both the intra- and extracellular currents are in reality volume currents in three-dimensional space. However, in computational models of neurons, the morphologies are typically described as branching cables of varying diameters, and the intracellular currents are assumed to run only in the longitudinal cable direction, having no radial components (Figure 2.6B). In practice, they are therefore computed using the 1D version of Ohm's law (equation (2.15)).

In contrast, tissue currents remain 3D volume currents. In reality, these will be currents that zig-zag their way through the mishmash of entangled neurites that make up neural tissue. However, for practical purposes, we typically describe tissue currents

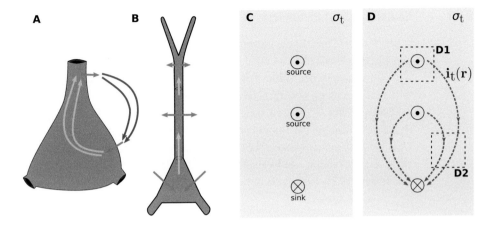

Figure 2.6 Current-source-density description. **A**: Closed current loops, involving intracellular volume currents (yellow), extracellular volume currents (red), and membrane currents (orange). The entire space will be filled with currents, and only a few paths are included in the illustration. **B**: Neuron approximated as a branching cable: intracellular currents run only longitudinally along the cable, and membrane currents are perpendicular to the cable. **C**: Membrane currents represented as a set of sources and sinks in a 3D tissue continuum. **D**: The sources/sinks give rise to tissue-volume currents $\mathbf{i}_t = -\sigma_t \nabla V_e$ in the tissue (illustrated with red, dashed arrows). If a tissue subvolume contains a neural source, a net tissue (volume) current equal to the source will leave this volume (D1). If a tissue subvolume does *not* contain a neural source, tissue currents entering and leaving this volume will sum to zero (D2).

on the coarse-grained tissue scale described in Section 2.3.2. We will denote tissue currents by

$$\mathbf{i}_t = -\sigma_t \nabla V_e \, , \tag{2.29}$$

where V_e is the extracellular potential, and we have used the index "t" to denote that $\mathbf{i}_t$ and σ_t represent the *tissue*-current density and conductivity, respectively.

2.6 Extracellular Potentials in the Brain

As indicated in Figure 2.6A, intracellular currents and currents in the surrounding tissue are coupled through the cellular membrane currents. Due to the principle of current conservation, the tissue currents (equation (2.29)), and thereby the extracellular potentials, can therefore be expressed as a function of the membrane currents.

   In Section 2.6.1, we formulate the standard theory for computing extracellular potentials, which is based on (1) describing neurons in terms of membrane current sources and (2) deriving an equation for the extracellular potentials resulting from these sources. A two-step framework for implementing this theory in computer simulations is outlined in Section 2.6.2, and some alternative modeling frameworks are briefly discussed in Sections 2.6.3 and 2.6.4.

2.6.1   Neurons as Current Sources

A set of neural membrane currents like that seen in Figure 2.6B can more generally be expressed in terms of a so-called *current-source density* (CSD). The CSD represents the membrane output per volume unit of the 3D tissue continuum (Figure 2.6C), and we will denote it by $C_t = C_t(\mathbf{r}, t)$ (with unit A/m^3).

The subscript "t" is used simply because the definition of the CSD is tightly linked to the definition of the tissue-current density $\mathbf{i}_t$ in equation (2.29). As we assume that C_t is the only source generating tissue currents in the surrounding tissue (Figure 2.6D), current conservation can be expressed mathematically through the continuity equation (Nicholson 1973, Nicholson & Freeman 1975)

$$\nabla \cdot \mathbf{i}_t = C_t \,. \tag{2.30}$$

Although the derivation of equation (2.30) is not trivial (see Section 4.3.2), it intuitively makes sense. It tells us that in a volume of space where there are no membrane sources ($C_t = 0$), the spatial rate of change of the tissue current ($\nabla \cdot \mathbf{i}_t$) must also be zero, so that no net tissue current will be entering or leaving such a volume.[2] In a volume that *does* receive membrane currents, equation (2.30) tells us that the current outputted from neurons into that volume (C_t) must leave the same volume in the form of a tissue current. As we will show in Chapter 4, this conservation law is the foundation for modeling extracellular potentials arising from neural activity.

Throughout most parts of this book, we assume that the tissue currents are given by Ohm's law (equation (2.29)). By that, we assume that brain tissue is a *linear* ohmic conductor, an assumption that we discuss in further detail in Chapter 5. When equation (2.29) is inserted into equation (2.30), we obtain a relationship between the neuronal current sources and the extracellular potential:

$$\nabla \cdot (\sigma_t \nabla V_e) = -C_t \,. \tag{2.31}$$

Generally, σ_t can be a position-dependent tensor, meaning that it can vary with position and be different in different spatial directions. If σ_t is the same in all directions and does not vary across space, equation (2.31) simplifies to the often-used form

$$\sigma_t \nabla^2 V_e = -C_t \,. \tag{2.32}$$

We note that although equation (2.32) resembles Poisson's equation,

$$\epsilon \nabla^2 V = -\rho \,, \tag{2.33}$$

from electrostatics, it represents an entirely different physical situation. While Poisson's equation from electrostatics determines the potential resulting from a fixed charge distribution in a dielectric medium, equation (2.32) determines the potential resulting from applied current sources in a conductive medium.

It may seem contradictory that V results from a current source in equation (2.32), when V according to equation (2.33) should follow from a charge distribution. The

---

[2] This does not mean that the tissue current itself must be zero; currents can still run *through* the volume.

response to this contradiction is that there indeed are charges also present when using equation (2.32), but they are implicit as a boundary condition accounted for by the source term. The sources are – for example, via capacitive membrane currents – responsible for setting up the charge distribution consistent with the electric potential. However, this charge distribution is not modeled explicitly, as the potential is instead computed from combining the principle of current conservation with Ohm's law for the (electroneutral) tissue medium embedding the sources.

Relating this to Figure 2.6D, the extracellular potential profile $V_e(\mathbf{r})$ computed with equations (2.31) or (2.32) can be interpreted as the profile that "is needed" in order to ensure current conservation in the tissue for a given source distribution $C_t(\mathbf{r})$. In general, $C_t(\mathbf{r})$ may be a continuous variable but was just three point sources in the illustration.

2.6.2 Two-Step (MC+VC) Scheme for Modeling Extracellular Potentials

To compute extracellular potentials using equations (2.31) or (2.32), we need to know the neuronal output, the current-source density C_t. Generally, the membrane currents in neurons depend on the membrane potential (V_m), defined as the difference between the intracellular (V_i) and the extracellular (V_e) potentials immediately inside and outside the membrane. Hence, membrane currents depend on extracellular potentials, while the extracellular potential in turn depends on the membrane currents of all nearby neurons.

Simulating all the coupled intra- and extracellular variables simultaneously in a self-consistent manner is computationally challenging. It is therefore common to uncouple them by making the simplifying assumption that the neurodynamics are independent of the dynamics of the extracellular potential. Since the range of the membrane potential ($\approx 100\,\mathrm{mV}$) tends to be much bigger than the range of the extracellular potential ($\approx 1\,\mathrm{mV}$), this is expectedly a reasonable approximation under many conditions (see Section 2.6.3 for a further discussion). One may then take a two-step approach to model the cellular and extracellular dynamics:

- Step 1: Compute the cellular dynamics (the intracellular currents, membrane currents, and membrane potential) using a separate framework (Figure 2.6B) assuming that these variables are independent of the effect that they have on extracellular potentials (and ion concentrations).
- Step 2: Use the membrane currents computed in Step 1 as "external" input currents (sinks and sources) C_t to the extracellular space (Figure 2.6C) and use equation (2.31) to compute $V_e(\mathbf{r})$ from the precomputed C_t (Figure 2.6D).

The standard framework for completing step 1 is presented in Chapter 3, and we will refer to it as the multicompartment (MC) framework, as it is based on the multicompartment modeling of neurons. The standard framework for completing step 2 is the volume-conductor (VC) theory presented in Chapter 4. Jointly, we will refer to the two-step scheme as the *MC+VC scheme*. The MC+VC scheme is the standard scheme for computing extracellular potentials and will be the default scheme used throughout this book. For completeness, a brief overview of alternative and more physically detailed schemes is given in Sections 2.6.3 and 2.6.4.

2.6.3   Alternative Schemes for Modeling Extracellular Potentials

There exist more comprehensive schemes for computing extracellular potentials than the MC+VC scheme, and an overview of alternative schemes is given in Figure 2.7. The figure includes four classes of schemes, distinguished in terms of (i) whether they couple the intracellular and extracellular dynamics and (ii) whether they are electrodiffusive. Schemes that do not fulfill (i) cannot account for so-called *ephaptic* effects – that is, the

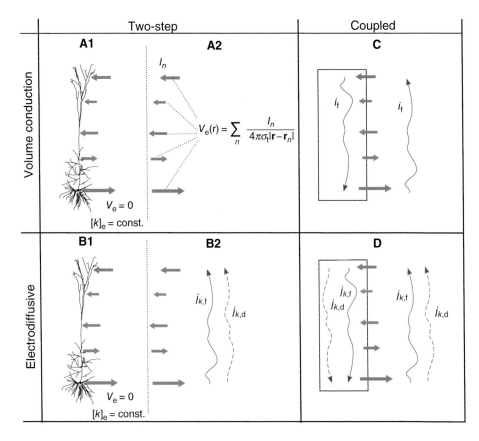

Figure 2.7 Overview of schemes for computing extracellular dynamics. The extracellular potential largely originates from neuronal membrane currents (orange arrows). **A**: Two-step MC+VC scheme. **A1**: Step 1 computes membrane neural currents in an independent simulation using multicompartment models (MC). **A2**: Step 2 uses the membrane output to compute the extracellular potential based on volume-conductor (VC) theory. **B**: Two-step MC+ED scheme. **B1**: Step 1 computes membrane currents and ion fluxes in an independent simulation based on MC models. **B2**: Step 2 uses the membrane output to compute the dynamics of extracellular ion concentrations and potentials using an electrodiffusive (ED) framework. **C**: All-VC scheme. Computes intracellular and extracellular electrodynamics simultaneously using a unified VC framework. Accounts for ephaptic effects but assumes ion concentrations to be constant. **D**: All-ED scheme. Computes intracellular and extracellular ion concentration and electrodynamics simultaneously using a unified electrodiffusive framework based on the Nernst-Planck equation (2.21). Accounts for ephaptic effects and ion-concentration dynamics.

non-synaptic effect that a neuron can have on its neighbors by causing changes in their shared extracellular environment. Schemes that do not fulfill (ii) cannot account for the effects that ion-concentration dynamics may have on the extracellular potential.

The (default) MC+VC scheme (Figure 2.7A) ticks neither of the (i) or (ii) boxes. It (i) does not couple the intra- and extracellular dynamics but computes the neurodynamics (step 1) under the assumption that it is independent of the resulting extracellular potential. In practice, this independence is normally implemented simply by setting the extracellular potential to be constant (typically zero) when the neurodynamics is computed. However, it is possible to combine the MC+VC scheme with a temporally varying extracellular potential, provided that it is imposed as an external input. This is, for example, relevant if one wishes to simulate extracellular electric stimulation (Appendix C). Also, the MC+VC scheme (ii) does not account for ion-concentration dynamics, since VC theory (presented in Chapter 4) assumes that all extracellular currents are ohmic drift currents.

The MC+ED scheme (Figure 2.7B), where "ED" stands for electrodiffusion, is a two-step scheme similar to the MC+VC scheme. Like in the MC+VC scheme, the neurodynamics is computed in a separate step 1 under the assumption of constant extracellular conditions. The MC+ED scheme differs from MC+VC in two main ways. Firstly, although the neural output (step 1) can be computed with the same type of multicompartmental neuron models as in the MC+VC scheme, the MC+ED scheme requires that all the neural output is made ion-specific. That is, the membrane neural currents must be converted to ion-specific fluxes (of Na^+, K^+, Cl^-, etc.), which in step 2 become the "input" sources to the extracellular medium. Secondly, the extracellular dynamics of ion concentrations and the resulting extracellular potential are computed in step 2 (the ED part) using an electrodiffusive framework, which accounts for the interplay between extracellular ion-concentration dynamics and extracellular potentials. A few implementations of this scheme exist (Halnes et al. 2016, Solbrå et al. 2018), and an electrodiffusive framework for use of this scheme is presented in some further details in Section 12.4 and Appendix I.

In the all-VC scheme (Figure 2.7C), the intra- and extracellular dynamics are modeled simultaneously on a unified framework based on volume conduction. Current conservation in the intra- and extracellular spaces is assured through suitable boundary conditions at the cellular membranes, constrained so that the current *entering* the membrane on one side is identical to the current *leaving* the membrane on the opposite side (Krassowska & Neu 1994). A few implementations of the all-VC scheme exist (Krassowska & Neu 1994, Agudelo-Toro & Neef 2013, Tveito et al. 2017) and has, in recent publications, been referred to as the extracellular-membrane-intracellular (EMI) framework (Tveito et al. 2017, Tveito et al. 2021). The all-VC scheme does not account for diffusive currents in the extra- and intracellular spaces.

The more comprehensive all-ED scheme (Figure 2.7D) models electrodiffusive processes both in the intra- and extracellular spaces using a unified framework, and it couples them through suitable boundary conditions at the cellular membrane. The all-ED scheme has been implemented in quite a variety of different versions, varying in terms of (i) the strategy chosen for solving the Nernst-Planck equation (equation (2.21)) for electrodiffusion, (ii) how boundary conditions are specified on cellular membranes, (iii) spatiotemporal resolution, and (iv) whether cells are modeled in a spatially explicit way. The topic of electrodiffusive modeling in neuroscience is revisited in Chapter 12, where we focus on models constructed to explain features of (macroscopic) extracellular potentials. More physically detailed models, tailored to describe microscopic electrodiffusive phenomena inside or near cell membranes will not be covered in this book, but see Savtchenko et al. (2017) or Jasielec (2021) for recent reviews of applications of electrodiffusive models in neuroscience.

Importantly, the fundamental equation in the MC+VC scheme is the CSD equation (2.31). As we will show in Chapter 4, this equation allows us to derive an analytical expression for the extracellular potential at all points in space once the neural current-source density C_t is known. In comparison, the alternative schemes are less "user friendly," as they all require that the extracellular variables are computed using numerical methods such as, for example, finite-element methods (Logg et al. 2012). This makes the alternative frameworks far more computationally demanding than the MC+VC framework.

As discussed further in Section 13.2, the aforementioned ephaptic effects are expected to be small under most conditions. Due to its computational simplicity (compared to the alternative schemes), the MC+VC scheme therefore remains the gold standard for computing extracellular electric signals generated by neurons and neural networks. Also, designated software such as our own LFPy (Lindén et al. 2014, Hagen et al. 2018) makes it easy to perform simulations using the two-step procedure.

Most of the simulations of extracellular potentials included in this book will be based on the MC+VC scheme. We note that, when using this scheme, the neglect of ephaptic effects is a concern only for the MC part (step 1) and not for the mapping from neural membrane currents to extracellular potentials (step 2: the VC part). This mapping does not in itself depend on whether ephaptic effects were accounted for when computing the neural dynamics. Put differently, ephaptic effects concern how the current-source density C_t in equation (2.31) is computed, but they do not affect the mapping from C_t to V_e.

2.6.4   Bi- and Tri-domain Models

A class of models that we did not include among the alternative frameworks earlier are the domain-type models, inspired by the bi-domain model (Eisenberg & Johnson 1970) and its successful applications to simulations of cardiac tissue (Henriquez 1993, Sundnes et al. 2006, Mori et al. 2008). For brain tissue, the most advanced models belonging to this category are the electrodiffusive tri-domain models (O'Connell & Mori 2016, Tuttle et al. 2019, Ellingsrud et al. 2022), composed of three domains representing (i) neurons, (ii) extracellular space, and (iii) a syncytium of gap-junction coupled glial cells.

Domain-type models do not account for any aspects of cellular geometries. Instead, they describe all constituents of tissue as a coarse-grained continuum where a set of domain-specific variables (domain-voltage, domain-ion concentrations, domain-volume fractions, etc.) and the interactions between them (membrane interactions) are defined at each point in space. This simplification allows for large-scale simulations of signals that propagate through tissue, such as waves of K^+, slow (DC-like) diffusion potentials, and glial buffering potentials during the pathological condition of spreading depression (Mori 2015, O'Connell & Mori 2016, Tuttle et al. 2019, Ellingsrud et al. 2022). Domain models are, however, not suited to model the faster fluctuations of extracellular potentials reflecting neurodynamics and synaptic integration, as these depend strongly on the morphologies of neurons (Einevoll, Kayser, et al. 2013).

2.7 Foundations of Electromagnetism

In classical physics, a complete description of the spatiotemporal development of electric and magnetic fields is given by the following physical laws:

- Maxwell's equations, which describe how electric charges and currents affect electric fields $\mathbf{E}$ (with unit V/m) and magnetic fields $\mathbf{B}$ (with unit tesla (T)).
- The Lorentz force law, which describes how electric and magnetic fields can accelerate charges (via Newton's law $\mathbf{F_q} = m_q \mathbf{a_q}$).

We end this basic theory chapter by briefly presenting Maxwell's equations (Section 2.7.1) and the Lorentz force (Section 2.7.3). Along the way, we introduce and discuss the main approximations that we make when we apply these physical laws in studies of brain tissue.

2.7.1 Maxwell's Equations

Electric and magnetic fields in biological tissue are well described by Maxwell's macroscopic equations for a linear medium, which are

$$\nabla \cdot \mathbf{E} = \frac{\rho_{\text{free}}}{\epsilon} , \tag{2.34}$$

$$\nabla \cdot \mathbf{B} = 0 , \tag{2.35}$$

$$\nabla \times \mathbf{E} = -\frac{\partial \mathbf{B}}{\partial t} , \tag{2.36}$$

$$\nabla \times \mathbf{B} = \mu \mathbf{i}_{\text{free}} + \mu \epsilon \frac{\partial \mathbf{E}}{\partial t} . \tag{2.37}$$

What we mean by "linear medium" in this context is explained in Section 2.7.1.2.

Equation (2.34) is Gauss's law (for electricity), with ϵ (with unit farad per meter (F/m)) being the electric permittivity of the medium and ρ_{free} being the free charge density. The law states that the divergence ($\nabla \cdot \mathbf{E}$) of the electric field is proportional to the local ρ_{free}. Integrating this over a closed surface, we get that the electric flux across this surface is proportional to the net charge enclosed by it. In general, ρ_{free} can be nonzero, although we argued earlier that brain tissue is practically electroneutral at the coarse-grained scale.

Equation (2.35) is Gauss's law for magnetism, and it is the magnetic equivalent to equation (2.34). Whereas equation (2.34) allows a nonzero divergence in the electric field due to a local free charge density, the divergence of the magnetic field must be zero because magnetic monopoles do not exist.

Equation (2.36) is the Maxwell-Faraday equation for electromagnetic induction. The equation tells us that a temporal variation in the magnetic field will generate an electric field perpendicular to the direction of the magnetic field variation (perpendicular because that is what the curl-operation ($\nabla \times$) tells us).

Equation (2.37) is Ampère's circuital law, with μ (with unit henries per meter (H/m)) being the magnetic permeability of the medium. According to equation (2.37), which is the magnetic equivalent to equation (2.36), a magnetic field can be induced either by an electric current of freely moving charges (first term on the right) or by a temporal variation in the electric field (last term on the right). The latter term does not represent an electric current of moving charges, but it has the same unit as a current and was called the *displacement current* by Maxwell. Displacement currents can exist in a vacuum, where they make up the electric component of electromagnetic radiation. In physical materials, displacement currents can contain additional contributions from the slight motion of charges bound in atoms, called dielectric polarization. The ohmic current introduced in equation (2.17) is an example of a current of free charges. The capacitive membrane current introduced in equation (2.20) is a 1D example of a displacement current.[3]

2.7.1.1   Current Conservation

The principle of current and charge conservation follows directly from Maxwell's macroscopic equations. To see this, we start by taking the divergence ($\nabla \cdot$) of both sides of equation (2.37). Since the divergence of a cross product is always zero, equation (2.37) then reduces to

$$\nabla \cdot \left(\mathbf{i}_{\text{free}} + \epsilon \frac{\partial \mathbf{E}}{\partial t} \right) = 0 . \tag{2.38}$$

This equation states that the total (free plus displacement) current density

$$\mathbf{i}_{\text{tot}} = \mathbf{i}_{\text{free}} + \epsilon \frac{\partial \mathbf{E}}{\partial t} \tag{2.39}$$

into any volume of space must be zero. To see the charge conservation explicitly, we write equation (2.38) as

$$\nabla \cdot \mathbf{i}_{\text{free}} = -\epsilon \frac{\partial (\nabla \cdot \mathbf{E})}{\partial t} \tag{2.40}$$

and insert equation (2.34) for $\nabla \cdot \mathbf{E}$ to obtain

$$\nabla \cdot \mathbf{i}_{\text{free}} = -\frac{\partial \rho_{\text{free}}}{\partial t} . \tag{2.41}$$

---

[3] "Capacitive membrane current" is the standard terminology in neuroscience and is a manifestation of the displacement current over the capacitive membrane.

Hence, if the left-hand side is nonzero, there is a net influx of free charges into a volume, and if that is the case, there must be an accumulation of charge there, as described by the right-hand side of the equation. This equation thus reveals the equivalence between the displacement current (last term on the right-hand side of equation (2.37)) and a local accumulation of free charge (right-hand side of equation (2.41)).

2.7.1.2 Constitutive Relations

A more general version of Maxwell's macroscopic equations contain the auxiliary displacement and magnetizing fields, $\mathbf{D}$ and $\mathbf{H}$, which include the effects of polarization and magnetization, respectively. As we will not encounter $\mathbf{D}$ and $\mathbf{H}$ later in this book, we do not define them in further detail here. However, we note that in order to solve the Maxwell equations, one must generally use constitutive relations that specify how $\mathbf{D}$ and $\mathbf{H}$ are related to $\mathbf{E}$ and $\mathbf{B}$. Otherwise, the system of equations will be underdetermined.

In equations (2.34)–(2.37), it is implicitly assumed that the medium is linear in the sense that we have used the linear constitutive relations $\mathbf{B} = \mu\mathbf{H}$ and $\mathbf{D} = \epsilon\mathbf{E}$. By this, we implicitly assume that polarizations and magnetizations are proportional to, and have the same direction as, $\mathbf{E}$ and $\mathbf{B}$, respectively.

The approximation that $\mathbf{D} = \epsilon\mathbf{E}$ is excellent for most materials. It holds as long as the external field is much weaker than the interatomic electric fields, which are of order 10^{11} V/m. This criterion requires very extreme conditions to be violated. In contrast, the approximation that $\mathbf{B} = \mu\mathbf{H}$ can break down in common magnetic materials. However, magnetization effects can be ignored in biological tissue, as its magnetic permeability is approximately equal to the vacuum permeability $\mu \approx \mu_0$ (Hämäläinen et al. 1993, Grodzinsky 2011). Hence, both linear constitutive relations are excellent approximations for brain tissue, and equations (2.34)–(2.37) apply.

Note that while μ is constant in brain tissue, ϵ may in general vary with space and depend on field frequency (more on this in Chapter 5).

2.7.2 Quasi-static Approximations of Maxwell's Equations

Maxwell's equations lay the theoretical fundament for understanding electromagnetic radiation, such as light. We can "see the light" by examining equation (2.36) and equation (2.37). Equation (2.36) tells us that a temporal change in the magnetic field induces an electric field perpendicular to the change, while equation (2.37) tells us that a temporal change in the electric field (last term on the right-hand side) induces a magnetic field perpendicular to the change. Electromagnetic radiation is due to such an interplay, where electric and magnetic fields act on each other through a periodic series of inductions and cancellations that propagate as a wave through a medium or free space (vacuum).

Although the occurrence of a bright idea is often illustrated in cartoons in the form of a light bulb appearing outside the protagonist's head, the generation of light – or other forms of electromagnetic waves – is in reality not a prominent feature of brain activity. For most brain phenomena, we are justified in using the so-called *quasi-static* approximations of Maxwell's equations. The term "quasi-static" is often used in a general fashion to refer to models of systems that do not generate electromagnetic waves. However, to be more precise, there are two different quasi-static approximations, the *quasi-electrostatic* and *quasi-magnetostatic*, and it is possible for one of them to be justified even when the other is not.

2.7.2.1   Quasi-electrostatic Approximation

The quasi-electrostatic approximation is that, in the calculation of $\mathbf{E}$, the source term associated with temporal changes in the magnetic field can be neglected. With $\partial \mathbf{B}/\partial t \approx 0$, equation (2.36) reduces to

$$\nabla \times \mathbf{E} = 0 . \tag{2.42}$$

This greatly simplifies the derivation of formulas for electric potentials in neural tissue and is a necessary criterion for expressing $\mathbf{E}$ as a gradient of a potential, like we did in equation (2.6). We used this approximation further in the CSD equation (equation (2.31)), where we assumed that volume currents in brain tissue depend solely on the gradients in the electric potential (and the conductivity). Thus, our theoretical foundation for computing extracellular potentials rests on the quasi-electrostatic approximation.

The quasi-electrostatic approximation is an excellent approximation for fields of physiological origin, which are normally of frequencies below a few thousand hertz (Plonsey & Heppner 1967, Hämäläinen et al. 1993, Bossetti et al. 2008, Gratiy et al. 2017). In this frequency range, neglecting the $\partial \mathbf{B}/\partial t$ term has been estimated to give a relative prediction error of only 10^{-7} for electric fields in neural tissue (Gratiy et al. 2017). The fulfillment of the quasi-electrostatic approximation alone is enough to exclude the mutual interplay between electric and magnetic fields that makes up electromagnetic radiation.

We note that the quasi-electrostatic approximation does not hold in the case of the strong and rapidly oscillating fields applied during transcranial magnetic stimulation (Roth & Basser 1990, Nagarajan & Durand 1996). After all, that is why transcranial magnetic stimulation has an effect.

2.7.2.2   Quasi-magnetostatic Approximation

The quasi-magnetostatic approximation is that, in the calculation of $\mathbf{B}$, the source term associated with temporal changes in the electric field can be neglected, so that equation (2.37) reduces to

$$\nabla \times \mathbf{B} = \mu \mathbf{i}_{\text{free}} . \tag{2.43}$$

The quasi-magnetostatic approximation is used in Chapter 11, where we neglect displacement currents $\epsilon \partial \mathbf{E}/\partial t$ when we compute magnetic fields.

As a related topic, we will in most parts of this book (with a few exceptions) also compute extracellular potentials under the assumption that tissue-volume currents are devoid of displacement currents. This assumption is implicit when we model tissue currents using Ohm's law $\mathbf{i}_t = -\sigma_t \nabla V_e$ with a real valued σ_t (equation (2.29)). By that, we assume a purely resistive medium, meaning that we must have $\mathbf{i}_{\text{free}} \approx -\sigma_t \nabla V_e$

and $\|\sigma_t \nabla V_e\| \gg \|\epsilon \partial \mathbf{E}/\partial t\|$ (Plonsey & Heppner 1967, Hämäläinen et al. 1993, Nunez & Srinivasan 2006) in the total current expressed in equation (2.39).

The question of whether a purely resistive brain tissue is a good approximation can be addressed without reference to magnetic fields or the Maxwell equations. It is ultimately a matter of determining (experimentally) whether Ohm's law applies. We revisit this topic in Section 5.4.1, where we review previous studies and provide support for the notion that brain tissue is predominantly resistive for physiological signals. We note, however, that the CSD equation (equation (2.31)) can be generalized to account for linear capacitive (displacement) effects by introducing a complex tissue conductivity $\hat{\sigma}_t$ (see Section 5.4.1).

Before moving on to other topics, we emphasize that displacement currents in the brain cannot be neglected altogether. The capacitive membrane currents make up an important fraction of the total membrane currents, and they are responsible for the temporal development of the membrane potential. Displacement currents may also become important on the tissue scale in the case of externally applied electric stimuli with fast-frequency oscillations (Bossetti et al. 2008).

2.7.3 The Lorentz Force

There are generally two ways through which magnetic fields can interact with electric currents. Firstly, a temporally varying magnetic field ($\partial \mathbf{B}/\partial t \neq 0$ in equation (2.36)) will generate an electric field ($\mathbf{E}$) and thus change the driving force for ohmic currents. However, for fields of physiological origin, such inductive effects can be neglected. This is what we do when making the quasi-electrostatic approximation (discussed earlier).

Secondly, even a static magnetic field ($\mathbf{B}$) can accelerate an electric charge via the Lorentz force,

$$\mathbf{F}_q = q(\mathbf{E} + \mathbf{u}_q \times \mathbf{B}), \tag{2.44}$$

which describes the collective electric and magnetic forces on a particle with charge q and velocity $\mathbf{u}_q$. Note that this equation is a generalization of equation (2.4) from the beginning of this chapter. The magnetic force is given by the cross product of the vectors $\mathbf{u}_q$ and $\mathbf{B}$, which means that it will act in the direction perpendicular to both of them.

The CSD equation (equation (2.31)) was derived under the assumption (stated in Section 2.4.4) that magnetic fields do not affect electric currents in brain tissue. In the beginning of this chapter (equation (2.4)), we thus excluded the magnetic component in the Lorentz force, assuming it to be negligibly small compared to the electric component. Here, we provide a justification for this assumption.

It follows from equation (2.44) that the magnetic component of the Lorentz force can be neglected if

$$\frac{uB}{E} \ll 1. \tag{2.45}$$

We are now thinking at the "tissue scale," so that u represents the averaged movement of charged particles in some preferred direction, whereas E and B represent typical electric and magnetic field strengths in coarse-grained brain tissue.

Geomagnetic fields near the surface of the earth ($\approx 50\,\mu\text{T}$) are about 10^8–10^9 times as strong as neuromagnetic signals ($\approx 100\,\text{fT}$ (Hämäläinen et al. 1993)). As the magnetic permeability (μ) of brain tissue is about the same as in free space (Hämäläinen et al. 1993), the magnetic field inside the brain will therefore be roughly the same as in the external world. In contrast, the electric conductivity of brain tissue ($\approx 0.3\,\text{S/m}$) is about 10^{11} times higher than for air. From the general requirement that the electric current density $\mathbf{i} = \sigma\mathbf{E}$ should be continuous on surfaces, it therefore follows that the atmospheric electric fields on the surface of the earth ($\approx 100\,\text{V/m}$) will be a factor 10^{11} lower in brain tissue than in air, giving a field contribution of approximately $10^{-9}\,\text{V/m}$ in the brain, which is much smaller than the fields of neuronal origin ($\approx 1\,\text{V/m}$ (Cordingley & Somjen 1978)).

Inserting the typical value of the endogenous electric field ($E = 1\,\text{V/m}$) and the external geomagnetic field ($B = 50\,\mu\text{T}$) into equation (2.45), we find that magnetic effects on the movement of charges are negligible provided that

$$u \ll 2 \times 10^4\,\text{m/s}. \tag{2.46}$$

As we explained in Section 2.4.1.1, the drift velocity of ions due to volume currents in brain tissue is on the order of nanometers per second, which definitely fulfills the criterion in equation (2.46). The maximal bulk flow of ions in the brain is therefore probably due to blood flow. Although blood is electroneutral and does not carry any net charge, it does carry positive and negative ions moving in a particular direction, and a magnetic field could in principle bend off oppositely charged ions in opposite directions. However, the maximal flow velocity in arteries has been estimated to be approximately 1 m/s (Bishop et al. 1986), which is by far slow enough for such magnetic effects to be safely neglected.

In fact, even the thermal velocities ($u \approx (k_B T/m)^{1/2} \approx 100$–$1000\,\text{m/s}$) of ions in a body-tempered brain would be too slow to violate the criterion in equation (2.46). We note, however, that the thermal velocity is not really a candidate for insertion into equation (2.46), since it represents the average absolute velocity of randomly walking particles, and therefore it does not give rise to any net movement of charges on the system level since magnetic forces acting on particles moving in one direction will be canceled by forces acting on particles moving in the opposite direction.

Hence, as we have assumed throughout this chapter, we can conclude that magnetic effects on the movement of charges in brain tissue are negligible. The reason why this is important is that $\mathbf{i}_{\text{free}}$ in equation (2.38) then becomes independent of $\mathbf{B}$, which we implicitly assumed when we used Ohm's law for volume conductors (equation (2.16)) to derive the CSD equation (equation (2.31)).

3 Neural Dynamics

Extracellular potentials measured in neural tissue are primarily generated by the electric activity of neurons, or more precisely, by neural membrane currents. To model extracellular potentials, we therefore first need to model neurons. Neural modeling is a core topic in computational neuroscience and has been covered in numerous textbooks in much greater detail than we intend to do here (Johnston & Wu 1994, Koch & Segev 1998, Koch 1999, Dayan & Abbott 2001, Hille 2001, Lytton 2002, Trappenberg 2002, Izhikevich 2007, De Schutter 2009, Ermentrout & Terman 2010, Gerstner et al. 2014, Miller 2018, Sterratt et al. 2023).

Extracellular potentials generally depend on the spatiotemporal distributions of currents over the somatodendritic membranes. We will often refer to these membrane currents as *sinks* and *sources*. As we explained in Section 2.6.1, a sink is a current that goes inward into the neuron (at a given location) and thus leaves the extracellular space, while a source is a current that goes outward from the neuron and thus enters the extracellular space. As current conservation (Section 2.4) requires that no net current can enter a neuron (as a whole), the sum of its source currents must equal the sum of its sink currents.

In large network simulations of neurons, it is common not to use spatially explicit neural models but to instead represent each individual neuron as a single compartment, meaning that all its dynamical variables are defined at a single point in space. Single compartment models, often called *point neurons*, can give accurate predictions of the times of action potentials (APs), but they do not generate extracellular potentials. The reason is that the sources and sinks in a point-model (by construction) are colocated in its one and only compartment, where current conservation requires that they cancel each other and amount to a zero net membrane current. When there is no membrane current anywhere, there is also no extracellular potential (Figure 3.1A).

To account for the spatial extension of neurons, we typically use *multicompartment* (MC) neuron models. Although the total sum of membrane currents is also zero in an MC model, the sources and sinks will generally be separated in space so that the membrane current is not zero locally. As depicted in Figure 3.1B–C, MC models therefore give rise to a spatial distribution of current sinks and sources.[1]

---

[1] The requirement that the membrane currents must sum to zero does not apply when the neuron receives input through a stimulus electrode or a gap junction synapse. These kinds of input are not membrane currents but are currents delivered into the intracellular space of the neurons (not crossing from the extracellular side of the membrane). In case of such inputs, the sum of membrane currents must equal the input current.

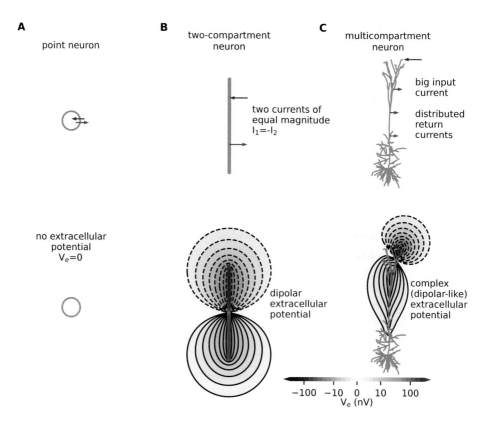

Figure 3.1 Extracellular potentials from neural activity. A: In point neurons, the membrane currents sum locally to zero (top), and they therefore do not generate extracellular potentials (bottom). **B**: In a two-compartment neuron, a current can enter one compartment while an identical current leaves the other compartment (top), causing a symmetric dipolar extracellular potential (bottom). **C**: In multicompartment models (here, the Hay model (Hay et al. 2011)), the spatial distribution of membrane currents (top) evokes complex extracellular potentials (bottom). Code available via www.cambridge.org/electricbrainsignals.

The most common MC modeling framework combines a so-called *Hodgkin-Huxley-type* description of the neural membrane mechanisms (presented in Section 3.1), with a cable-theory description of how currents spread spatially in dendrites and axons (presented in Section 3.2). This framework has become the gold standard for biophysically detailed neural simulations on the cellular and network level, and it has been used to simulate the dynamics of large neural networks (see e.g. Traub et al. (2005), Markram et al. (2015), Billeh et al. (2020)). We will here mainly focus on this standard framework, and we will refer to it simply as the MC framework.

An MC model is characterized by its morphology and its membrane mechanisms, and its key dynamical variable is the membrane potential V_m. As we explain further in Section 3.2.1, the morphology of the real neuron – which may look something like Figure 3.1C – is represented as a discretized set of compartments connected by resistors.

This structure houses two kinds of physiological currents: (i) currents that run intracellularly between compartments (along dendrites or axons) and (ii) the membrane currents in each compartment (the sinks or sources). Both kinds of currents depend (among other things) on the membrane potential in the various compartments, and together they determine how the membrane potential develops over time. In addition to the physiological currents, external stimuli (iii) can be delivered to the model, representing for example electrode current injections.

In the following, we first present a framework for modeling the membrane currents in a single compartment in Section 3.1, and then we go on to show how a number of such compartments can be connected together to create an MC model in Section 3.2. Together, these two sections provide a theoretical framework for modeling neurons that should be sufficient for most practical applications. The cable theory, presented in Section 3.3, allows us to derive analytical formulas for how electric signals spread in simplified (passive) neural dendrites. Although these formulas may not be accurate for real neurons, they still offer useful, qualitative insights into the spatiotemporal aspects of neural signaling. Readers that crave further biophysical insight into the ionic movements that mediate the membrane currents will get an introduction to this in Section 3.4. In Section 3.5, we give a short introduction on how one can construct neural network models at various levels of biophysical detail. Finally, Section 3.6 ends this chapter by summarizing what we bring along from the MC models when we set out to model extracellular potentials in chapters to follow.

3.1   Membrane Currents

A fundament for biophysics-based neural modeling was laid by Hodgkin and Huxley (1952a), who developed a mathematical model for the AP generation and propagation in the squid giant axon. The same kind of mathematical formalism has later been successfully used to model a diversity of membrane mechanisms in many different neuron types (see e.g. Koch (1999)), and models based on this type of formalism are often referred to as Hodgkin-Huxley (HH) type models.

In HH-type models, the membrane typically includes three classes of intrinsic membrane currents, normally represented as current densities (current per membrane area). These are (i) the capacitive current density i_c, (ii) the leakage current density i_L, and (iii) the current densities through active ion channels i_w, of which there are several different kinds (w indexes an arbitrary active ion channel). In addition, a neuron may receive stimuli, normally represented as total currents rather than current densities. These can, for example, be (iv) a synaptic input I_{syn} or (v) external current injections I_{ext} by an electrode.

The membrane currents (i–iii) depend on membrane properties and are often called distributed mechanisms, as they scale with the membrane area and are proportional to (i) the specific capacitance and (ii–iii) the specific conductance per membrane area. In contrast, the stimuli (iv–v) are often called point processes as they are delivered at a singular point on the membrane in the case of synapses, as well as at a single point within the cell (cytosol) in case of external stimulation currents.

If the neuron is modeled as a single compartment, the currents entering that compartment must sum to zero, so that

$$i_c + i_L + \sum_w i_w + \frac{I_{syn}}{A} - \frac{I_{ext}}{A} = 0, \tag{3.1}$$

where the point currents must be divided by the compartment area A. By convention, the first four of these currents are defined as positive when going outwards (from the intra- to the extracellular space). These are rightfully referred to as membrane (or transmembrane) currents as they go through the membrane from the intra- to the extracellular space. In contrast, the external stimulus is defined as positive when injected *into* the neuron. Also, it is conceptually different from the other currents in that it is injected with an electrode placed inside the cell. This is important to keep in mind for the later chapters of this book because it means that, unlike the membrane currents, the injected current does not (directly) evoke extracellular potentials because it does not originate from the extracellular space. This also means that when a current is injected into a neuron by an electrode, the membrane currents no longer sum to zero but instead sum to the injected current. Below, we define the various currents that go into equation (3.1).

3.1.1   Capacitive Current

The capacitive current density,

$$i_c = c_m \frac{dV_m}{dt}, \tag{3.2}$$

describes the charging of the membrane to a potential V_m due to a charge density accumulating on the outside and inside of the capacitive membrane. Here, c_m is the *specific membrane capacitance* defined as capacitance per membrane area. An illustration of how to interpret the capacitive current is given in Figure 3.2.

If we insert equation (3.2) into equation (3.1), we get

$$c_m \frac{dV_m}{dt} = -i_L - \sum_w i_w - \frac{I_{syn}}{A} + \frac{I_{ext}}{A}, \tag{3.3}$$

which gives us an intuitive understanding of neurodynamics: if the sum of ionic currents entering the neuron (right-hand side) is nonzero, it will lead to a charging (left-hand side) of the membrane.

In a study by Gentet et al. (2000), a value $c_m \approx 0.9\,\mu F/cm^2$ was found in several classes of neurons, suggesting that c_m to a good approximation may be treated as a "biological constant" of neural membranes. However, other studies have come up with other estimates, and the constancy of c_m is debated (see e.g. Eyal et al. (2016)). In the original HH model, c_m had the value $1\,\mu F/cm^2$, and this value – which is close to that found by Gentet et al. (2000) – is often used as a default value by modelers.

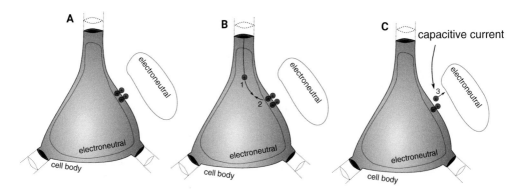

Figure 3.2 Capacitive currents ensure current conservation. A: The extracellular and intracellular bulk solutions are essentially electroneutral, and the only region where there is a nonzero charge density is in thin Debye layers on either side of the capacitive membrane. The membrane separates a surface charge density η on one side from a surface charge density $-\eta$ on the opposite side, giving rise to a membrane potential of $V_m = \eta/c_m$. This charge separation process can be described in terms of a capacitive current: a membrane current that is not due to ions crossing the membrane but is due to ions piling up on either side of it. **B, C**: An outward capacitive current could correspond to an anion (blue) leaving the membrane on the inside, which will coincide with a cation (red) leaving the membrane on the outside. Thus, the capacitive current ensures current conservation and gives rise to electric ionic volume currents both in the intra- and extracellular space.

3.1.2 Leakage Current

The leakage current density is given by

$$i_L = \overline{g}_L (V_m - E_L), \tag{3.4}$$

where $\overline{g}_L$ is the leak conductance, and the overline is used to indicate that it is a constant. The factor $(V_m - E_L)$ is often called the *driving force*, and E_L is the *leak reversal potential*. The biophysical origin of the reversal potential is explained later (see Section 3.4). For now, we may simply think of E_L as the "target potential" that the leakage current will strive to drive the membrane potential towards.

The neural membrane possesses an orchestra of mechanisms, including ion pumps, cotransporters, and a series of different active ion channels. Only a small subset of these mechanisms are typically included explicitly in a computational model. The reason may partly be that the modeler wants to keep things simple and partly that not all membrane mechanisms have been identified or been characterized experimentally in the first place. In reality, i_L does not represent a single, particular type of membrane mechanism but should rather be seen as a phenomenological way of summarizing the effect of all otherwise unaccounted-for membrane mechanisms.

Together, the capacitive current and the leakage current determine the passive properties of the membrane. If the neuron were to include only these two currents, it could be well modeled as a resistor-capacitor circuit (RC) circuit (Figure 3.3), and such RC-neuron models are often used to simulate the subthreshold dynamics of neurons (Lapicque 1907, Brunel & van Rossum 2007, Sterratt et al. 2023).

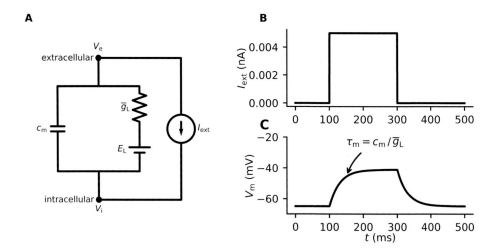

Figure 3.3 RC neuron and responses to step input current. A: A neuron model containing only a capacitive and a leakage current can be represented as an RC circuit with specific membrane resistivity $r_m = 1/\overline{g}_L$ and specific capacitance c_m. The RC model receives an electrode injection I_{ext} (converted to current density by division by membrane area A). **B,C:** A step-pulse injection charges the membrane, and when the input is terminated, the membrane potential returns to the value E_L. The membrane's time constant $\tau_m = r_m c_m$. Here, the single-compartment model is a cylinder with length 20 µm, diameter 10 µm, $c_m = 1\,\mu F/cm^2$, $\overline{g}_L = 33.3\,\mu S/cm^2$, and $E_L = -65\,mV$ so that $\tau_m \approx 30\,ms$. Code available via www.cambridge.org/electricbrainsignals.

In the RC model, E_L is the resting potential – that is, the potential that the membrane settles on when it does not receive any input. If active ion channels are added to the model, they may in principle have a nonzero activity at rest. They will then affect the resting potential so that it differs from E_L. We note that $\overline{g}_L$ represents the *specific conductance per unit area*, but we will simply refer to it as a *conductance*. In neuroscience applications, it is normally specified in units of millisiemens per square centimeter (mS/cm^2).

3.1.3   Active Ion Channels

In addition to i_c and i_L, biophysical neural models include a number of active ion channels. These account for the fancy aspects of neurodynamics, and a main legacy of Hodgkin and Huxley (HH) is the mathematical model they proposed for the kinetics of these (Hodgkin & Huxley 1952a). In the HH formalism, ionic membrane currents through active channels are modeled as

$$i_w = g_w(V_m - E_w),\tag{3.5}$$

where the conductance $g_w(V_m, t)$ is a function of the membrane potential and time, and E_w is the reversal potential of the current that travels through the channel type (see Section 3.4). In analogy with the leak reversal potential, we may think of E_w as the target potential that the current through ion channel w will strive to drive the membrane potential towards.

The mathematical model Hodgkin and Huxley proposed for the kinetics of the active conductances is

$$g_w = \bar{g}_w m_w^{\alpha_w} h_w^{\beta_w} . \tag{3.6}$$

Here, the constant $\bar{g}_w$ represents the conductance when all channels of type w are fully open, while the degree of openness varies with the so-called *gating variables*, m_w and h_w. The gating variables determine how the ion channels activate or deactivate (open or close), and m_w and h_w represent two types of gates differing in terms of their opening/closing kinetics. The exponents α_w and β_w represent the number of gates of each type within channel w.

At the level of a single ion channel, the openings or closings occur as stochastic events, and the channel will either be in the closed or open state (meaning that it cannot be partially open). The values of m_w and h_w then indicate the *probability* that a given gate is in the open state, and their values are numbers between 0 (definitely in the closed state) and 1 (definitely in the open state). For a channel to be open, all gates must be open, and the probability for this is given by the product $m_w^{\alpha_w} h_w^{\beta_w}$.

Importantly, when the HH formalism is applied on the cellular level, i_w in equation (3.5) with g_w as in equation (3.6) is not the current density through a single ion channel but rather the total current density through a large number of channels of the same type w. The probability that a given single gate (m_w or h_w) is open will then equal the *fraction* of gates of that type that is open. On the cellular level, the product $m_w^{\alpha_w} h_w^{\beta_w}$ thus corresponds to the fraction of ion channels in which all gates are open so that they can conduct current. Consequently, the product $\bar{g}_w m_w^{\alpha_w} h_w^{\beta_w}$ corresponds to the ion-channel conductance on the cellular level.

For voltage-gated ion channels, the dynamics of the gating variables can be described by kinetics equations on the form

$$\frac{\xi(V_m, t)}{dt} = \frac{\xi_\infty(V_m) - \xi}{\tau_\xi(V_m)}, \text{ for } \xi \in \{m, h\} . \tag{3.7}$$

Here, the steady-state activation $\xi_\infty(V_m)$ represents the fraction of gates that will end up in the open state if the cell is clamped at a given potential V_m for a sufficiently long time. However, the process of opening the gates takes time, as accounted for by the activation time constant $\tau_\xi(V_m)$. Both $\xi_\infty(V_m)$ and $\tau_\xi(V_m)$ are functions of the membrane potential. These are typically determined experimentally for each individual ion-channel type (e.g. as in Hodgkin & Huxley 1952a). We will think of $\xi_\infty(V_m)$ and $\tau_\xi(V_m)$ as known functions, and we will not go into the experimental challenges involved in determining them.

If one prefers to think of neurons in terms of circuit diagrams, current densities through active ion channels are simply added in parallel to the passive i_c and i_L current densities. As an example, the electric circuit representation of the original HH model is depicted in Figure 3.4, and the full set of equations for the HH model is summarized in

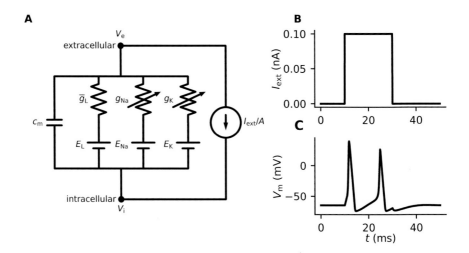

Figure 3.4 Hodgkin-Huxley model and response to step input current. A: In addition to a capacitive and a leakage current, active neuron models contain a number of active ion channels. The example diagram shows the original Hodgkin-Huxley model, with an active Na^+ and an active K^+ channel. The arrows through the active conductances indicate that they are variable. An electrode injection I_{ext} has been converted to current density by division by membrane area A. **B, C**: Model response to step input current. Here, the single-compartment neuron model is a cylinder with length 20 μm, diameter 10 μm, $c_m = 1 \mu F/cm^2$, and conductance values $(\bar{g}_L, \bar{g}_{Na}, \bar{g}_K)$ as summarized in Box 3.1. Code available via www.cambridge.org/electricbrainsignals.

Box 3.1. Compared to modern biophysically detailed neuron models, the original HH model is relatively simple in that it only contains two voltage-gated active ion channels: a Na^+ channel with three activation gates (m^3) and one inactivation gate (h) as well as a K^+ channel with four activation gates n^4. The two active ion channels are together responsible for AP generation.

Note that there exits ion channels whose activation or inactivation do not depend on V_m but instead on some other variable, such as the concentration of some ion species or ligand (Hille 2001, Sterratt et al. 2023). An HH-type formalism can in many cases be applied also to these kinds of ion channels, provided that $\xi_\infty(V_m)$ and $\tau_\xi(V_m)$ in equation (3.7) can be replaced with experimentally determined functions of the relevant variables. An important example would be the Ca^{2+}-activated ion channels, briefly introduced in Section 3.4.6.

Like all models, the HH formalism for ion-channel gating is based on a set of simplifying approximations. Firstly, the currents described by equation (3.5) are sometimes

---

Box 3.1 Hodgkin-Huxley Equations for Squid Axon at 6.3 °C.

$$c_\text{m}\frac{dV_\text{m}}{dt} = -\overline{g}_\text{L}(V_\text{m} - E_\text{L}) - \overline{g}_\text{Na}m^3h(V_\text{m} - E_\text{Na}) - \overline{g}_\text{K}n^4(V_\text{m} - E_\text{K})$$

$$\frac{d\xi(V_\text{m},t)}{dt} = \frac{\xi_\infty(V_\text{m}) - \xi}{\xi(V_\text{m})}, \text{ for } \xi \in \{m,h,n\}$$

$$\xi_\infty(V_\text{m}) = \frac{\alpha_\xi(V_\text{m})}{\alpha_\xi(V_\text{m}) + \beta_\xi(V_\text{m})}, \text{ for } \xi \in \{m,h,n\}$$

$$\tau_\xi(V_\text{m}) = \frac{1}{\alpha_\xi(V_\text{m}) + \beta_\xi(V_\text{m})}, \text{ for } \xi \in \{m,h,n\}$$

$$\alpha_n = 0.01\frac{V_\text{m} + 55}{1 - e^{-(V_\text{m}+55)/10}}$$

$$\beta_n = 0.125e^{-(V_\text{m}+65)/80}$$

$$\alpha_m = \frac{0.1V_\text{m} + 40}{1 - e^{-(V_\text{m}+40)/10}}$$

$$\beta_m = 4e^{-(V_\text{m}+65)/18}$$

$$\alpha_h = 0.07e^{-(V_\text{m}+65)/20}$$

$$\beta_h = \frac{1}{1 + e^{-(V_\text{m}+35))/10}}$$

$$c_\text{m} = 1.0\,\mu\text{F/cm}^2$$

$$\overline{g}_\text{Na} = 120 \text{ mS/cm}^2$$

$$\overline{g}_\text{K} = 36 \text{ mS/cm}^2$$

$$\overline{g}_\text{L} = 0.3 \text{ mS/cm}^2$$

$$E_\text{Na} = 50 \text{ mV}$$

$$E_\text{K} = -77 \text{ mV}$$

$$E_\text{L} = -54.4 \text{ mV}$$

Equations were taken from Sterratt et al. (2023). V_m must be inserted with unit mV. Time constants (τ_ξ) come out in units of ms. Rate constants (α's and β's) come out in units of ms^{-1}.

called *quasi-ohmic*, as they depend linearly on the driving force ($V_\text{m} - E_w$), when they in reality are nonlinear functions of both the ionic concentrations in the intra- and extracellular space and the membrane potential V_m (for more on this, see Section 3.4.2). Secondly, since the HH-type formalism is based on deterministic equations, it cannot account for ion-channel noise, which can stem from the stochastic gating at the single-channel level. Thirdly, when channels open and close, movement of charges in channel proteins gives rise to a tiny current called the *gating current*, which is not included in the HH formalism. Fourthly, the relatively simple gating kinetics described by HH-type channels is not accurate for all ion channels, and other ion-channel models have been developed based on a formalism more compatible with statistical physics and thermodynamics or based on highly detailed Markov-type formalisms based on single-channel recordings. More comprehensive overviews of alternative ion-channel models can be found in other textbooks (Koch 1999, Hille 2001, De Schutter 2009, Sterratt et al. 2023). Despite its limitations, the HH formalism (equations (3.5) and (3.6)) has become a standard choice when modeling neurodynamics on the cellular level.

3.1.4   Synapses

In an in vivo context, a neuron is typically driven by the synaptic input it receives from other neurons. In this section, introduce some synapse models that are commonly used in neural simulations.

3.1.4.1   Conductance-Based Synapses

A conductance-based chemical synapse can be modeled analogously to how ion channels are modeled:

$$I_{\text{syn}}(t) = G_{\text{syn}}(t)\big(V_{\text{m}}(t) - E_{\text{syn}}\big), \tag{3.8}$$

where $G_{\text{syn}}(t)$ is the synaptic conductance and E_{syn} is the reversal potential of the synapse.

Whereas ion channels are normally described as distributed mechanisms, giving rise to a current density (per membrane area), synapses are normally described as point processes: they exist at a singular location and give rise to a total input current at this particular location. The synaptic conductance is therefore denoted with a capital G (typically given in units of millisiemens (mS)) to distinguish it from the specific conductances per unit membrane area (g) of ion channels.

The reversal potential E_{syn} determines whether the synapse is *excitatory*, which means that it will depolarize the membrane potential of the postsynaptic neuron and bring it closer to the threshold for generating APs, or *inhibitory*, which means that it will hyperpolarize or oppose depolarization of the membrane potential, keeping it below the AP-firing threshold.

Examples of excitatory synapses are AMPA and NMDA synapses, in which E_{syn} ($\approx$ 0–10 mV) is high above the action-potential firing threshold. The most common inhibitory synapse is the GABA$_\text{A}$ synapse, in which E_{syn} ($\approx$ −70 to −60 mV) is close to the resting membrane potential of the neuron.

The synapse model described by equation (3.8) is said to be *conductance based* because the activation of the synapse is modeled as a transient change in the synaptic conductance. Unlike ion channels, which are often voltage- or calcium-activated, the chemical synapse is activated by neurotransmitters received from a presynaptic cell. However, in MC-type models, the transmitter-activation processes occurring in the synaptic cleft are normally not explicitly described. Instead, one normally uses a phenomenological synapse model, where the synaptic conductance $G_{\text{syn}}(t)$ is represented as a constant $\overline{g}_{\text{syn}}$ multiplied with a temporal kernel determining the opening and closing of postsynaptic neurotransmitter-gated ion channels associated with the synapse:

$$I_{\text{syn}}(t) = \overline{g}_{\text{syn}} f(t)\big(V_{\text{m}}(t) - E_{\text{syn}}\big), \tag{3.9}$$

where we have assumed that the synapse was activated at $t = 0$. Typical choices for $f(t)$ are

(i) exponential decay: $\qquad f(t) = e^{-t/\tau}\,\Theta(t)$ $\qquad\qquad$ (3.10)

(ii) α-function: $\qquad f(t) = \dfrac{t}{\tau}e^{1-t/\tau}\,\Theta(t)$ $\qquad\qquad$ (3.11)

(iii) β-function: $\qquad f(t) = \left(\dfrac{e^{-t/\tau_1} - e^{-t/\tau_2}}{e^{-\tau_{\text{peak}}/\tau_1} - e^{-\tau_{\text{peak}}/\tau_2}}\right)\Theta(t)$ $\qquad$ (3.12)

$$\text{with:}\quad \tau_{\text{peak}} = \frac{\tau_2\tau_1}{\tau_2 - \tau_1}\log\left(\frac{\tau_2}{\tau_1}\right),$$

where $\Theta(t)$ is the (Heaviside) unit step function $\Theta(t \geq 0) = 1$, $\Theta(t < 0) = 0$ and where the time constants τ, τ_1, and τ_2 are tuned to experimental data from the particular synapse type that one wants to model. The waveforms (i)–(iii) have been normalized so that their peak amplitude is 1, meaning that $\overline{g}_{\text{syn}}$ in equation (3.9) can be interpreted as the peak conductance of the synapse, analogous to how $\overline{g}_w$ is the peak conductance for fully open ion channels in Section 3.1.3.

The simple waveforms (i)–(iii) are typically used for AMPA and GABA$_\text{A}$ synapses. NMDA synapses have an additional dependency on membrane voltage and the concentration of extracellular Mg^{2+}, and they are commonly modeled as (Koch 1999, Sterratt et al. 2023)

$$I_{\text{NMDA}}(t) = \frac{\overline{g}_{\text{NMDA}}}{1 + \chi[\text{Mg}^{2+}]e^{-\zeta V_m}}\,f(t)\left(V_m(t) - E_{\text{syn}}\right),\qquad (3.13)$$

where χ and ζ are parameters that determine the Mg^{2+} and voltage dependencies, respectively, and that must be tuned to the data from the NMDA synapse being modeled. The temporal kernel $f(t)$ can in principle be any of the waveforms (i)–(iii) defined earlier, but it is commonly taken to be a β-function (Koch 1999, Sterratt et al. 2023). Due to the dependence on V_m and Mg^{2+}, the conductance of the fully open NMDA synapse is not $\overline{g}_{\text{NMDA}}$, but $\overline{g}_{\text{NMDA}}/(1 + \chi[\text{Mg}^{2+}]e^{-\zeta V_m})$.

The conductance $\overline{g}_{\text{syn}}$ (or $\overline{g}_{\text{NMDA}}$) determines the efficacy of the synaptic transmission and must be tuned by the modeler or determined experimentally to get the right peak amplitude of the synaptic current. We note that the synaptic strength can change over time through various processes that are often grouped together under the term "synaptic plasticity." Synaptic plasticity is involved in learning and memory formation in the brain. We will not go further into this topic here, but for more sophisticated synapse models including plasticity rules, see for example the book chapter by Roth & van Rossum (2009) or the review by Magee & Grienberger (2020).

3.1.4.2 Current-Based Synapses

Unlike conductance-based synapses (equation (3.8)), which depend on the driving force $(V_m(t) - E_{\text{syn}})$, current-based synapses assume that the synaptic current is independent of the state $(V_m(t))$ of the neuron. The relationship between $\overline{g}_{\text{syn}}$ and I_{syn} then becomes linear so that the synapse can be modeled as

$$I_{\text{syn}}(t) = \overline{I}_{\text{syn}}f(t).\qquad (3.14)$$

We can think of the current-based synapse as a conductance-based synapse where the driving force $(V_m(t) - E_{syn})$ has been approximated by a constant (cf. equation (3.9)) so that

$$\overline{I}_{syn} = \overline{g}_{syn}(\langle V_m(t)\rangle_t - E_{syn}),\qquad(3.15)$$

where $\langle V_m(t)\rangle_t$ denotes the temporal mean (Appendix E.1) of the membrane potential. Approximating the time-dependent membrane potential with the average membrane potential is expectedly most appropriate for excitatory synapses where the reversal potential E_{syn} of the synapse is close to 0 mV. With a membrane potential that most of the time varies between, say, –60 and –70 mV, the relative variation of the synaptic drive $V_m(t) - E_{syn}$ is then on the order of only 10 percent, and the error introduced when replacing $V_m(t)$ with $\langle V_m(t)\rangle_t$ in equation (3.15) is thus modest. For inhibitory synapses with E_{syn} close to the resting membrane potential of the neuron, the relative variation in the synaptic drive will be much greater, and the validity of the approximation is less clear.

The temporal kernel $f(t)$ can have the same exponential waveforms (i)–(iv) as for the conductance-based synapses. Current-based synapses have been popular in point-neuron network simulations in particular, as these simplified synapse models make such networks amenable to rigorous mathematical analysis (Brunel 2000, Deco et al. 2008, Tetzlaff et al. 2012, Helias et al. 2013, de Kamps 2013, Schuecker et al. 2015, Bos et al. 2016) and also computationally efficient (Brette et al. 2007, Plesser et al. 2007, Helias et al. 2012, Kunkel et al. 2014, Jordan et al. 2018). For point-neuron networks, effects of conductance-based synapses versus current-based synapses on network activity have been investigated (Richardson 2004, Cavallari et al. 2014). In the context of this book, current-based synapses are useful because they (given some additional approximations) can be used to obtain a linear relationship between the synaptic input to neurons and the extracellular potential (see Chapter 6).

3.1.4.3   Gap Junctions

Some neurons (and astrocytes) form electrical synapses called *gap junctions*, which are different from the chemical synapses discussed earlier. A gap junction is basically an electric coupling between a branch of one neuron (with membrane potential V_{m1}) and a branch of another neuron (with membrane potential V_{m2}). Gap junctions are formed by ion-permeable channel proteins that connect the cytosol of the two cells at the junction.

The current through the gap junction is given by the synaptic conductance multiplied by the voltage difference across the junction,

$$I_{gap} = G_{gap}(V_{m2} - V_{m1}).\qquad(3.16)$$

3.1.5   External Current Injections

In experimental patch clamp recordings, neurons receive electrode injections, often given as step-current injections, ramp-current injections, or sinusoidally varying current injections. In MC models, an electrode injection is (like synaptic input) usually described as a point process – that is, as a total current injected at a singular location or, in practice, into a single compartment of the model. In this way, it differs from the capacitive- and ion-channel-current densities, which are referred to as distributed mechanisms as they scale with the membrane area of the compartment.

While capacitive currents, ion-channel currents, or currents through chemical synapses represent charge transfers from the immediate outside to the immediate inside of the membrane, electrode injections are injected directly into the intracellular space and should not be considered as membrane currents. As mentioned earlier, this is important to keep in mind when we set out to model the extracellular potential: injected currents do not contribute directly to the extracellular potential, only indirectly via the membrane (return) currents that they trigger.[2] We recall that current conservation requires that the sum of the currents into a neuron (as a whole) must be zero (Figure 3.1). With no electrode injection present, this implies that the sum of the membrane currents must be zero. With an electrode injection present, the sum of the membrane (outward) currents must instead equal the injected current.

3.2 Multicompartment Models

When we introduced the various membrane currents earlier, we mostly pictured the neuron as being represented as a single compartment. In doing that, we implicitly assumed that the whole neuron was isopotential (same V_m everywhere). However, in neurons with long and branchy dendrites, V_m in the soma can be completely different from V_m in the tip of a distal dendrite, and the spatial variability in V_m can be accounted for by the use of multicompartmental (MC) models. The neural morphology is then typically represented as isopotential cylindrical compartments connected with resistors, and a single value of V_m is computed for each individual compartment (Figure 3.5A).

3.2.1 Multicompartment Formalism

As a simple introduction to the formalism used in MC models, let us consider a subset of three connected cylindrical compartments of a non-branching dendritic cable, which we number $n-1$, n, and $n+1$ (Figure 3.5B). For simplicity, we assume that the three cylinders have the same length L and diameter d. The three types of currents that might be present in the middle compartment n are (i) the axial currents $I_{n-1,n}$ and $I_{n,n+1}$ from/to the neighboring compartments, (ii) the membrane currents $I_{m,n}$ (including synaptic currents), and (iii) possibly an external stimulus $I_{ext,n}$.

The membrane currents are those we know from Section 3.1, but to keep notation short, we have here grouped them together into a total membrane current,

---

[2] The same holds for currents through gap junctions, as they pass directly from one cell to another without entering the extracellular space.

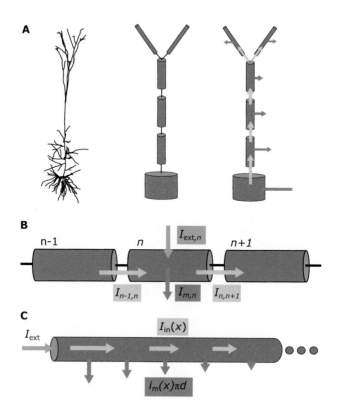

Figure 3.5 Multicompartment model. A: Representations of a neural morphology as a number of interconnected compartments. For illustrational purposes, only six compartments were included, but many MC models have several hundred compartments. **B**: Subset of interconnected (trunk) compartments. The currents involved in MC modeling include the sum of membrane currents ($I_{m,n}$) in a compartment n, the intracellular currents running between compartments ($I_{n-1,n}$ and $I_{n,n+1}$), and (if present) an external stimulus $I_{ext,n}$. **C**: Semi-infinite cable receiving external input in the finite end.

$$I_{m,n} = \pi d L \left(i_{c,n} + i_{L,n} + \sum_w i_{w,n} \right) + I_{syn,n}\,, \tag{3.17}$$

where the current densities in equation (3.1) have been converted to total currents through multiplication with the membrane area $A = \pi d L$ (cylinder side wall) of compartment n.

The external stimulus current $I_{ext,n}$ affects the neurodynamics in the same way as the membrane currents. For the purpose of computing the neurodynamics, we could therefore have included it as part of the general membrane current ($I_{m,n}$). However, as discussed in Section 3.1.5, $I_{ext,n}$ differs from the other membrane currents as it enters the inside cytosol of the neuron directly and not from the extracellular space. We have therefore chosen to keep $I_{m,n}$ and $I_{ext,n}$ separate.

The axial current between two compartments is proportional to the intracellular voltage difference between the compartments, as determined by Ohm's law,

$$I_{n-1,n} = \frac{V_{i,n-1} - V_{i,n}}{4r_a L/(\pi d^2)} \, ,$$

$$I_{n,n+1} = \frac{V_{i,n} - V_{i,n+1}}{4r_a L/(\pi d^2)} \, . \tag{3.18}$$

Here, the denominators represent the axial resistance between two compartments, defined in terms of the axial resistivity r_a ($\Omega\,$cm), a material property of the cytosol solution, the cross-section area $\pi d^2/4$, and the compartment length or travel distance L.

With all currents specified, the dynamics of the system in Figure 3.5B is computed using Kirchhoff's current law, which demands that the sum of the currents into a given compartment (n) is zero,

$$I_{n-1,n} - I_{n,n+1} - I_{m,n} + I_{ext,n} = 0 \, . \tag{3.19}$$

The equations for the axial currents (equation (3.18)) become more complicated when the connected cylinders are of different lengths and diameters, especially at branch points. However, the theory for computing the dynamics in branching structures with varying diameters is well established (Rall 1977, Rall 1989), and designated software such as NEURON (Hines & Carnevale 1997, Hines et al. 2009) and Arbor (Akar et al. 2019) offers solutions for compartmentalizing neurons once the morphology is specified. Different strategies for determining a suitable spatial discretization of cable models such as the so-called *d_lambda rule* are discussed in Hines & Carnevale (2001). For the remainder of this chapter, we limit ourselves to consider the simplified, unbranched scenario (Figure 3.5B), as we deem this as sufficient for establishing a basic understanding of MC modeling.

We note that the axial currents (equation (3.18)) depend on the intracellular potential $V_{i,n}$, while the membrane currents (equation (3.17)) depend on the membrane potential $V_{m,n} = V_{i,n} - V_{e,n}$, where $V_{e,n}$ is the extracellular potential immediately outside the membrane. The system of equations in equation (3.19) is therefore underdetermined. In most MC models, this challenge is overcome by assuming that the extracellular potential V_e is constant in time and space. One can then substitute $V_{i,n} = V_{m,n} + V_e$ in equation (3.18) with $V_{m,n}$ because the constant appears in all the terms and can be eliminated.

The assumption of an isopotential extracellular space, which corresponds to assuming an infinite extracellular conductivity, may appear peculiar, especially within a book where the dynamics of the extracellular potential is the main topic. We comment further on this assumption and its consequences in Section 13.2. For now, we simply note that, although the extracellular potential is not zero in reality, variations in V_e are generally much smaller than variations in V_m so that effects from variations in V_e on V_m are likely to be small. This is the justification for neglecting the impact of V_e in most MC models.

In the following, we thus replace V_i in equation (3.18) with V_m. If we then insert equation (3.18) and the full expression for the membrane currents (equation (3.17)) into Kirchhoff's current law (equation (3.19)), we get the fundamental equation for MC models:

$$c_m \frac{dV_{m,n}}{dt} = -i_{L,n} - \sum_w i_{w,n} - \frac{I_{syn,n}}{\pi d L} + \frac{I_{ext,n}}{\pi d L} + \frac{d}{4r_a}\left(\frac{V_{m,n+1} - V_{m,n}}{L^2} - \frac{V_{m,n} - V_{m,n-1}}{L^2}\right).$$

$$\tag{3.20}$$

3.2.2   Endpoint Boundary Conditions

Equation (3.20) assumes that a compartment n has two neighbors ($n-1$ and $n+1$). This will not be true for the endpoint compartments (let us call them compartment 1 and N), and we must therefore select appropriate boundary conditions for these compartments. The most commonly used boundary condition is the sealed-end condition, meaning that no axial currents leave at cable endpoints, so that $I_{0,1} = 0$ in one end of an unbranched cable and $I_{N,N+1} = 0$ in the other end. This is, for example, the default boundary condition in the NEURON simulator (Hines & Carnevale 1997, Hines et al. 2009). At the cable endpoints, equation (3.20) is then replaced with

$$c_m \frac{dV_{m,1}}{dt} = -i_{L,1} - \sum_w i_{w,1} - \frac{I_{syn,1}}{\pi d L} + \frac{I_{ext,1}}{\pi d L} + \frac{d}{4r_a} \frac{V_{m,2} - V_{m,1}}{L^2} \tag{3.21}$$

and

$$c_m \frac{dV_{m,N}}{dt} = -i_{L,N} - \sum_w i_{w,N} - \frac{I_{syn,N}}{\pi d L} + \frac{I_{ext,N}}{\pi d L} - \frac{d}{4r_a} \frac{V_{m,N} - V_{m,N-1}}{L^2}. \tag{3.22}$$

3.2.3   Passive Multicompartment Models

The neural response to small perturbations from the resting potential often has a close-to-linear relationship between the membrane potential and membrane currents (Mauro et al. 1970, Koch 1984). When modeling subthreshold neural dynamics, a common simplification is therefore to assume that the ionic membrane current density has the linear form

$$i_{m,n} = g_m(V_{m,n} - E_m), \tag{3.23}$$

where E_m is the membrane resting potential and g_m the specific membrane conductance. This is essentially the same as the passive leakage current density defined earlier (equation (3.4)). However, g_L and E_L have been replaced with g_m and E_m, respectively, as we may think of equation (3.23) as comprising not only the leakage current, but also additional contributions from active currents that are passive-like and partially open near the resting potential. In certain simplified point-neuron models, such a linear relationship between membrane currents and membrane potential is assumed to hold all the way up to the action-potential firing threshold (Sterratt et al. 2023).

Models that approximate the membrane ionic current density with equation (3.23) are often called passive. In passive models, it is customary to replace the specific membrane conductance g_m with the specific membrane resistance $r_m = 1/g_m$ ($\Omega\,\mathrm{cm}^2$). For MC models, the passive version of equation (3.20) then becomes

$$c_m \frac{dV_{m,n}}{dt} = \frac{E_m - V_{m,n}}{r_m} - \frac{I_{syn,n}}{\pi d L} + \frac{I_{ext,n}}{\pi d L} + \frac{d}{4r_a} \left(\frac{V_{m,n+1} - V_{m,n}}{L^2} - \frac{V_{m,n} - V_{m,n-1}}{L^2} \right). \tag{3.24}$$

Passive MC models (equation (3.24)) are commonly used as an approximation for signaling in neural dendrites, which tend to have a lower density of active mechanisms compared to the soma and axon (Koch 1999). Sometimes, the passive approximation is used for all compartments in the MC model – for example, to simulate responses to subthreshold synaptic input. This is a common approximation used in simulations of the local field potential (see Chapter 8). As we shall see later, the passive model is also useful in that it allows us to derive analytical results for signal propagation in neural dendrites.

3.2.4 Two-Compartment Model

As we explained in the beginning of this chapter (Figure 3.1), the simplest model that can produce extracellular potentials is a two-compartment model. In this limiting case, the dynamics of compartments $n = 1$ and $n = 2$ are described by the edge-compartment formulas, equations (3.21) and (3.22), respectively, so that the two-compartment model becomes

$$c_m \frac{dV_{m,1}}{dt} = -i_{L,1} - \sum_w i_{w,1} - \frac{I_{syn,1}}{\pi d L} + \frac{I_{ext,1}}{\pi d L} + \frac{d}{4r_a}\frac{V_{m,2}-V_{m,1}}{L^2}, \tag{3.25}$$

$$c_m \frac{dV_{m,2}}{dt} = -i_{L,2} - \sum_w i_{w,2} - \frac{I_{syn,2}}{\pi d L} + \frac{I_{ext,2}}{\pi d L} - \frac{d}{4r_a}\frac{V_{m,2}-V_{m,1}}{L^2}. \tag{3.26}$$

The passive version of the two-compartment model is

$$c_m \frac{dV_{m,1}}{dt} = \frac{E_m - V_{m,1}}{r_m} - \frac{I_{syn,1}}{\pi d L} + \frac{I_{ext,1}}{\pi d L} + \frac{d}{4r_a}\frac{V_{m,2}-V_{m,1}}{L^2}, \tag{3.27}$$

$$c_m \frac{dV_{m,2}}{dt} = \frac{E_m - V_{m,2}}{r_m} - \frac{I_{syn,2}}{\pi d L} + \frac{I_{ext,2}}{\pi d L} - \frac{d}{4r_a}\frac{V_{m,2}-V_{m,1}}{L^2}. \tag{3.28}$$

Although the two-compartment model fails to account for many important features of the extracellular potential, its simplicity gives it pedagogical value, and we will in later chapters use it to establish some qualitative insights into the relationship between neural activity and extracellular potentials.

3.2.5 Biophysically Detailed Cell Models

The term *biophysically detailed* cell models usually refers to multicompartment models with a detailed morphology and many different active ion channels. Such models can be constructed by combining the different principles introduced so far in this chapter.

When constructing biophysically detailed cell models, the aim is typically to reproduce a specific experiment – for example, one where current is injected into a neuron while the response (ideally at several locations along the morphology) in the membrane potential is measured. The modeling typically involves the reconstruction of the neural morphology as well as an optimization procedure where parameters of the passive and active ion channels are adjusted until the model reproduces the experimental data as close as possible (Hay et al. 2011, Almog & Korngreen 2014, Van Geit et al. 2016).

As we will demonstrate in coming chapters, extracellular signals depend on how membrane currents are distributed over the neural morphology. Hence, there are salient features in such signals that can only be captured with biophysically detailed neural models, as simpler models lack the necessary complexity.

3.2.6   The Hay Model

In many of the figures and simulation examples in this book, we have used a bio-physically detailed model of a layer 5 pyramidal cell as our "default" biophysically detailed cell model. The layer 5 pyramidal cells are large, well studied, and abundant, and they serve as important input–output units in the cortex. Their dendrites span the entire column, thus receiving input from all cortical layers, and they are the main output units from the cortex to various other parts of the brain. Furthermore, because of their geometrical alignment, large size, and abundance, they are expected to be important contributors to different electric brain signals.

The model that we use represents an adult rat cortical layer 5b pyramidal cell. It was developed by Hay et al. (2011), and we refer to it simply as the Hay model. The Hay model was fitted to experimental data using multiobjective optimization with an evolutionary algorithm and has 10 active ionic conductances, and the model can exhibit both dendritic Ca^{2+} spikes and back-propagating action potentials. It has become an often-used cell model in computational neuroscience and is available from ModelDB (Hines et al. 2004, Hay et al. 2011) (parameters from the Hay et al. (2011) Table 3, accession #139653, "cell #1"). We encountered the Hay model already in Figure 1.1 of this book.

3.3   Cable Theory

As we indicated in Section 3.2.3, the passive model allows us to derive analytical results for how signals spread in cellular structures, such as neural dendrites. To derive the analytical formulas, we envision that we have an (infinitely long) chain of identical and interconnected cylinders, and we let each of them become infinitesimally short so that they form a continuum (Figure 3.5C). Mathematically, that means that we let $L \to \delta x$ in equation (3.24) and take the limit $\delta x \to 0$. From this, we can derive the cable equation (see e.g. Sterratt et al. (2023)):

$$c_{\mathrm{m}} \frac{\partial V_{\mathrm{m}}}{\partial t} = \frac{E_{\mathrm{m}} - V_{\mathrm{m}}}{r_{\mathrm{m}}} + \frac{d}{4r_{\mathrm{a}}} \frac{\partial^2 V_{\mathrm{m}}}{\partial x^2} + \frac{\mathcal{I}_{\mathrm{ext}} - \mathcal{I}_{\mathrm{syn}}}{\pi d} , \qquad (3.29)$$

where we have introduced the synaptic and external stimulus current per unit length (with unit mA/cm),

$$\mathcal{I}_{\mathrm{ext}}(x,t) = \lim_{\delta x \to 0} I_{\mathrm{ext}}(x,t)/\delta x ,$$
$$\mathcal{I}_{\mathrm{syn}}(x,t) = \lim_{\delta x \to 0} I_{\mathrm{syn}}(x,t)/\delta x . \qquad (3.30)$$

To improve our analytical understanding of dendritic integration, it is useful to reformulate the cable equation as

$$\tau_{\mathrm{m}} \frac{\partial V_{\mathrm{m}}}{\partial t} = E_{\mathrm{m}} - V_{\mathrm{m}} + \lambda^2 \frac{\partial^2 V_{\mathrm{m}}}{\partial x^2} + \frac{r_{\mathrm{m}}}{\pi d} \left(\mathcal{I}_{\mathrm{ext}} - \mathcal{I}_{\mathrm{syn}} \right) , \qquad (3.31)$$

where we have multiplied all terms with r_m and introduced the *length constant*

$$\lambda = \sqrt{\frac{dr_m}{4r_a}} \tag{3.32}$$

and the membrane time constant

$$\tau_m = r_m c_m . \tag{3.33}$$

Here, τ_m is the typical time scale (dimensionless time: t/τ_m), while λ is the typical length scale (dimensionless length: x/λ) of electric signals in the cable. If a certain point along the cable is perturbed so that the potential is shifted from rest to a predetermined value V_m, τ_m will determine how fast the local potential falls back towards rest, while λ will determine how far the local perturbation will spread along the cable. For example, an apical dendrite of a cortical neuron with diameter $d = 4\,\mu m$, membrane resistivity $r_m = 20\,k\Omega\,cm^2$, and axial resistivity $r_a = 200\,\Omega\,cm$ (Koch 1999) will, according to equation (3.32), have a length constant of $\lambda = 1\,mm$.

The cable equation is a continuous version of equation (3.24). Whereas MC models generally must be solved numerically, the cable equation allows the spatiotemporal evolution of the membrane potential to be solved analytically for some idealized scenarios. Below, we consider a couple of scenarios that might make the interpretation of τ_m and λ clearer.

3.3.1 Steady-State Solutions of the Cable Equation

We first consider the steady-state solution for a semi-infinite cable, containing no synapses but receiving a constant external current injection at a sealed end in $x = 0$ (Figure 3.5C). Since only the end point receives the stimulus, we may attack this problem by solving equation (3.31) for all other points $x > 0$ and then account for the effects of the stimulus current afterwards as a boundary condition.

At steady state, $\partial V_m/\partial t = 0$, and equation (3.31) becomes the ordinary differential equation

$$0 = E_m - V + \lambda^2 \frac{d^2 V_m}{dx^2}, \quad \text{for } x > 0 . \tag{3.34}$$

If we introduce the new variable $\Delta V_m = V_m - E_m$, equation (3.34) simplifies to

$$\frac{d^2 \Delta V_m}{dx^2} - \frac{1}{\lambda^2} \Delta V_m = 0 , \tag{3.35}$$

which has the solution

$$\Delta V_m(x) = \Delta V_m(0) e^{-x/\lambda} \tag{3.36}$$

or

$$V_m(x) = E_m + (V_m(0) - E_m) e^{-x/\lambda} . \tag{3.37}$$

We note that the general solution to equation (3.35) also has a term containing the positive exponential $e^{x/\lambda}$, but this term diverges when $x \to \infty$ and is excluded on the count of being unphysical.

To obtain the final solution to the problem, we need to specify the boundary value $V_m(0)$, which will depend on the external stimulus. Since the stimulus is injected in a singular point ($x = 0$), we go back to expressing it in terms of a total current I_{ext} and not as a current per unit length $\mathcal{I}_{ext}$. The stimulus can be introduced through a boundary condition demanding that the injected current must be identical to the axial current in the cable at $x = 0$:

$$I_{ext} = -\frac{\pi d^2}{4} \frac{1}{r_a} \frac{dV_m}{dx}\Big|_{x=0}. \tag{3.38}$$

If we insert for dV_m/dx (calculated from equation (3.37)) and insert equation (3.32) for λ, we find that

$$V_m(0) = E_m + R_\infty I_{ext}, \tag{3.39}$$

where we have defined the steady-state input resistance of the semi-infinite cable as

$$R_\infty = \sqrt{\frac{4 r_m r_a}{\pi^2 d^3}}. \tag{3.40}$$

If we now insert equation (3.39) into equation (3.37), we obtain our final solution

$$V_m(x) = E_m + R_\infty I_{ext} e^{-x/\lambda}. \tag{3.41}$$

Some insights that can be derived from the steady-state solution are:

- The steady-state input resistance of the semi-infinite cable R_∞ is proportional to $1/d^{3/2}$ (see equation (3.40)) and is thus higher the thinner the dendrite.
- The amplitude of the membrane potential decays exponentially from the injection site and outwards, and the amplitude is reduced by a factor $1/e$ over the length λ (equation (3.37)). Neurons with dendrites much shorter than λ are said to be electrotonically compact, meaning that a potential shift generated at some location spreads with relatively little attenuation to all parts of the dendrites.
- Since $\lambda \propto \sqrt{d}$ (equation (3.32)), voltage deflections will spread further the thicker the dendrite.
- Since $\lambda \propto \sqrt{r_m/r_a}$ (equation (3.32)), voltage deflections spread more easily when the membrane resistance is high (makes cable less leaky) and the axial resistivity low.

3.3.2   Ball-and-Stick Neuron Model

The cable equation has played an important role in computational neuroscience, much due to the work of Wilfrid Rall, who showed that certain classes of dendritic trees can be reduced to a single, equivalent cylinder preserving both the membrane area and the input conductance of the original dendritic tree (Rall 1959, Koch 1999). For this reduction to be perfect, the dendritic tree must meet a set of geometrical conditions:

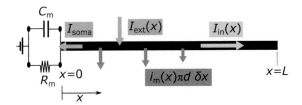

Figure 3.6 Ball-and-stick neuron. Illustration of a ball-and-stick neuron model receiving external input. In the illustration, the stick (a cylinder with diameter d and length L) receives a total input current (green) in a specific point. This causes intracellular (yellow) currents and distributed membrane-return currents (orange), represented as current densities (per membrane area) times the membrane area of an infinitesimal cylinder segment, $\pi d \delta x$. A single-compartment soma (with total membrane capacitance C_m and resistance R_m) determines the boundary condition in the end ($x = 0$) of the cable, which in turn will determine the current leaving (pink) in the endpoint.

1. All branches have the same r_a and r_m.
2. All terminal branches must end with the same boundary conditions (e.g. the sealed end condition).
3. All terminal branches must end at the same *electrotonic distance* from the origin of the main branch. This (terminal) electrotonic distance is found by summing the electrotonic distances of all branches along the path, where the electrotonic distance of a single branch n equals its length l_n divided by its length constant (λ_n).
4. In every branch point, the diameter of the parent branch d_0 and its two daughters d_1 and d_2 must satisfy the so-called $d^{3/2}$ *rule*, stating that $d_0^{3/2} = d_1^{3/2} + d_2^{3/2}$.
5. In the case of dendritic input, the input must be identical in all locations at the same electrotonic distance.

Although real dendritic trees will never meet all the conditions listed, modeling the entire dendritic tree as a cylinder has become a popular approximation since it allows for dendritic signaling to be studied analytically by solving the cable equation. Analytical solutions can also be derived for a system where one end of the finite cylinder is connected to a single-compartment soma, as shown for example by Rall (1959) or Johnston & Wu (1994, section 4.5). This phenomenological neuron, commonly referred to as the ball-and-stick model (Figure 3.6), is a much-used simplified neuron model.

3.3.3   Intrinsic Dendritic Filtering

The cable properties of dendrites cause a low-pass filtering of extracellular signals commonly referred to as *intrinsic dendritic filtering* (Pettersen & Einevoll 2008, Lindén et al. 2010). Understanding dendritic filtering is important for understanding how neural events, such as an AP (Section 7.3.1) or a synaptic input (Section 8.4.1), are reflected in the extracellular potential.

As we will see in Chapter 4, it is the membrane currents that set up the extracellular potentials. The physical origin of the intrinsic dendritic filtering effect is the frequency-dependence of capacitive membrane currents. It can be illustrated by considering a sinusoidal current injected into one end of a compartmentalized passive dendritic stick (Figure 3.7A). From Kirchhoff's current law, we know that the sum of the currents into a neural compartment must always be zero. The injected current I_{inj} will therefore at all times be equal to the sum of the membrane current in the first compartment $I_{m,1}$ and the axial current $I_{a,1}$, which becomes the input to the next compartment (compartment 2). Likewise, this input will leave compartment 2 partly as membrane currents and partly as axial currents into compartment 3, and so on. The generated extracellular signal will be determined by the magnitudes and spatial distribution of the membrane-return currents along the stick.

For a constant current injection into a semi-infinite cable, the steady-state distribution of return currents is determined by the distribution of the membrane potential, which in turn is determined by the length constant λ (equation (3.32)). By definition, λ is the length over which the potential falls to a fraction $1/e$ of its boundary value when the finite end of a semi-infinite cable is fixed at a constant potential or, equivalently, if the end of a semi-infinite cable receives a constant direct-current (DC) input. Therefore, λ is often referred to as the DC length constant.

If the end of the cable instead receives an alternating current injection, the distribution of membrane-return currents will also depend on the frequency of the input current (Figure 3.7C). This is again a consequence of the capacitive properties of the membrane: when the frequency increases, the capacitive component of the membrane current increases, resulting in the membrane becoming effectively leakier so that currents will travel a shorter distance along the cable before returning to the extracellular space. Hence, the average separation distance between input and return currents will decrease with the frequency of the injected current.

In the context of the ball-and-stick neuron model in Figure 3.7, this dendritic filtering effect will be quantified below, where we define the frequency-dependent length constant for the case of a semi-infinite and a finite stick.

3.3.3.1   Semi-infinite Stick

In the case of a semi-infinite cable (or semi-infinite stick), the frequency-dependent length constant is given by (see Appendix A)

$$\lambda_{AC}(f) = \lambda \sqrt{\frac{2}{1 + \sqrt{(2\pi f \tau_m)^2 + 1}}} . \qquad (3.42)$$

Here, τ_m is the membrane time constant (equation (3.33)), λ is the DC length constant (equation (3.32)), and f is the frequency of the external current injection to the finite end of the cable (or soma in the ball-and-stick model).[3]

---

[3] Instead of using a current injection, the equivalent boundary condition can be obtained by imposing an oscillating potential with frequency f in the finite end of the cable.

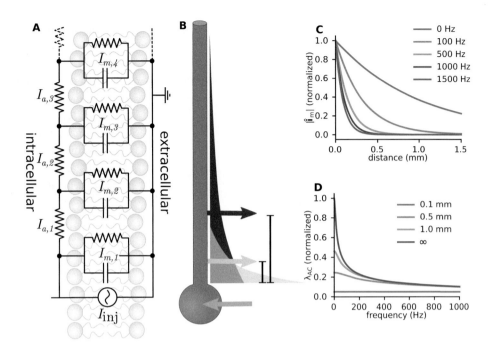

Figure 3.7 Intrinsic dendritic filtering. A sinusoidal current injected into the soma of a ball-and-stick model enters the dendrite as an axial current and returns to the extracellular space via the dendrite membrane. **A**: Circuit diagram. **B**: Illustration of the distribution of dendritic return currents in cases of slow (dark grey) and fast (light grey) oscillations. The higher the frequency, the closer the return currents will be to the soma. The arrows illustrate the weighted mean of return-current positions. **C**: Distribution of return currents for a selection of stimulus frequencies, where $\hat{i}_m$ is the membrane current using a formulation with complex numbers (Pettersen & Einevoll 2008). **D**: The frequency-dependence of the length constant $\lambda_{AC}(f)$ in sticks of various lengths. The average vertical position of return currents in (B) can be quantified by $\lambda_{AC}(f)$, which in turn reflects the weighted mean of the return-current positions (C). Code available via www.cambridge.org/electricbrainsignals.

In the limiting case of $f = 0$, $\lambda_{AC}(0)$ in equation (3.42) becomes identical to the DC length constant λ. In general, $\lambda_{AC}(f)$ decreases with f (Figure 3.7B, D), and in the limit of very high frequencies ($f \gg 1/2\pi\tau_m$),

$$\lambda_{AC}(f) \sim \frac{\lambda}{\sqrt{\pi f \tau_m}} \,. \tag{3.43}$$

Thus, low-frequency current inputs will tend to spread further in dendritic structures compared to high-frequency current inputs. As we will see later in this book, the decrease of $\lambda_{AC}(f)$ with frequency implies that low-frequency components in the membrane currents tend to give larger contributions to extracellular potentials than high-frequency components.

We note here that MC models rest on the assumption that neural morphology is well represented by a one-dimensional, branching cable, meaning that radial components of intracellular currents are neglected. A criterion for neglecting the radial dimension of the cable is that the length constant is much greater than the cable diameter (Koch 1999). Since the DC constant λ is in the order of 1 mm and dendritic diameters in the order of 1 μm, equation (3.43) suggests that this approximation will be acceptable for frequencies below the megahertz range.

3.3.3.2   Finite Stick

For a stick of finite length, $\lambda_{AC}(f)$ is not given by equation (3.42). However, for a stick receiving an input with frequency f in one end (soma-end), an AC length constant can instead be defined as the weighted mean value of the position of the return currents.[4] In the limit of an infinitely long stick, it can be shown that this return-current-based definition of $\lambda_{AC}(f)$ becomes identical to the membrane-potential-based definition in equation (3.42) (Pettersen & Einevoll 2008).

Numerical examples of the value of $\lambda_{AC}(f)$ for a finite stick are shown in Figure 3.7D.

3.3.3.3   Return-Current Approximation

To get a qualitative understanding of the implications of intrinsic dendritic filtering, it is convenient to assume as an approximation that all return currents from a given oscillatory component leave the dendrite at a single point located at a distance $\lambda_{AC}(f)$ away from the soma (Pettersen & Einevoll 2008). If we make this approximation, each oscillatory component gives rise to a current dipole with a dipole length given by $\lambda_{AC}(f)$. In Chapter 7, we use this approximation to obtain qualitative insights into how the shape of extracellular spikes (signatures of neural APs) varies depending on position relative to the soma and dendrites.

3.4   Ion-Concentration Dynamics and Reversal Potentials

Up until this point, we have focused on electric currents in neurons but have talked little of their biophysical origin – that is, the ionic movements that mediate these currents. Ionic membrane fluxes are largely driven by diffusion and thus depend on the intra- and extracellular solutions having different ionic compositions.

In the current section, we go further into how neural activity is rooted in ionic movements. Before we proceed, we note that the concentrations of the main charge carriers in brain tissue tend to remain approximately constant under normal working conditions. Under such circumstances, the MC models defined in Section 3.2.1, with membrane mechanisms as specified in Section 3.1, give a complete and operational framework for modeling the electric activity and distribution of membrane currents in neurons. We will

---

[4] Computing $\lambda_{AC}(f)$ is easiest if we use a complex notation (defined in Appendix F.2). For a complex stimulus, return currents $\hat{i}_m$ are complex, and $\lambda_{AC}(f) = (\int_0^L x|\hat{i}_m| dx)/(\int_0^L |\hat{i}_m| dx)$ where x is the distance from the input-receiving soma-end of the stick and L is the length of the stick (Pettersen & Einevoll 2008).

consider the effects of concentration changes in Chapter 12, but we will neglect them in most parts of this book. Some readers may therefore choose to skip the rest of this section.

As APs are generated by Na^+ and K^+ ions flowing across neural membranes, it may seem like a rather bold modeling assumption that the concentrations of these ions remain constant. However, under many circumstances, the assumption is well founded. One reason is that the number of ions crossing the membrane during a brief period of activity, such as the generation of an AP, is so small that it only leads to tiny changes in the ion concentrations. To see why, let us consider an example with a spherical single-compartment neuron with radius r. This neuron contains a number $N_0 = (4/3)\pi r^3 N_A [Na^+]$ of Na^+ ions, where N_A is Avogadro's number and $[Na^+]$ is the intracellular Na^+ concentration. Let us next assume that Na^+ ions enter the neuron and increase the membrane potential by ΔV_m. The total charge needed for this will be $\Delta Q = 4\pi r^2 c_m \Delta V_m$, which corresponds to a number of $N_{new} = \Delta Q/e$ of ions, where e is the elementary charge. From this, we may compute the ratio between the number of Na^+ ions that entered the neuron and the number of ions that were already there:

$$\frac{N_{new}}{N_0} = \frac{3 c_m \Delta V_m}{F r [Na^+]}. \tag{3.44}$$

Here, we have used Faraday's constant $F = e N_A = 96\,458.3\,C/mol$. If we insert the values $\Delta V_m = 100\,mV = 0.1\,V$ (typical AP amplitude), $[Na^+] = 10\,mM$ (typical intracellular Na^+ concentration), and $c_m = 1\,\mu F/cm^2 = 10^{-2}\,F/m^2$. The ratio then becomes

$$\frac{N_{new}}{N_0} = \frac{3.1 \times 10^{-9}\,m}{r}. \tag{3.45}$$

Hence, even in a small compartment with radius $r = 10^{-6}\,m$, the number of ions that enters during an AP will only change the concentration by a fraction: 3.1×10^{-3} (i.e. about 0.3 percent). This suggests that concentration changes on a short time scale are small enough to be neglected.

Since neurons possess homeostatic machinery of ion pumps and cotransporters that strive to maintain the membrane ion-concentration gradients, the assumption of constant ion concentrations also tends to hold on a longer time scale (Hille 2001, Somjen 2004). Perhaps the most important player in this machinery is the Na^+/K^+-pump discovered by Skou (1957), which uses the energy of 1 ATP molecule to pump 2 K^+ ions into the neuron and 3 Na^+ ions out, thus reversing the ionic exchange that occurs during AP generation. As a result of this pump, the intracellular space tends to remain comparatively rich in K^+ while the extracellular space tends to remain comparatively rich on Na^+. Typical values of the ion concentrations of the main charge carriers inside and outside neurons are given in Table 3.1.

Later, we will explain how the ionic concentrations are implicitly present in the HH formalism as they determine the ionic reversal potentials (E_w in equation (3.5)). We will also comment briefly on the cases where the assumptions of constant ion concentrations are not applicable.

	K$^+$	Na$^+$	Mg^{2+}	Cl$^-$	Ca^{2+}	HCO3$^-$
Intracellular (mM)	125	10	0.5	6.6	6×10^{-5}	18
Extracellular (mM)	2.9	147	0.7	119	1	23.3
Nernst potential (mV)	-100	72	4.5	-77	129	-6.9

Table 3.1. Major charge carrier concentrations inside and outside a typical mammalian neuron. Concentration values were taken from Table 2-1 in Somjen (2004) for neurons in the central nervous system (intracellular) and human cerebrospinal fluid (extracellular). Values vary with species and brain regions. Nernst potentials were computed from equation (3.50) assuming a body temperature of 309.15 K.

3.4.1   Nernst Potential

Ion channels are proteins coiled up inside the membrane in a way that makes them form pores through the membrane. Some of these pores are selectively permeable only to specific ions. The ion flux through an open ion channel will be mediated by a combination of (i) diffusion and (ii) electric drift as predicted from the Nernst-Planck equation (equation (2.21)).

The ionic reversal potential is defined as the membrane potential at which the diffusive and drift currents of a given ion species are in an equilibrium so that they are equal in magnitude but oppositely directed. If we approximate the membrane currents as one-dimensional (in the x-direction) and perpendicular to the membrane, the Nernst-Planck equation for an ion species k is

$$j_k = j_{k,\mathrm{diff}} + j_{k,\mathrm{drift}} = -D_{\mathrm{m},k}\left(\frac{d[k]}{dx} + \frac{Fz_k}{RT}[k]\frac{dV}{dx}\right), \tag{3.46}$$

where $[k]$ denotes the concentration of ion k (mM). The first term inside the parenthesis of equation (3.46) is the diffusive flux density along the concentration gradient, and the second term is the electric drift flux density along the voltage gradient. Note that the effective diffusion constant for transport across the membrane $D_{\mathrm{m},k}$ is not equal to the diffusion constant D_k in dilute saline solutions.

The reversal potential (also called the Nernst potential) is the equilibrium potential for which there is no net flux so that $j_{k,\mathrm{diff}} = -j_{k,\mathrm{drift}}$. According to equation (3.46), this occurs when

$$\frac{1}{[k]}\frac{d[k]}{dx} = -\frac{Fz_k}{RT}\frac{dV}{dx}. \tag{3.47}$$

We may multiply both sides by dx and rearrange this to get

$$-dV = \frac{RT}{Fz_k}\frac{d[k]}{[k]}. \tag{3.48}$$

If we integrate this from the intracellular to the extracellular side of the membrane, we get

$$-\int_{V_i}^{V_e} dV = \frac{RT}{Fz_k}\int_{[k]_i}^{[k]_e}\frac{d[k]}{[k]}. \tag{3.49}$$

The integral on the left-hand side equals $V_i - V_e$, where V_e is the extracellular potential immediately outside the membrane. This corresponds to the membrane potential for which the net flux in equation (3.46) is zero and is, by definition, the reversal potential E_k. Hence, according to equation (3.49), the reversal potential is given by

$$E_k = \frac{RT}{Fz_k} \ln \frac{[k]_e}{[k]_i}. \tag{3.50}$$

Equation (3.50) was used to compute the reversal potentials listed in Table 3.1.

To establish an intuitive understanding of what the reversal potential means, let us use a K^+ channel as an example. For simplicity, we assume that it opens at a time when the neuron is resting ($V_m = V_i - V_e \approx -70\,\text{mV}$) and that it is the only ion channel that opens. Since the intracellular space is more K^+-rich than the extracellular space, diffusion will drive K^+ out from the neuron, and since V_m is negative, electric drift will drive K^+ into the neuron. When the concentration difference is as large as it is over the neural membrane, the diffusive process will dominate and give a net efflux of K^+, leading to a hyperpolarization of the neuron. V_m will then gradually drop towards the K^+ reversal potential $E_K = -100\,\text{mV}$ (Table 3.1). As V_m becomes more negative, the drift component becomes stronger, and when V_m reaches E_K, the drift component becomes equal in magnitude to the diffusive component so that the net K^+ current becomes zero. E_K is called the reversal potential because if V_m were to drop below this value, the direction of the current would reverse, meaning that the drift component would become dominant over diffusion and result in an influx of K^+ ions into the neuron (against the concentration gradient).

Since ion concentrations are generally assumed to remain constant in MC models, the reversal potentials are also assumed to be constant. As a consequence, one typically does not worry about ionic concentrations when constructing such models but simply uses values for E_k based on experimental measurements of the potentials at which the various membrane currents reverse.

3.4.2 Goldman-Hodgkin-Katz Current Equation

The reversal potential in equation (3.50) is the potential for which the net ion flux in equation (3.46) is zero. Finding the relationship between the ion flux (or current) and the membrane potential for other values of the potential is less trivial. However, if one makes the assumptions that (i) ions cross the membrane independently and (ii) the electric field within the membrane is constant, one can derive from equation (3.46) the Goldman-Hodgkin-Katz (GHK) current equation for the membrane currents (see e.g. Hodgkin & Katz (1949) or Hille (2001, chapter 14)):

$$i_k = p_k z_k F \frac{z_k F V_m}{RT} \left(\frac{[k]_i - [k]_e e^{-z_k F V_m/RT}}{1 - e^{-z_k F V_m/RT}} \right). \tag{3.51}$$

Here, the specific permeability p_k (with unit $\text{cm}^{-1}\,\text{s}^{-1}$) is defined as the constant of proportionality between the concentration difference and molar flux density of ion k per membrane area. The permeability is proportional to the diffusion constant $D_{m,k}$ in equation (3.46), and the relationship between the two is discussed in more detail in Hille (2001, chapter 14). Solving equation (3.51) for when $i_k = 0$ gives us back the reversal potential in equation (3.46).

Looking at equation (3.51), we see that the membrane currents are nonlinear functions of both the ionic concentrations $[k]$ in the intra- and extracellular space and the membrane potential V_m. In comparison, the membrane current densities $i_k = g_k(V_m - E_k)$ used in the HH-type formalism (equation (3.5)) are linearized (sometimes called quasi-ohmic) versions of the Goldman-Hodgkin-Katz equation. That is, they are linearly proportional to the deviance from the reversal potential E_k, with conductances g_k proportional to p_k in equation (3.51). Such a linear current–voltage relationship is for many ion species a good approximation within the voltage range that neurons normally operate.

3.4.3   Reversal Potential of a Non-specific Ion Channel

To find the reversal potential of an ion channel that is permeable to several ion species simultaneously, the criterion that $j_{k,\text{diff}} = -j_{k,\text{drift}}$, which led to equation (3.47), must be replaced with the criterion that $\sum_k i_{k,\text{diff}} = -\sum_k i_{k,\text{drift}}$. This gives us the equilibrium potential E_{rev} for when the currents mediated by the individual ion species jointly sum to a zero net current. For example, if an ion channel is permeable to the three most abundant charge carriers, K^+, Na^+, and Cl^- with permeabilities p_K, p_{Na}, and p_{Cl}, it can be shown that the reversal potential will be

$$E_{\text{rev}} = \frac{RT}{F} \ln \frac{p_K[K^+]_e + p_{Na}[Na^+]_e + p_{Cl}[Cl^-]_i}{p_K[K^+]_i + p_{Na^+}[Na]_i + p_{Cl}[Cl^-]_e}. \tag{3.52}$$

This equation in known as the Goldman equation, but it is also sometimes referred to as the Goldman-Hodgkin-Katz (GHK) voltage equation and can be derived from the GHK current equation (equation (3.51)). Note that, since the permeabilities occur in both the numerator and denominator, the value of E_{rev} depends only on their relative differences.

If one uses the quasi-ohmic linear form for the current densities in equation (3.51), one can derive as a counterpart to equation (3.52) that

$$E_{\text{rev}} = \frac{g_K E_K + g_{Na} E_{Na} + g_{Cl} E_{Cl}}{g_K + g_{Na} + g_{Cl}} \tag{3.53}$$

for the compound current density

$$i = g(V_m - E_{\text{rev}}), \tag{3.54}$$

with

$$g = g_K + g_{Na} + g_{Cl}. \tag{3.55}$$

3.4.4   Leakage-Reversal Potential

If the membrane's (passive) leak conductances for K^+, Na^+, and Cl^- are denoted $\overline{g}_K$, $\overline{g}_{Na}$, and $\overline{g}_{Cl}$, we can define a total non-specific leakage current $i_L = \overline{g}_L(V_m - E_L)$, with E_L and $\overline{g}_L$ computed from equations (3.53) and (3.55), respectively. However, while the electric leakage current is zero when $V_m = E_L$, the fluxes of the respective ion species are not zero individually. For the concentrations to remain constant over time, this non-equilibrium in terms of fluxes must be counterbalanced by homeostatic machinery involving the electrogenic Na^+/K^+-pump (discussed earlier in this section), as well as possibly a number of cotransporters and other electrogenic exchanger pumps.

There exist computational models that explicitly account for the homeostatic processes. These models contain a combination of active and passive ion channels, ion pumps, and ion cotransporters, which are tuned so that they together drive the model into the true resting state where the membrane potential and the intra- and extracellular ion concentrations remain constant (see e.g. Kager et al. (2000), Øyehaug et al. (2012), Wei et al. (2014), or Sætra et al. (2020)).

While the above-mentioned ion-conserving models are useful for studying cellular activity under circumstances involving large deviances from baseline concentrations, many relevant problems can be addressed without modeling the homeostatic processes in such detail. In most MC models, the ensemble of processes that work to maintain baseline conditions are simply assumed to do their job in the background and are not explicitly modeled. The leakage current is then defined pragmatically as a linear approximation of all membrane currents that are not accounted for by other terms in the model. E_L should then be interpreted more cautiously as accounting for an unknown combination of passive leakage, electrogenic pump activity, and possible contributions from active ion channels that are not explicitly included in the model (see Offner (1991) for a critical study of this approximation). When constructing models, E_L is normally not measured separately or computed from equation (3.52), but it is instead introduced as a free parameter, adjusted to the value that gives the model neuron the correct resting membrane potential. This was, for example, done in the work by Hodgkin & Huxley (1952*b*).

3.4.5   Resting Membrane Potential

The GHK voltage equation (equation (3.52)) is sometimes used to describe the resting membrane potential of neurons. That is, it is used to estimate the relationship between the resting membrane potential, the intra- and extracellular ion concentrations, and the membrane's permeability to the ions during rest. For example, Hodgkin & Katz (1949) found by trial and error that the membrane potential at and near rest in the squid giant axon was well fitted with permeability ratios $p_K : p_{Na} : p_{Cl} = 1 : 0.04 : 0.45$. Although such fittings can be valuable for model construction, it should be noted that it is not entirely correct to view the resting state as an electrodiffusive equilibrium determined by the GHK voltage equation. As we mentioned when we discussed the leakage-reversal potential earlier, the reason is that the resting state is bound to involve some contribution from the electrogenic Na^+/K^+-pump (see e.g. Anderson et al. (2010)) and possibly other electrogenic exchanger pumps. The exchanger pumps consume energy to perform work against concentration gradients, and their activity is not described by the GHK voltage equation.

The equilibrium between passive leakages and pump activity during rest was simulated in the computational model by Sætra, Einevoll & Halnes (2020, supplementary material, S2). The leakage-reversal potential E_L in this model can be computed to be -62.5 mV, while the model's resting potential was approximately -68 mV. The deviation between E_L and the resting potential was mainly due to the electrogenic Na^+/K^+-pump. Hence, the model suggests that the electrogenic pump current in a resting neuron is big enough to shift the membrane potential by a few millivolts.[5]

3.4.6   Intracellular Calcium Dynamics

While most ion concentrations in MC models are assumed to be constant, it is common to make an exception for the intracellular Ca^{2+} concentration $[Ca^{2+}]_i$. The main reason is that $[Ca^{2+}]_i$ is orders of magnitude lower than the concentrations of the other main ion species (Table 3.1). Unlike for the other ion concentrations, quite-dramatic relative changes in $[Ca^{2+}]_i$ can therefore occur on a short time scale – for example, during the opening of Ca^{2+} channels – and will lead to subsequent changes in the Ca^{2+} reversal potential.

Keeping explicit track of $[Ca^{2+}]_i$ allows the modeler to recalculate Ca^{2+} reversal potentials under the time course of simulation. This allows Ca^{2+} currents to be modeled more accurately, and many MC models even employ the GHK equation (equation (3.51)) to describe Ca^{2+} currents even more accurately (Hille 2001, Destexhe & Huguenard 2009).

Perhaps the most obvious need for tracking $[Ca^{2+}]_i$ comes when the model neuron contains Ca^{2+}-gated ion channels – that is, ion channels that open or close as a function of $[Ca^{2+}]_i$. These ion channels are often modeled with the same kind of kinetics scheme as voltage-gated ion channels (equation (3.7)) but with the voltage variable replaced with $[Ca^{2+}]_i$ so that

$$\frac{d\xi([Ca^{2+}]_i, t)}{dt} = \frac{\xi_\infty([Ca^{2+}]_i) - \xi}{\tau_\xi([Ca^{2+}]_i)}, \text{ for } \xi \in \{m, h\}. \qquad (3.56)$$

$[Ca^{2+}]_i$ may change due to influx or efflux over the membrane through ion channels, pumps, or cotransporters, due to a cascade of chemical Ca^{2+} buffering processes where Ca^{2+} ions bind to other molecules with different capacities and kinetics or due to the release or uptake of Ca^{2+} from intracellular stores (Sterratt et al. 2023). Due to the many factors affecting the intracellular Ca^{2+}, the concentration may often vary spatially within the cell and can, for example, be higher near the membrane than in the average bulk solution, at least for a transient period during Ca^{2+} channel influx. Some modelers therefore account for intracellular concentration gradients by, for example, using a model consisting of spherical or cylindrical co-centric shells (Sterratt et al. 2023). However, we will here describe only a simple model for the Ca^{2+} dynamics, based on a framework on the form

---

[5] This means that if the pump is turned off, the membrane resting potential will initially be shifted by a few millivolts. Long-term effects will be more dramatic: without pump activity, the neuron will gradually dissipate its concentration gradients and eventually enter depolarization block.

$$\frac{d[\text{Ca}^{2+}]_i}{dt} = \gamma i_{\text{Ca}} - \frac{[\text{Ca}^{2+}]_i - [\text{Ca}^{2+}]_{i,0}}{\tau_{\text{Ca}}}. \tag{3.57}$$

Here, the first term on the right is an inward membrane current that causes $[\text{Ca}^{2+}]_i$ to increase, and the last term represents the summed activity of an unspecified orchestra of mechanisms that jointly make $[\text{Ca}^{2+}]_i$ decay towards some baseline value $[\text{Ca}^{2+}]_{i,0}$ with a time constant τ_{Ca}. One might expect that the conversion factor γ should be determined by the surface-to-volume ratio of the neural compartment where $[\text{Ca}^{2+}]_i$ is defined so that $\gamma = \text{area}/(\text{volume} \times 2F)$, with F being the Faraday constant. However, in practice, a large fraction of the Ca^{2+} entering through the membrane will not proceed inwards into the cell but will almost immediately bind to other molecules, so that only a small fraction of the entering Ca^{2+} ends up as free Ca^{2+} inside the cytosol (see e.g. Sala & Hernández-Cruz (1990) or Destexhe et al. (1994)). In a modeling context, one may therefore interpret equation (3.57) more phenomenologically and consider γ and τ_{Ca} as tuning parameters, specified to obtain a good match between the model and experimental data on $[\text{Ca}^{2+}]_i$. A simple framework, as in equation (3.57), is often deemed sufficient and is used in many MC models.

In addition to the modeling of Ca^{2+} channels and Ca^{2+}-dependent ion channels, track-keeping of $[\text{Ca}^{2+}]_i$ could be motivated by the aim to reproduce data from Ca^{2+} imaging experiments or, inversely, to use Ca^{2+} imaging data to fit parameters in MC models (see e.g. Taylor et al. (2009)). Furthermore, Ca^{2+} acts as a second messenger in neurons and can trigger a number of intracellular signal transduction cascades (Sterratt et al. 2023, chapter 6), a topic that we do not go into here.

3.4.7 Intracellular Dynamics of Other Ion Species

Although it might be useful for modeling Ca^{2+} dynamics, a simple model as in equation (3.57) might not be meaningful for modeling changes the more abundant ion species (K^+, Na^+, and Cl^-). One reason is that equation (3.57) only considers the intracellular concentration. This makes sense for Ca^{2+}: since the intracellular concentration is much smaller than the extracellular concentration, dramatic relative changes of $[\text{Ca}^{2+}]_i$ can occur without requiring correspondingly dramatic relative changes in the extracellular concentration. The same does not hold for the more abundant species. Since their concentrations are of the same order of magnitude on both sides of the membrane, accurate book-keeping would imply the modeling of both intra- and extracellular concentration dynamics. Modeling of extracellular concentrations will in turn require a compartmentalization of the extracellular space and a suitable numerical scheme for computing the extracellular dynamics.

Another reason is that equation (3.57) is local in the sense that the concentration is affected exclusively by membrane ionic currents. In reality, the axial intracellular currents are also carried by ions, especially by the more abundant ones (Qian & Sejnowski 1989). A complete and consistent model of ion-concentration dynamics and its effect on intra- and extracellular potentials would require that an electrodiffusive framework be used to describe the dynamics both in the intra- and extracellular space (see Figure 2.7D). Such frameworks exist (see e.g. Ellingsrud et al. (2020) and Sætra et al. (2020)), but they are generally computationally heavy and not based on the MC framework that we have presented here.

Possible effects of extracellular ion-concentration dynamics on extracellular potentials are discussed further in Chapter 12.

3.5   Neural Network Models

So far in this chapter, we have introduced the core principles governing the dynamics of individual neurons and their synapses. However, a neuron does not make a brain, and the human brain contains by estimate an average of 86 billion neurons, whereof about 16 billion are in the cerebral cortex (Azevedo et al. 2009, Herculano-Houzel 2009). These neurons are heterogeneous in terms of their biophysical properties and can be classified in terms of their gene expression, brain region, morphological and electrical features, and so on. (Markram et al. 2004, Harris & Shepherd 2015, Zeng & Sanes 2017). Moreover, an average neuron typically forms thousands of synapses with other neurons. The total number of synapses in the human brain is not precisely known, but estimates go as high as 1,000 trillion (10^{15}), and the number of synapses in the human neocortex is believed to be about 150 trillion (Pakkenberg et al. 2003). With this neuron count, this corresponds to roughly 10,000 synapses per neuron. Although 10,000 synapses may sound like a lot, these numbers indicate that the probability that two given neurons (among the 86 billion) are connected is relatively low. As such, the brain is on average a sparsely connected network.

In order to explain different dynamical phenomena and the processing of information that occurs in the circuitry of the brain, a whole host of biologically inspired recurrent (as well as non-recurrent, e.g. feed-forward) neural network models has been developed in the past.[6]

Here, we shall briefly introduce the main types of neural network models accounting for various degrees of biophysical detail and scale, from highly detailed models down to phenomenological population models, as illustrated in Figure 3.8. The main motivation in the context of this book is that, in order to explain extracellular signals generated by groups of neurons, one must also account for the population activity, not only the activity of individual neurons isolated from the network in which they are embedded.

3.5.1   Networks of Biophysically Detailed Neurons

Network models that aim to simultaneously account for the geometrical and electrical properties of neurons – that is, biophysically detailed network models – are illustrated at the top level in Figure 3.8. These networks are typically composed of geometrically detailed MC models instrumented with a mix of linear (e.g. passive leakage) ion

---

[6] A recurrent network refers to a network that involves a loop or feedback mechanism, so that the output of a process is fed back into the input of the same process in a subsequent fashion.

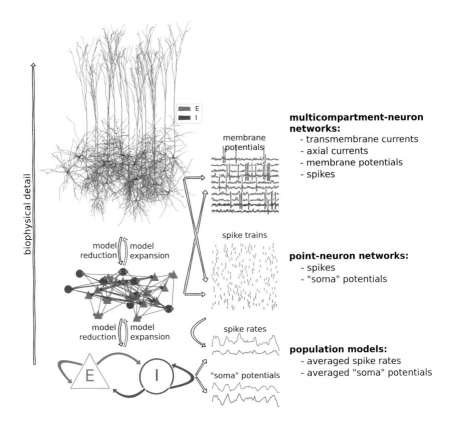

Figure 3.8 Illustration of two-population neural network models at various levels of biophysical detail as well as the typical signals that may be extracted at each level. Less detail typically allows for simulating larger networks on the same computational resources. The downward arrows bridging the different levels of detail denote "model reduction," a systematic approach for reducing the level of biophysical detail while preserving some features of the more detailed model (e.g. spike statistics). Conversely, the upward arrows denote "model expansion," a systematic approach to increase the level of detail (e.g. to compute transmembrane currents). Figure adapted from Hagen et al. (2022) (CC-BY 4.0 International license).

channels, nonlinear (e.g. voltage- or $[Ca^{2+}]_i$-dependent) ion channels, event-activated synaptic currents, and stimulation currents. They account for the membrane-voltage dynamics throughout the neural geometries.

At the technical level, one main complication of implementing an MC-neuron network model of N neurons and simulating it, versus executing N-independent MC neuron simulations, is that one must define the pairwise synaptic connectivity between pre- and postsynaptic neurons. Triggers for synaptic activation are typically APs generated by the presynaptic neuron, which propagates along the axon and, after a certain time delay, activates the synaptic conductance or current in the postsynaptic neuron. In biological neural circuits, this time delay reflects both the axonal transmission time from the site of the AP initialization (typically the soma or axon hillock) to the presynapse as well as the dynamics of neurotransmitter release, diffusion, and uptake, along with possible activation delays on the postsynaptic side. However, as the transmission time for a particular synaptic connection is expected to be fixed, it remains customary to not explicitly model the axon or synaptic interaction in detail. Instead, a membrane-voltage threshold detection step on the upslope of the membrane potential to determine the AP time (or "spike" time) of the presynaptic neuron can be implemented straightforwardly. The corresponding synapse conductances on the postsynaptic neurons are then activated at the time corresponding to the AP time plus an appropriate delay time representing the axonal delay.

Software for simulating biologically inspired network models allows users to not deal directly with the low-level implementation details outlined above. Thus, in terms of constructing biophysically detailed neural network models, the significant effort can then rather be spent on determining the single-neuron and connectivity parameters. These parameters can be set either via anatomical or biophysical constraints from experimental data or via different iterative schemes for parameter optimization, aiming for predictions that explain experimental observations (e.g. sets of spike trains resembling experimental data) or certain mechanisms. The availability of detailed MC-neuron-model implementations representing different neuron classes is continuously growing as quite-streamlined in vitro pipelines for whole-cell patch clamp recordings, cell-geometry reconstruction, and corresponding fitting of conductance parameters to the measured intracellular recordings have been established – for instance, in the context of the Allen Cell Type Database and the Blue Brain Project (Hay et al. 2011, Markram et al. 2015, Ramaswamy et al. 2015, Gouwens et al. 2018). Similarly, paired whole-cell patch clamp recording methods can be used to extract synapse connectivity parameters and rules (pairwise connection probabilities, temporal dynamics, weight dynamics, and so on) for different classes of neurons (Markram et al. 2015, Campagnola et al. 2022).

Another main complicating factor when determining the connectivity parameters in MC-neuron networks is the detailed connectome itself. For MC-neuron networks, the connectivity can be represented by a 3D array where the first dimension represents presynaptic neurons indices, the second the postsynaptic neuron indices, and the third the indices of the compartments corresponding to the synapse locations. Three-dimensional reconstructions of smaller samples of neural tissue are possible using electron microscopy (reviewed by Kubota et al. (2018)). However, instantiating the connectivity in models of larger networks is usually procedural (Reimann et al. 2015, Billeh et al. 2020, Hjorth et al. 2021). Thus, due to the arduous team effort required to determine suitable parameters for use in biophysically detailed network models as well as the computational requirements for running them, the number of published biophysically detailed neural network models remains comparatively low. Some examples of biophysically detailed neural network models are Traub et al. (2005), Markram et al. (2015), and Billeh et al. (2020).

3.5.2   Spiking Point-Neuron Network Models

The most common observables used to experimentally assess neural activity have historically been timings of somatic APs or voltage traces as shown in Figure 3.8. Many (if not most) studies aiming to explain neural population dynamics have therefore deliberately chosen to use simple mathematical models for individual neurons, accounting only for the somatic membrane potential dynamics, or even simpler models, accounting only for the timing of somatic APs. Such models include no spatial information about neural geometry nor on how the membrane potential varies throughout the dendrites and axons. This family of neuron models is often referred to as point-neuron models. The individual point neurons and other network elements (for stimuli, recording, etc.) may, however, be assigned locations corresponding to brain area, cortical layer, feature space (e.g. desired orientation tuning), or similar, which may facilitate connection routines that depend on location and distance between the neurons.

Neuron models that either produce APs or predict the exact timing of APs are commonly referred to as "spiking neuron models," and their sequences of APs are commonly referred to as "spike trains." Since this terminology is prevalent in the neural-network literature, we will adopt it here but note that the term "spike" in this context is used synonymously to "AP." We have, in other parts of the book, used the term "spike" to refer exclusively to the extracellular signature of the AP. However, since APs are indeed the source of extracellular spikes, the reference to a spiking neuron should hopefully not cause any confusion.

The difference between point-neuron models and the previously introduced single-compartment neuron models is mainly a technicality. Single-compartment models contain some structural information concerning compartment size and are typically parameterized in terms of specific membrane capacitance and conductances (per membrane area). In contrast, point-neuron models have no spatial extension and are parameterized in terms of absolute capacitance and conductances. It is, however, straightforward to compute the absolute capacitance and conductances for a single-compartment model and use it as a point-model. A spiking point-neuron model could in this way be obtained from the single-compartment Hodgkin-Huxley (HH) model illustrated in Figure 3.4.

As the nonlinear AP-generating mechanisms in the HH model are computationally expensive, much-simpler spiking point-neuron models have been developed. Typically, these simpler models do not account for the shape of the AP itself. Instead, they simply predict the time of its occurrence by combining a model of the subthreshold voltage dynamics with a threshold value at which the AP is postulated to occur. Next, the membrane voltage is simply reset to a predetermined subthreshold value and clamped for some duration in order to mimic the AP refractory period. In the simplest models of this kind, the subthreshold dynamics are given by the passive model in equation (3.20), which for a point neuron becomes

$$\tau_m \frac{dV_m}{dt} = -(V_m - E_L) + R_m I_{syn} + R_m I_{ext}, \qquad (3.58)$$

where $\tau_m = C_m R_m$. More sophisticated spiking point-neuron models exist, introducing for instance nonlinearities in subthreshold dynamics or spike threshold dynamics that are dependent on the spiking history or stochastic processes (Lapicque 1907, Brunel & van Rossum 2007, Brunel 2013, Gerstner et al. 2014). Interested readers are advised to consult the more comprehensive reviews of these types of point-neuron models by Brunel (2013) and Gerstner et al. (2014). We note that describing the somatic voltage (or only the subthreshold part of it) is not a strict requirement for point neurons – they may very well be represented by a statistical model for spike emission as a function of the synaptic input current (e.g. McCulloch & Pitts (1943)).

Their mathematical simplicity lends point neurons to be implemented and simulated very efficiently in comparison to biophysically detailed MC-neuron models. In the case of linear subthreshold dynamics, membrane potential values and spike times of point neurons may also be solved for exactly on temporally discrete time grids (Rotter & Diesmann 1999). As a consequence, the dynamics of recurrently connected networks of spiking point neurons can be studied at a much greater scale and pace than similarly sized networks of MC-neuron models. As typical point neurons have very few intrinsic parameters compared to MC-neuron models, the use of point neurons in network simulations has allowed effects of network connectivity to be separated from cellular dynamics. This separability has in turn allowed for a rigorous mathematical analysis of observed phenomena such as asynchronous or synchronous, regular, and irregular spiking activity (see e.g. Brunel (2000)).

Point-neuron network models can accurately capture observations present in experimental data gathered in vivo, including irregular firing patterns (Softky & Koch 1993, Amit & Brunel 1997, van Vreeswijk & Sompolinsky 1997, Shadlen & Newsome 1998), fluctuations of the membrane potential (Destexhe et al. 1999), asynchronous firing patterns (Ecker et al. 2010, Renart et al. 2010, Ostojic 2014), correlations in neural activity (Okun & Lampl 2008, Gentet et al. 2010, Helias et al. 2013), self-sustained activity (Kriener et al. 2014), and realistic firing rates across cortical lamina (Potjans & Diesmann 2014, Billeh et al. 2020). Similarly, reduced models derived from biophysically detailed MC-neuron models have shown that point-neuron network models can accurately capture spiking statistics of the biophysically detailed model variants (see e.g. Rössert et al. (2016), Billeh et al. (2020)). In the case of Rössert et al. (2016), the effect of dendritic filtering on synaptic input currents is captured by computing an equivalent filter for each possible synapse location and type using a Green's function formalism (Wybo et al. 2013). The point neurons are modeled using a generalized integrate-and-fire formalism (Jolivet et al. 2004, Pozzorini et al. 2015), and the linear filters applied to the synaptic currents result in well-preserved postsynaptic potentials and output spike trains in the point neurons corresponding to the biophysically detailed neuron models.

3.5.3   Neural Population Models

As indicated in the previous subsection, the spiking dynamics of large-scale MC-neuron networks can sometimes be preserved in simplified networks of few-compartment or spiking point neurons. However, when describing brain dynamics on the very large scale, simulating each individual spike of each individual neuron becomes computationally expensive even if one uses the simplest available model for the individual spiking neurons. Further steps of simplification may therefore be made in order to describe neural dynamics at the level of populations, as illustrated at the bottom end of the chart in Figure 3.8. Such models are commonly referred to as *rate models* and are typically used to model brain activity at the *mesoscopic* (local populations, brain areas) and *macroscopic* (whole brain) levels.

Some rate models describe the mean activity of subsets of neurons ("populations") as a function of time alone, but other rate-based formalisms exist that describe activity both in space and time. The former type is referred to as *neural-mass* or *population-rate* models, while the latter type is referred to as *neural-field* models. Neural-field models have been used extensively for describing phenomena such as stationary activity bumps, traveling activity waves, and advancing activity wavefronts (see e.g. Amari (1977), Deco et al. (2008), Coombes et al. (2014)). Some rate-model frameworks have been used to describe the activity of individual neurons (Ermentrout & Terman 2010, chapter 11). Other more mathematically comprehensive frameworks allow for a more direct mapping of model parameters between spiking neural-network models and population-rate models (Cain et al. 2016, Schwalger et al. 2017, Senk et al. 2020).

Here, we will not delve into much mathematical detail as the body of literature on both rate models and neural-field models and their derivations is vast – see, for example Ermentrout & Terman (2010) and Gerstner et al. (2014). To give a simple example, the firing rate $v(t)$ in a single-population rate model may be described by (Gerstner et al. 2014, equation 15.17)

$$\tau_m \frac{\partial v(t)}{\partial t} = -v(t) + G\left(I(t)\right) , \qquad (3.59)$$

where τ_m is the time constant of the effective low-pass filter characterizing the system, $I(t)$ describes the input current (from external and recurrent sources), and G is a stationary gain function here (see Gerstner et al. (2014) for details).

In the case of neural fields, the rate model for a single layer in a recurrently connected structure may take the form (Amari 1977):

$$\tau_m \frac{\partial u(x,t)}{\partial t} = -u(x,t) + \int w(x-y)f\left(u(y,t)\right) \, dy + s(x,t) , \text{ where} \qquad (3.60)$$

$$f(u) = \begin{cases} 0, u \leq 0 \\ 1, u > 1 , \end{cases} \qquad (3.61)$$

$$w(x-y) = w(x,y) . \qquad (3.62)$$

Here $u(x,t)$ is a dynamical variable representing the net input to neurons located at point x on a line so that the population firing rate of these neurons is given by $f(u)$, while $w(x)$ denotes a distance-dependent weighting function (representing, for example, a Mexican hat local connectivity pattern), and $s(x,t)$ represents the stimulus. Amari (1977) investigated such neural-field equations to demonstrate mechanisms of dynamical patterns like traveling waves and global oscillations. Frameworks like The Virtual Brain (Ritter et al. 2013, Sanz-Leon et al. 2013, Sanz-Leon et al. 2015) use related formalisms to account for population-averaged whole-brain activity across parcellations (brain areas).

3.6   Discussion: From Neurodynamics to Electric Brain Signals

The MC modeling framework is a well-established and well-tested formalism. However, like any modeling framework, it is a simplified representation of the real biological system and rests on a series of simplifying assumptions. Since MC-type models have become a pillar in modern computational neuroscience, their weaknesses and general applicability have been the subject of numerous critical studies (see e.g. Rinzel (1990), Koch (1999, chapter 2), Meunier & Segev (2002), Lindsay et al. (2004), Catterall et al. (2012), Almog & Korngreen (2016), Drukarch et al. (2018)). We will not give a exhaustive review of the concerns that may be raised with MC models, but we will discuss some of their limitations in Section 13.2.

The main focus of this book is not on the modeling of neural dynamics as such, but on how various aspects of neural dynamics (once modeled) give rise to extracellular potentials. Before we take the leap into the extracellular space, we briefly summarize what it is that we bring along from the neurodynamics described in the current chapter.

As we explained in Section 2.6.2, we will in most parts of this book use a two-step MC+VC scheme, which combines multicompartment models of neurons (the MC part) with volume-conductor theory (the VC part) to model the extracellular potential V_e. The MC framework presented in the current chapter is a recipe for accomplishing step 1. The only thing we need from the MC simulation to compute the extracellular potential is the distribution of membrane currents. This distribution can be abstractly represented in terms of a current-source density C_t, as in equation (2.31). However, in practice, it is often described in terms of a set of point sources – that is, one membrane-current source per neural compartment in the MC model. In the next chapter, we will present the volume-conductor (VC) theory for modeling V_e, which is the recipe for accomplishing step 2, where the extracellular potentials are computed from the known distribution of membrane currents.

Although glial cells are less electrically active than neurons, glial membrane currents will in principle contribute to V_e also. The MC framework, which we presented as a way to model neurons, is equally well suited to model the electric activity of glial cells. From a modeling point of view, the main difference between these two cell types is the membrane mechanisms that they possess, but the same Hodgkin-Huxley-type formalism should be able to capture the dynamics of glial membrane mechanisms also.

In cases where the neurodynamics is not computed using the MC framework but with more abstract models, such as point neurons (as in Section 3.5.2) or firing-rate models (as in Section 3.5.3), the mapping from neurodynamics to extracellular potentials is less trivial. The reason is that the spatial distribution of membrane currents, and thus the biophysical origin for the extracellular potentials, are missing from such abstract models. Several mapping strategies that allow such abstract models to be used to predict extracellular potentials have nevertheless been developed, and some of these are presented in Section 6.4.

4 Volume-Conductor Theory

As we saw in the previous chapter, multicompartment (MC) models of morphologically detailed neurons allow us to compute the membrane currents at various locations on the neural membrane. This chapter is about how we, when we know the distribution of neural membrane currents, can use volume-conductor (VC) theory to predict the resulting extracellular potential V_e at any given point in space.

VC theory, which treats brain tissue at the macroscopic scale as a continuous volume conductor, is the foundation of the forward modeling of extracellular potentials inside and outside the brain, from extracellular spikes and LFPs to ECoGs and EEGs. VC theory is also used in the forward modeling of magnetic fields, but we will postpone the magnetic theory to Chapter 11.

A general starting point for VC theory is the current-source density (CSD) equation that we introduced in Section 2.6.1. However, starting out with the CSD equation can appear a bit abstract and mathematical and it is not always necessary. Before we revisit and discuss the CSD equation thoroughly in Section 4.3, we will instead establish a basic and intuitive understanding of VC theory by following a simpler path, using a few simple idealized examples as starting points. Before we begin, let us emphasize that all computations of V_e in the current chapter will be based on the following assumptions:

1. Neural membrane currents are known from a previous independent simulation, meaning that we are using the two-step (MC+VC) scheme defined in Section 2.6.2. When doing this, we implicitly assume that changes in V_e evoked by neurodynamics do not (ephaptically) act back on neurodynamics (see Section 13.2 for further discussion).
2. Brain tissue can be represented as a continuous medium, as argued in Section 2.3.2 and defined in further detail in Chapter 5.
3. The current density in the tissue depends linearly on the electric field and is given by Ohm's law for volume conductors:

$$\mathbf{i}_t(\mathbf{r}) = -\sigma_t \nabla V_e(\mathbf{r}), \qquad (4.1)$$

where the subscript "t" indicates that $\mathbf{i}_t$ is a macroscopic current density through *tissue*, defined as current per tissue cross-section area, and σ_t is the macroscopic tissue conductivity.[1]

---

[1] For more detailed definitions of $\mathbf{i}_t$ and σ_t, see Chapter 5.

4.   The effective tissue conductivity (σ_t) is constant (does not vary with space or time), the same in all spatial directions (isotropic medium), and frequency-independent (no capacitive effects on tissue currents).

We note that while the three first assumptions (1–3) are prerequisites for using VC theory in the standard form presented here, the assumption (4) that σ_t is a constant is just a simplification that we make in the current chapter. Ways to apply VC theory with an inhomogeneous, anisotropic, or frequency-dependent σ_t are discussed in Chapter 5.

4.1   Point-Source Approximation

Currents that come out from MC model simulations are typically represented as a set of discrete current sources (one source per neural compartment), as we illustrated previously in Figure 2.6. Let us therefore start as simple as possible and derive the contribution to V_e from a single point source (I_0) at position $\mathbf{r}_0 = 0$ (Figure 4.1A).

A simple way to derive an expression for V_e in this example is to first exploit the spherical symmetry of the problem. Due to the symmetry, we know that the tissue current density $\mathbf{i}_t$ evoked by I_0 must be radially directed, so its magnitude is solely a function of distance ($r = \|\mathbf{r}\|$) from the source. This allows us to drop the vector notation in equation (4.1) and express the current density as the scalar function

$$i_t(r) = -\sigma_t \frac{dV_e(r)}{dr} \; , \tag{4.2}$$

where the radial directionality is kept implicit.

Next, we consider a spherical surface with the source in its center. The net current leaving this surface must equal the current density $i_t(r)$ on the surface multiplied with the surface area $4\pi r^2$. Due to the principle of current conservation, we also know that the net current leaving any surface enclosing the source must equal I_0 because otherwise charge would pile up inside the surface. Hence, we can equate

$$I_0 = 4\pi r^2 i_t = -4\pi \sigma_t r^2 \frac{dV_e(r)}{dr} \; , \tag{4.3}$$

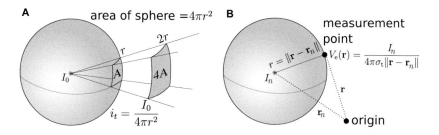

Figure 4.1 Extracellular potential from a single neural-point source current. A: Point source I_0 in the origin, $r = 0$. B: Point source I_n at position $\mathbf{r}_n$.

which we can rearrange to get

$$\frac{dV_e(r)}{dr} = -\frac{I_0}{4\pi\sigma_t r^2} .$$ (4.4)

To obtain the final solution for V_e, we define our reference point (ground) for the potential to be infinitely far away ($V_e(\infty) = 0$) and integrate equation (4.4) from ∞ to r:

$$\int_\infty^r \frac{dV_e(r')}{dr'} dr' = \int_\infty^r -\frac{I_0}{4\pi\sigma_t r'^2} dr' .$$ (4.5)

This leads to the final expression

$$V_e(r) = \frac{I_0}{4\pi\sigma_t r} ,$$ (4.6)

where r is the distance from the source.

In the example above, we made things mathematically transparent by assuming that the current source was placed in the origin $r = 0$. For a point source I_n located at an arbitrary point $\mathbf{r}_n$ (Figure 4.1B), the corresponding expression for the extracellular potential is

$$V_e(\mathbf{r}) = \frac{I_n}{4\pi\sigma_t\|\mathbf{r} - \mathbf{r}_n\|} .$$ (4.7)

We note that the single point source that we have looked at so far cannot represent a neuron since the sum of the currents crossing a neural membrane at various locations is always zero. However, the activity of a neuron can be represented as an assembly of such point sources, $I_1, I_2, I_3, \ldots$ in locations $\mathbf{r}_1, \mathbf{r}_2, \mathbf{r}_3, \ldots$, which collectively sum up to a zero net membrane current. As we stated in the beginning of this chapter, we assume that the extracellular medium is linear. This means that the contributions from the individual point sources add up linearly, and the potential at position $\mathbf{r}$ is given by

$$V_e(\mathbf{r}) = \frac{I_1}{4\pi\sigma_t\|\mathbf{r} - \mathbf{r}_1\|} + \frac{I_2}{4\pi\sigma_t\|\mathbf{r} - \mathbf{r}_2\|} + \frac{I_3}{4\pi\sigma_t\|\mathbf{r} - \mathbf{r}_3\|} + \ldots = \sum_n \frac{I_n}{4\pi\sigma_t\|\mathbf{r} - \mathbf{r}_n\|} .$$
(4.8)

Equation (4.8) is referred to as the point-source approximation (Holt & Koch 1999) since it approximates the neuron as a set of point-current sources. That is, each compartment in a multicompartment neuron model is assumed to deliver a singular current to the extracellular space, typically chosen to be located at the compartment midpoint.

4.2 Line-Source Approximation

A more sophisticated alternative to making the point-source approximation is to assume that the membrane current is evenly distributed over the compartment axis, a choice that is referred to as the *line-source approximation* (Holt 1998, Holt & Koch 1999, Lindén et al. 2014).

The idea behind the line-source approximation is to approximate a straight line segment (think "cylinder axis of a neural compartment") as a uniformly distributed continuum of infinitesimal point sources. The contribution to the extracellular potential

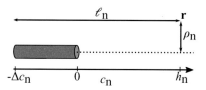

Figure 4.2 Line-source approximation. Definitions of the symbols used in the line-source approximation.

from a current I_n distributed over a compartment n can then be found analytically by integrating equation (4.7) over the center-line axis of the compartment (Holt 1998, appendix C).

If we consider a compartment of length Δc_n with a net membrane current I_n, the current from an infinitesimal part dc_n of the compartment is given by $I_n dc_n/\Delta c_n$. Using the coordinate system illustrated in Figure 4.2, the integral to solve is then

$$V_e(\mathbf{r}) = V_e(h_n, \rho_n, \Delta c_n) = \frac{1}{4\pi\sigma_t} \frac{I_n}{\Delta c_n} \int_{-\Delta c_n}^{0} \frac{dc_n}{\sqrt{(h_n - c_n)^2 + \rho_n^2}}. \qquad (4.9)$$

Here, ρ_n is the distance perpendicular to the compartment's center axis, and h_n is the longitudinal distance from the end of the compartment. The solution to this integral is

$$V_e(\mathbf{r}) = \frac{1}{4\pi\sigma_t} \frac{I_n}{\Delta c_n} \log \left| \frac{\sqrt{h_n^2 + \rho_n^2} - h_n}{\sqrt{\ell_n^2 + \rho_n^2} - \ell_n} \right|, \qquad (4.10)$$

where $\ell_n = \Delta c_n + h_n$ is the longitudinal distance from the start of the compartment to $\mathbf{r}$.

Note that equation (4.10) is only valid outside the neuron, which puts a lower bound on ρ_n. In MC models, neural compartments are normally represented as cylinders, and equation (4.10) is then valid for ρ_n greater than the cylinder radius. Outside the cylinder, it has been shown that the line-source approximation gives essentially the same result as a model where the effects of the cylindrical geometry on V_e is explicitly accounted for (see Holt (1998, appendix C)).

Equations (4.10) and (4.7) are analogous expressions for the cases where the membrane currents in a single neural compartment are treated as a line source or point source, respectively. Like for the point-source approximation, the contributions from individual line sources add up linearly. If we have multiple compartments n, the extracellular potential can therefore be computed as

$$V_e(\mathbf{r}) = \frac{1}{4\pi\sigma_t} \sum_n \frac{I_n}{\Delta c_n} \log \left| \frac{\sqrt{h_n^2 + \rho_n^2} - h_n}{\sqrt{\ell_n^2 + \rho_n^2} - \ell_n} \right|. \qquad (4.11)$$

We note that the different compartments here in general will be oriented at different angles. The line-source approximation can be expected to give better predictions of V_e than the point-source approximation at points in space that are very near neural membranes (Holt & Koch 1999). At points further away from membranes, the two approximations give converging predictions (Figure 4.3).

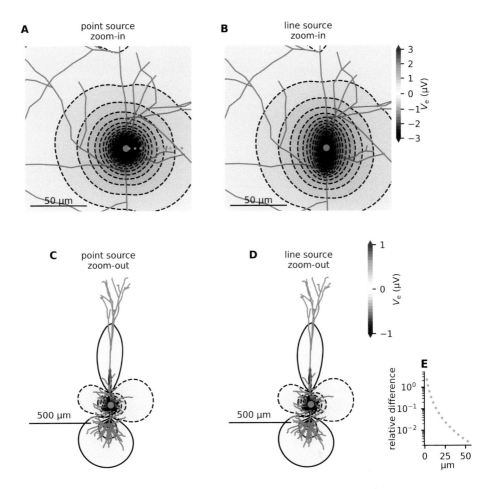

Figure 4.3 Point-source versus line-source approximation. A: Instantaneous extracellular potential following a single excitatory synaptic input to the Hay model, calculated with the point-source approximation. **B**: Same as panel A, using the line-source approximation. **C, D**: Same as panels A and B, respectively, but zoomed out showing the entire Hay model and extracellular potential. **E**: The relative difference between the results from the two approximations as a function of distance (along the orange dotted line in panel A) from the synapse location (blue dot). The relative difference was less than 10 percent for distances greater than ∼13 μm and less than 1 percent for distances greater than ∼35 μm. In all cases, V_e is shown at the time point when the maximum absolute value was observed. Code available via www.cambridge.org/electricbrainsignals.

4.3   Current-Source Density Description

Instead of representing neural output as a set of point or line sources, we can represent it more generally in terms of a *current-source density* (CSD), which we denote as C_t (Figure 4.4). The CSD features in the general current-conservation equation,

$$\nabla \cdot \mathbf{i}_t = C_t \tag{4.12}$$

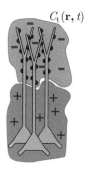

$C_t(\mathbf{r}, t)$

Figure 4.4 Neural current sources. When simulated using a numerical scheme, here illustrated by a population of pyramidal cells receiving excitatory apical synaptic input, the membrane currents are known at a discrete set of neural compartments and represent a set of point sources. As a mathematical generality, we can describe neural membrane currents as a continuous CSD distribution $C_t(\mathbf{r}, t)$. Note also that the sum of the currents entering/leaving every single neuron is always zero, so the spatial integral over $C_t(\mathbf{r}, t)$ must be zero at all times.

for coarse-grained neural tissue, which we have seen earlier as equation (2.30) and refer to as the CSD equation. The CSD equation determines the relationship between any distribution of sources (represented by C_t) and the currents in the surrounding tissue (represented by $\mathbf{i}_t$).

In Section 4.3.1, we first focus on the use of the CSD equation. In Section 4.3.2, we show how this equation can be derived, and in Section 4.3.3, we provide some further discussion on its interpretation. Parts of this section is quite mathematical, and some readers may choose to skip it since forward modeling in practice is normally based on using either the point-source or line-source formalism presented in the preceding sections.

4.3.1   Use of the Current-Source Density Equation

The CSD is what one traditionally has estimated when doing inverse modeling of the local field potential (LFP) – that is, when one aims to infer the underlying distribution of cellular membrane sources explaining recorded extracellular potentials. Defining neural activity in terms of a CSD can also be useful within the forward modeling context and has, for example, been used to compute extracellular potentials from assumed CSD distributions (Pettersen et al. 2006, Einevoll et al. 2007, Cserpan et al. 2017). However, forward modeling is more commonly based on the MC+VC framework, where the neural output is represented as a set of point sources (Section 4.1) or line sources (Section 4.2).

The CSD equation essentially just constitutes a more general starting point for VC theory than the point-source or line-source representations. The CSD is a density distribution (with unit A/m^3) and thus a spatially continuous function, but as we show later, discontinuities – such as point sources – can be represented as a CSD with the use of the Dirac delta function.

Since VC theory is based on the assumption that i_t is ohmic (equation (4.1)), and since we in the current chapter assume that σ_t is constant and isotropic, equation (4.12) reduces to a direct relationship between the Laplacian (∇^2) of the extracellular potential and the CSD:

$$\sigma_t \nabla^2 V_e = -C_t, \qquad (4.13)$$

which we have seen earlier as equation (2.32). The essential idea behind VC theory is that we can solve equation (4.13) for V_e once C_t is known.

A standard mathematical strategy for solving linear partial differential equations, such as equation (4.13), is to (i) first find its Green's function G and then to (ii) express the general solution as a convolution of a general source C_t with G. By definition, G is the so-called *impulse response* to a unitary singular source $C_t(\mathbf{r}) = \delta^3(\mathbf{r} - \mathbf{r}_n)$ at some position $\mathbf{r}_n$, where $\delta^3(\mathbf{r})$ is the Dirac delta function in three dimensions. Also, G should have the same boundary condition as V_e at infinity, where both should be zero. Hence, we seek the solution to

$$\sigma_t \nabla^2 G(\mathbf{r}) = -\delta^3(\mathbf{r} - \mathbf{r}_n), \qquad (4.14)$$

with the boundary condition $G(\infty) = 0$. We note that an impulse response given by $C_t(\mathbf{r}) = I_n \delta^3(\mathbf{r} - \mathbf{r}_n)$ is the equivalent CSD description of the singular current source we considered earlier. That means that we already have the solution to equation (4.14): it is given by equation (4.7), with $I_n = 1$. However, we provide here an alternative derivation of the same result by integrating over the delta function.

To proceed, we divide both sides of equation (4.14) with σ_t and integrate over an arbitrary 3D volume containing the impulse source:

$$\iiint \nabla^2 G(\mathbf{r}) \, dv = -\frac{1}{\sigma_t} \iiint \delta^3(\mathbf{r} - \mathbf{r}_n) \, dv = -\frac{1}{\sigma_t}, \qquad (4.15)$$

where the final step follows from the definition of the 3D Dirac delta function. Using Gauss's theorem, we can convert the volume integral on the left-hand side of equation (4.15) to an integral over the surface (s) that encloses our arbitrary volume, so that equation (4.15) becomes

$$\oiint \nabla G(\mathbf{r}) \cdot d\mathbf{s} = -\frac{1}{\sigma_t}. \qquad (4.16)$$

To solve the surface integral, it is convenient to choose our arbitrary volume to be a sphere centered at the source location $\mathbf{r}_n$ and with radius $R = \|\mathbf{r} - \mathbf{r}_n\|$ (Figure 4.1B). Due to the symmetry of the problem, we then know that G must be the same over the whole spherical surface so that $G(\mathbf{r}) = G(R)$. We also know that its gradient is perpendicular to the surface-area increment $d\mathbf{s} = \mathbf{e}_r ds$ and constant over the surface, so that

$$\nabla G(\mathbf{r}) = \frac{dG(R)}{dR} \mathbf{e}_r. \qquad (4.17)$$

If we use this, equation (4.16) simplifies to

$$\frac{dG(R)}{dR} \oiint d\mathbf{s} = -\frac{1}{\sigma_t}, \qquad (4.18)$$

which has the solution

$$4\pi R^2 \frac{dG(R)}{dR} = -\frac{1}{\sigma_t} .$$ (4.19)

We may now integrate this from ∞ to R and use $G(\infty) = 0$ to get

$$G(R) = \frac{1}{4\pi\sigma_t R} .$$ (4.20)

Finally, we use $G(R) = G(\mathbf{r})$ and $R = \|\mathbf{r} - \mathbf{r}_n\|$ to obtain the desired Green's function

$$G(\mathbf{r}) = \frac{1}{4\pi\sigma_t\|\mathbf{r} - \mathbf{r}_n\|} .$$ (4.21)

As earlier stated, the solution of equation (4.13) for a general function $C_t(\mathbf{r}')$ can be expressed as a convolution of $C_t(\mathbf{r}')$ with G, meaning that

$$V_e(\mathbf{r}) = \frac{1}{4\pi\sigma_t} \iiint \frac{C_t(\mathbf{r}')}{\|\mathbf{r} - \mathbf{r}'\|} dv' ,$$ (4.22)

where the volume integral runs over an arbitrary volume containing all the sources – that is, it must run over all $\mathbf{r}'$ where $C_t(\mathbf{r}') \neq 0$. Equation (4.22) is the continuous counterpart to equation (4.8), allowing the current-source density (C_t) to be any function. If we describe the CSD as a sum of point sources, so that $C_t(\mathbf{r}) = \sum_n I_n \delta^3(\mathbf{r} - \mathbf{r}_n)$, equation (4.22) reduces to equation (4.8).

4.3.2   Derivation of the Current-Source Density Equation

So far, we have used the CSD equation (equation (4.12)) more or less as a postulate. CSD analysis was introduced by Pitts (1952), and its theoretical and practical basis was further established in the early works by Nicholson and coworkers (Nicholson & Llinas 1971, Nicholson 1973, Freeman & Nicholson 1975, Nicholson & Freeman 1975). A more detailed derivation was presented later by Gratiy et al. (2017), taking a more microscopic starting point. This detailed derivation is presented here.

4.3.2.1   Conservation of Microscopically Defined Current

The starting point is a total, microscopically defined, current density $\mathbf{i}_\mu$.[2] By "microscopic," we mean that it can be an intracellular, extracellular, or membrane-current density, depending on its position in space. It can generally consist of all sorts of currents (resistive, diffusive, displacement, advective). Since it contains all currents (including displacement currents), we know that it is conserved (see equation (2.38)), so that

$$\nabla \cdot \mathbf{i}_\mu = 0$$ (4.23)

should be satisfied at all points in space.

---

[2] $\mathbf{i}_\mu$ is still macroscopic enough for a continuity description to be warranted (see Section 2.3.1).

4.3.2.2   Definition of Coarse-Grained Current

The microscopic $\mathbf{i}_\mu$ is impractical for describing large-scale tissue processes as this would force us to map out the detailed structure of tissue with extremely fine resolution. We therefore define a coarse-grained version of this current density,

$$\langle \mathbf{i}_\mu(\mathbf{r}) \rangle = \iiint \mathbf{i}_\mu(\mathbf{r}')w(\mathbf{r} - \mathbf{r}')\, dv' \tag{4.24}$$

that is applicable on a larger spatial scale (Section 2.3.2). We have introduced here an averaging kernel $w(\mathbf{r})$, which defines the coarse-graining scale. The exact shape of $w(\mathbf{r} - \mathbf{r}')$ is not critical for the derivation of the CSD equation. Its essence is to put weight on the currents present in a basin around position $\mathbf{r}$, and we may picture it to have a constant plateau-value within some range around position $\mathbf{r}$ while it decays smoothly to zero at a distance $\|\mathbf{r} - \mathbf{r}'\| = R$. As we wish to average over a volume large enough to not critically depend on microstructures in the tissue, R should be much larger than the typical diameter of a dendrite and should be at least some tens of micrometers (see Section 2.3.2). The kernel has unit m^{-3} and is normalized to unity:

$$\iiint w(\mathbf{r} - \mathbf{r}')\, dv' = 1 . \tag{4.25}$$

4.3.2.3   Conservation of Coarse-Grained Current

The coarse-grained current is also conserved. To show that this is true, we compute the averaging

$$\nabla \cdot \langle \mathbf{i}_\mu(\mathbf{r}) \rangle = \iiint \mathbf{i}_\mu(\mathbf{r}')\nabla w(\mathbf{r} - \mathbf{r}')\, dv' , \tag{4.26}$$

where the nabla-operator could be moved inside the integration since it operates on $\mathbf{r}$ and not the integration variable $\mathbf{r}'$. Since $\nabla w(\mathbf{r} - \mathbf{r}') = -\nabla' w(\mathbf{r} - \mathbf{r}')$, we may rewrite this integral as

$$\nabla \cdot \langle \mathbf{i}_\mu(\mathbf{r}) \rangle = - \iiint \mathbf{i}_\mu(\mathbf{r}')\nabla' w(\mathbf{r} - \mathbf{r}')\, dv' . \tag{4.27}$$

Further, using the vector identity

$$\mathbf{i}_\mu \cdot \nabla w = \nabla \cdot (w\mathbf{i}_\mu) - w\nabla \cdot \mathbf{i}_\mu, \tag{4.28}$$

we can write equation (4.27) as

$$\nabla \cdot \langle \mathbf{i}_\mu(\mathbf{r}) \rangle = \iiint \nabla' \cdot \left(w(\mathbf{r} - \mathbf{r}')\mathbf{i}_\mu(\mathbf{r}')\right)\, dv' - \iiint w(\mathbf{r} - \mathbf{r}')\nabla' \cdot \mathbf{i}_\mu(\mathbf{r}')\, dv'. \tag{4.29}$$

The last term in this equation is by definition $\langle \nabla \cdot \mathbf{i}_\mu \rangle$, which according to equation (4.23) must be zero. The second term can be converted to the surface integral by use of the divergence theorem, so that equation (4.29) becomes

$$\nabla \cdot \langle \mathbf{i}_\mu(\mathbf{r}) \rangle = \oiint \mathbf{i}_\mu(\mathbf{r}')w(\mathbf{r} - \mathbf{r}')ds' . \tag{4.30}$$

Since the averaging was performed over the entire space, the integral is taken over a surface infinitely removed from the position $\mathbf{r}$, and since the averaging kernel w is nonzero only in the local region surrounding $\mathbf{r}$, the integral must be zero. Hence, a conservation law

$$\nabla \cdot \langle \mathbf{i}_\mu(\mathbf{r}) \rangle = 0 \tag{4.31}$$

also applies for the coarse-grained current.

4.3.2.4 Splitting into Cellular and Extracellular Domains

To proceed with the derivation of the CSD equation, we formally perform the spatial averaging separately over the cellular (c) and extracellular (e) domains of the reference volume, so that we can write:

$$\langle \mathbf{i}_\mu \rangle = \langle \mathbf{i}_\mu \rangle_c + \langle \mathbf{i}_\mu \rangle_e \,. \tag{4.32}$$

The conservation law (equation (4.31)) then becomes

$$\nabla \cdot \langle \mathbf{i}_\mu \rangle_c + \nabla \cdot \langle \mathbf{i}_\mu \rangle_e = 0 \,. \tag{4.33}$$

The advantage with this coarse-grained resolution is that it introduces a *bi-domain representation*, where we may think of the cellular and extracellular spaces as two coexisting and interacting domains that are both defined continuously over the whole tissue space rather than within their respective subspaces (Gratiy et al. 2017).

4.3.2.5 Definition of Extracellular Volume-Current Density

In the splitting, we let the boundary of the cellular domain(s) run on the external surface of the cellular membranes so that $\langle \mathbf{i}_\mu \rangle_c$ is the cellular current density averaged over the cytoplasm *and* membrane. This is important because it leaves us with an extracellular domain (e) that is electroneutral since it contains no membrane-current contributions, thus no capacitive (displacement) currents.[3] For brevity, we may define

$$\mathbf{i}_e = \langle \mathbf{i}_\mu \rangle_e \,, \tag{4.34}$$

where we let it be implicit that $\mathbf{i}_e$ is the extracellular volume-current density defined on a coarse-grained scale. We typically assume that $\mathbf{i}_e$ is purely ohmic, so that

$$\mathbf{i}_e = -\sigma_e \nabla V_e \,, \tag{4.35}$$

although it in principle could contain additional contributions from diffusion and advection. Here, σ_e is the conductivity of the extracellular medium (see Section 5.2 for the definition).

4.3.2.6 Definition of Current-Source Density

Unlike the extracellular domain, the cellular domain is (microscopically) electroneutral only internally in the cytoplasm, while the membrane currents at the cellular boundaries include capacitive currents that represent charge piling up on either side of the membranes when membrane potentials vary. This does not cause us any trouble, as we

---

[3] With the assumption that the extracellular medium is purely ohmic, all capacitive currents in brain tissue are membrane currents.

do not need to explicitly express the currents on the cellular domain to derive the coarse-grained CSD equation.

To obtain the CSD, we instead start by considering the averaging expression over the cellular domain:

$$\langle \mathbf{i}_\mu(\mathbf{r})\rangle_c = \iiint_{v_c} \mathbf{i}_\mu(\mathbf{r}')w(\mathbf{r}-\mathbf{r}')\,dv' \,. \tag{4.36}$$

This integral runs over all cellular volumes (v_c) within the (infinite) averaging volume but puts weight only on contributions inside the kernel (where $\|\mathbf{r}-\mathbf{r}'\| < R$). We take the divergence of this expression to obtain

$$\nabla \cdot \langle \mathbf{i}_\mu(\mathbf{r})\rangle_c = \iiint_{v_c} \mathbf{i}_\mu(\mathbf{r}')\nabla w(\mathbf{r}-\mathbf{r}')\,dv' \,, \tag{4.37}$$

where the nabla-operator could be moved inside the integration since it operates on $\mathbf{r}$ and not the integration variable $\mathbf{r}'$.

Since $\nabla w(\mathbf{r}-\mathbf{r}') = -\nabla' w(\mathbf{r}-\mathbf{r}')$, equation (4.37) can be rewritten as

$$\nabla \cdot \langle \mathbf{i}_\mu(\mathbf{r})\rangle_c = - \iiint_{v_c} \mathbf{i}_\mu(\mathbf{r}')\nabla' w(\mathbf{r}-\mathbf{r}')\,dv' \,. \tag{4.38}$$

The chain rule allows us to rewrite the integrand as

$$- \mathbf{i}_\mu(\mathbf{r}')\nabla' w(\mathbf{r}-\mathbf{r}') = -\nabla' \cdot \big(w(\mathbf{r}-\mathbf{r}')\mathbf{i}_\mu(\mathbf{r}')\big) + w(\mathbf{r}-\mathbf{r}')\nabla' \cdot \mathbf{i}_\mu(\mathbf{r}') \,. \tag{4.39}$$

With this, equation (4.37) becomes

$$\begin{aligned}\nabla \cdot \langle \mathbf{i}_\mu\rangle_c &= - \iiint_{v_c} \nabla' \cdot \big(w(\mathbf{r}-\mathbf{r}')\mathbf{i}_\mu(\mathbf{r}')\big)\,dv' + \iiint_{v_c} w(\mathbf{r}-\mathbf{r}')\nabla \cdot \mathbf{i}_\mu(\mathbf{r}')\,dv' \\ &= - \iiint_{v_c} \nabla' \cdot \big(w(\mathbf{r}-\mathbf{r}')\mathbf{i}_\mu(\mathbf{r}')\big)\,dv' + \langle \nabla \cdot \mathbf{i}_\mu(\mathbf{r})\rangle_c \,, \end{aligned} \tag{4.40}$$

where the last step follows from the definition of the coarse graining.

Since current conservation requires that $\nabla \cdot \mathbf{i}_\mu(\mathbf{r}) = 0$, we lose the last term on the right, and the equation reduces to

$$\nabla \cdot \langle \mathbf{i}_\mu\rangle_c = - \iiint_{v_c} \nabla' \cdot \big(w(\mathbf{r}-\mathbf{r}')\mathbf{i}_\mu(\mathbf{r}')\big)\,dv' \,. \tag{4.41}$$

Using the divergence theorem, we can convert the volume integral on the right-hand side to a surface integral

$$\nabla \cdot \langle \mathbf{i}_\mu\rangle_c = - \oiint_{S_c} w(\mathbf{r}-\mathbf{r}')\mathbf{i}_\mu \cdot \mathbf{ds} \,, \tag{4.42}$$

which runs over all the cellular surfaces (membranes) within the averaging volume.

We recognize that, on the cellular surfaces, the component of $\mathbf{i}_\mu$ that is normal to the cellular surface is the membrane-current density i_m:

$$\mathbf{i}_\mu \cdot \mathbf{ds} = i_m ds \,. \tag{4.43}$$

With that, equation (4.42) becomes

$$\nabla \cdot \langle \mathbf{i}_\mu \rangle_\mathrm{c} = - \oiint_{S_\mathrm{c}} w(\mathbf{r} - \mathbf{r}') i_m ds \ . \tag{4.44}$$

This last integral thus gives us the total current that comes out from the membranes inside the coarse-graining volume (weighted with distance from position $\mathbf{r}$). This works as a good formal definition of the CSD:

$$\nabla \cdot \langle \mathbf{i}_\mu \rangle_\mathrm{c} = -C_\mathrm{all} \ . \tag{4.45}$$

We have let the subscript "all" indicate that the CSD computed in this way includes all membrane sources present in the tissue. The reason for introducing this subscript will become clear in Section 4.3.3, where we compare this theoretically derived CSD equation with the version used in forward modeling based on the MC+VC scheme.

4.3.2.7   Arriving at the Current-Source Density Equation

Inserting equations (4.45) and (4.34) into equation (4.32), we obtain the CSD equation:

$$\nabla \cdot \mathbf{i}_\mathrm{e} = C_\mathrm{all} \ . \tag{4.46}$$

Importantly, $\mathbf{i}_\mathrm{e}$ is the average current density running *extracellularly* through tissue, defined as extracellular current per unit *tissue* cross-section area and not per *extracellular* cross-section area, although $\mathbf{i}_\mathrm{e}$ by definition only runs through the extracellular parts of this cross-section area. Likewise, C_all is the source-current per total tissue volume.

Assuming that the extracellular current density is ohmic (equation (4.35)) and that the extracellular conductivity σ_e is constant, equation (4.46) becomes

$$\sigma_\mathrm{e} \nabla^2 V_\mathrm{e} = -C_\mathrm{all} \ . \tag{4.47}$$

Equations (4.46) and (4.47), respectively, closely resemble the CSD equations (4.12) and (4.13), defined earlier, but as we discuss below, the two versions have some subtle differences in terms of their interpretation.

4.3.3   Tissue Currents versus Extracellular Currents

If we compare equation (4.46) to the similar equation (4.12) used in VC theory,

$$\nabla \cdot \mathbf{i}_\mathrm{t}(\mathbf{r}) = C_\mathrm{t}(\mathbf{r}) \ , \tag{4.48}$$

we see that they differ in that the former contains the pair $\mathbf{i}_\mathrm{e}$ and C_all where the latter contains the pair $\mathbf{i}_\mathrm{t}$ and C_t.

As illustrated in Figure 4.5, the difference between the two scenarios is that, in the theoretical derivation of equation (4.46), C_all by definition includes absolutely all the membrane-current sources present in the tissue. Since all cellular currents thereby are accounted for in C_all, $\mathbf{i}_\mathrm{e}$ on the right-hand side of the equation is exclusively extracellular, as we have indicated by giving it the subscript "e" (Figure 4.5B).

In contrast, when performing forward modeling using the two-step MC+VC scheme, we generally have knowledge only of the sources (C_t) that we ourselves have simulated

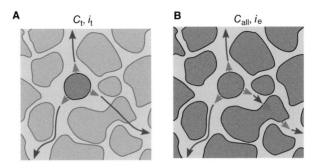

Figure 4.5 Illustration of the difference between C_t and C_{all}. A: Simulated cellular sources C_t (dark grey) evoke volume currents $\mathbf{i}_t$ in surrounding tissue. Tissue currents can partly take intracellular pathways through cells not explicitly simulated (light grey). **B**: Theoretical scenario when all cellular sources C_{all} are known. All currents crossing membranes are accounted for in C_{all}. Remaining volume currents $\mathbf{i}_e$ are then purely extracellular.

and impressed upon the tissue. In real tissue, it seems likely that the tissue currents evoked by a given assembly of sources (C_t) at least to some degree will interact with cellular membranes in the surrounding tissue and thereby induce additional membrane currents not included in C_t (Figure 4.5A). To use VC theory and still allow for this possibility, we may accept that some (unknown) fraction of the tissue-current density may (i) cross membranes and evoke (unknown) current sources additional to those simulated ($C_t \neq C_{all}$) and (ii) travel through intracellular pathways ($\mathbf{i}_t \neq \mathbf{i}_e$).

By assumption, the sources not accounted for by C_t in equation (4.48) are then accounted for by allowing $\mathbf{i}_t$ to be partially intracellular, which amounts to allowing σ_t to be greater than σ_e. The remaining question is then whether these tissue currents of unknown intra- and extracellular composition are well approximated by the ohmic relation, $\mathbf{i}_t = -\sigma_t \nabla V_e$. As we discuss further in Section 5.3, there is experimental evidence that they are.

For now, we conclude that we use equation (4.48) instead of equation (4.46) in forward modeling to account for the (related) facts that (i) the simulated C_t does not by necessity include absolutely all the membrane currents present in the tissue, and (ii) the tissue currents surrounding C_t may not exclusively travel along extracellular pathways.

4.4   Dipole Approximation

When calculating V_e at large distances from its underlying current sources, it can be convenient to use the *dipole approximation*. The justification of this approximation is that a neuron's contribution to V_e becomes increasingly dipolar with distance. The dipole approximation is commonly used when simulating EEG or MEG signals, as we will speak more of in Chapters 9 and 11. By "dipole," we here refer to a "current dipole," not a "charge dipole." Charge dipoles will not be encountered in this book.

If the distance from the center of a volume containing a set of current sources to the measurement point is larger than the maximal distance from the volume center to any source (Jackson 1998), as illustrated in Figure B.1, we can reformulate equation (4.8) by using the multipole expansion (Nunez & Srinivasan 2006):

$$V_e(\mathbf{r}) = V_e(R) = \frac{C_{\text{monopole}}}{4\pi\sigma_t R} + \frac{C_{\text{dipole}}}{4\pi\sigma_t R^2} + \frac{C_{\text{quadrupole}}}{4\pi\sigma_t R^3} + \frac{C_{\text{octopole}}}{4\pi\sigma_t R^4} + \dots ,\qquad (4.49)$$

where $R = \|\mathbf{r} - \mathbf{r}_c\|$ is the distance between the measurement point and the center $\mathbf{r}_c$ of the source distribution (defined in Section B.1).

Here the terms C_{monopole}, C_{dipole}, and so on represent the monopolar, dipolar, etc. contributions to V_e, and these terms can in general be quite complex as they depend on the angular and radial coordinates of the recording electrode and all the different current sources. For their full specification, see the derivation in Appendix B.

As we explained in Section 3.1, the net sum of currents into the neuron as a whole is always zero. Provided that the neuron does not receive an internal current injection, this means that the sum of the currents over the neural membrane is zero and that the monopolar contribution (the first term in equation (4.49)) vanishes ($C_{\text{monopole}} = 0$). Furthermore, the quadrupole, octopole, and higher-order contributions to V_e decay more rapidly with distance R than the dipole contribution. When we are sufficiently far away from the source distribution, V_e can therefore be well approximated by the dipole contribution alone:

$$V_e(\mathbf{r}) \approx \frac{1}{4\pi\sigma_t} \frac{\mathbf{P} \cdot \mathbf{R}}{R^3} = \frac{1}{4\pi\sigma_t} \frac{P\cos\theta}{R^2} ,\qquad (4.50)$$

where we have expressed C_{dipole} in terms of the dipole moment $\mathbf{P}$ (with unit Am) and the angle θ between the dipole moment and the distance vector $\mathbf{R} = \mathbf{r} - \mathbf{r}_c$, as explained in Appendix B. Equation (4.50) is known as the dipole approximation and is an alternative to equation (4.8) for predicting the extracellular potential.

The relationship between $\mathbf{P}$ and a set of (neural) current sources I_n in positions r_n is (Nunez & Srinivasan 2006, Pettersen et al. 2008, Lindén et al. 2010, Pettersen et al. 2014, Næss et al. 2021)

$$\mathbf{P} = \sum_{n=1}^{N} I_n \mathbf{r}_n .\qquad (4.51)$$

In the simplest case of a two-compartmental neuron with a current sink $-I$ at location $\mathbf{r}_1$ and a current source I (of the same magnitude) at location $\mathbf{r}_2$, the dipole moment is $\mathbf{P} = -I\mathbf{r}_1 + I\mathbf{r}_2 = I(\mathbf{r}_2 - \mathbf{r}_1) = I\mathbf{d}$, where $\mathbf{d}$ is the distance vector between the sink and source, determining both the length d and the direction of the dipole (see Figure 3.1B).

The dipole approximation (equation (4.50)) is good provided that the measurement point for V_e is far away from the source-sink distribution that it originates from, like in the case of an EEG. As a rule of thumb, Nunez & Srinivasan (2006) suggest that the approximation is applicable when R is three to four times greater than the dipole length: $R > 3d$ or $R > 4d$. For a numerical example evaluation of this rule of thumb, see Section B.5.

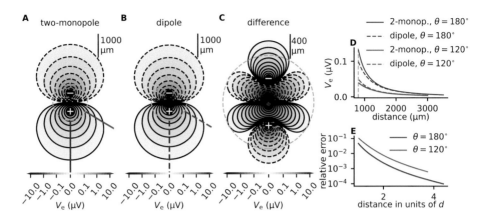

Figure 4.6 **Illustration of dipole approximation.** **A**: The extracellular potential in the vicinity of two current sources with identical magnitude but opposite polarity (a two-monopole). The separation between the sources is $d = 800\,\mu\text{m}$, and the extracellular potential is calculated with equation (4.8). **B**: The extracellular potential from the dipole approximation applied to the example in panel A, through the use of equation (4.50). **C**: The difference between results from using the two approaches described in panels A and B. **D**: The amplitude decay with distance for two different angles, marked by the red and blue lines in panel A and B. The minimum distance from the center is $800\,\mu\text{m}$, marked by the cyan circle in panels A–C. **E**: The relative error introduced by the dipole approximation as a function of the distance in units of the dipole length $d\ (= 800\,\mu\text{m})$. Code available via www.cambridge.org/electricbrainsignals.

A simple illustration of how the accuracy of the dipole approximation depends on distance is found in Figure 4.6, which compares a two-monopole (like what we would get from a two-compartment neuron model, Figure 3.1) to a dipole. As can be seen, the V_e from the two cases is very similar far away from the center, but the difference is substantial when very close.

More complex MC neuron models can be expected to give rise to source/sink patterns that are substantially more complex than the simple two-monopole, but seen from a sufficiently large distance, any nonzero dipole component will eventually dominate the higher-order terms. This is illustrated in Figure 4.7, which shows a simulation of the extracellular potential evoked by the Hay neuron receiving a single synaptic input, when the detailed spatial distribution of membrane currents is explicitly accounted for. The figure shows how the complex patterns in the extracellular potential in the vicinity of the neuron (panels A–C) look like the potential from a dipole when seen from a larger distance (panels D–F). For more examples of the dipole approximation applied to calculate V_e from MC neurons, see Lindén et al. (2010), Næss et al. (2021), and Chapter 9.

4.4.1 Distance-Decay of Potential from Dipole

If we move away from a dipole in a direction radially from the center of the dipole, it follows from equation (4.50) that the dipole's contribution to V_e decays with distance

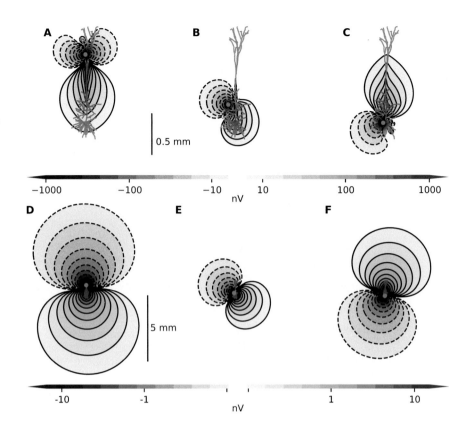

Figure 4.7 Seen from a distance, electric potentials from synaptic inputs resemble electric potentials from single dipoles. **A–C**: Extracellular potential, calculated with equation (4.8), in the vicinity of a neuron receiving a single synaptic input for three different synaptic input locations. Because of the complex distribution of the membrane currents, the extracellular potentials will have a complex but mostly dipole-like shape. **D–F**: Same as panel A–C, but seen from a greater distance so that the dipole contribution is dominating the extracellular potential. Code available via www.cambridge.org/electricbrainsignals.

as $1/r^2$, except when $\theta = 90°$ where $V_e = 0$ (Figure 4.8). This $1/r^2$ dependency is commonly stated as a rule of thumb for dipoles, but it is generally true only if we move along an axis that is radial to the dipole center. If we move along an axis that is not radial to the dipole center, we will not get a strict $1/r^2$ dependence. However, V_e will still decay as $1/r^2$ in the far-field limit ($r \to \infty$) except in the special case where the axis of movement is perfectly perpendicular to the dipole orientation.

To show this, we consider the general case of a dipole tilted an arbitrary angle α from the z-axis, and we examine how $V_e(x)$ varies along a horizontal axis in a plane at a distance z_0 below the dipole position (Figure 4.8A). Realizing that $\theta = 90° + \beta - \alpha$, trigonometric identities allow us to express

$$\cos \theta = \sin(\alpha - \beta) = \sin \alpha \cos \beta - \cos \alpha \sin \beta \,. \tag{4.52}$$

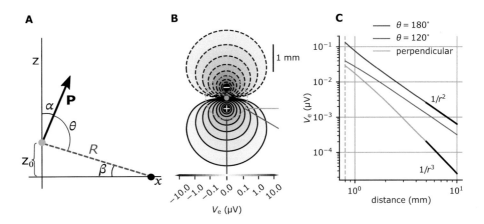

Figure 4.8 Distance-decay of potential from dipole is direction-dependent. A: Illustration of the model used to explore how the extracellular potential varies along a horizontal axis a distance z_0 below a dipole **P** tilted an angle α from the vertical axis. **B**: The extracellular potential in the vicinity of two current sources with identical magnitude but opposite polarity (a two-monopole). The separation between the sources is $d = 800$ μm, and the extracellular potential is calculated through equation (4.8). **C**: The amplitude of the potential as a function of distance for three different scenarios: the direction is directly downwards from the center of the dipole (blue), the direction is radially to the center of the dipole at an angle of $\theta=120°$ (red), and lastly, the direction is perpendicular to the dipole direction, horizontally from one pole of the two-monopole (orange). The black curves overlaying the blue and orange curves depict true $1/r^2$- and $1/r^3$-decays, respectively. Code available via www.cambridge.org/electricbrainsignals.

Insertion this into equation (4.50) gives us

$$V_e(x) \approx \frac{1}{4\pi\sigma_t} \frac{P}{R^2} \left(\frac{x}{R} \sin\alpha - \frac{z_0}{R} \cos\alpha \right), \tag{4.53}$$

where we have used $\cos\beta = x/R$ and $\sin\beta = z_0/R$.

We next use

$$R = \sqrt{x^2 + z_0^2} = x\sqrt{1 + \left(\frac{z_0}{x}\right)^2}. \tag{4.54}$$

In the far-field limit where $(z_0/x)^2 \to 0$, we find through the use of the Taylor expansion that equation (4.53) can be written as

$$V_e(x) \approx \frac{P}{4\pi\sigma_t} \left(\frac{\sin\alpha}{x^2} - \frac{z_0 \cos\alpha}{x^3} \right), \tag{4.55}$$

which thus contains a term that decays as $1/x^2$ and a term that decays as $1/x^3$. As long as $\alpha \neq 0$, it will always be the case that the first term will dominate after a certain distance $x \gg z_0$ so that the far-field decay will be

$$V_e(x) \approx \frac{P}{4\pi\sigma_t} \frac{\sin\alpha}{x^2}. \tag{4.56}$$

Since $\sin\alpha$ is a constant, this corresponds to a $1/x^2$-decay.

Only in the special case where the dipole is perfectly vertical ($\alpha = 0$) will the second term in equation (4.55) dominate in the far-field limit, and we get the $1/x^3$-decay

$$V_e(x) \approx \frac{P}{4\pi\sigma_t} \frac{z_0}{x^3} . \tag{4.57}$$

The special $1/r^3$ scenario is illustrated in Figure 4.8B–C, for the case where a vertical dipole is constituted by a two-compartment neuron model oriented in the z-direction, with a current source in the soma and a current sink in the dendrite. The dipole is in this case positioned midways between the somatic and dendritic source, and if we move horizontally ($r = x$) along an axis in the soma-plane, θ will continuously change (from $\theta = 180°$ towards $\theta = 90°$). As a consequence, V_e approaches the $1/r^3$ decay (orange curve) at far distances, as predicted by equation (4.57). In comparison, if we move away from the dipole in directions radially to the dipole center, V_e will approach a $1/r^2$-decay in the far-field limit (blue and red curves).

As a sanity check, in the case where $z_0 = 0$, we are moving (radially and horizontally) away from the dipole center so that $R = x$ and $\alpha = 90° - \theta$. Since we then have $\sin\alpha = \cos\theta$, equation (4.53) reduces to equation (4.50) for radial decay, and V_e decays as $1/x^2$ as it should. In the special case where the dipole is vertical, $\alpha = 0$, $\theta = 90°$, and $V_e(x) = 0$.

The take-home message from this analysis is that the V_e far away from a dipole will always decay as $1/r^2$, unless we are moving along an axis that is (i) perfectly perpendicular to the dipole orientation and (ii) not radial to the dipole center.

4.4.2   Multi-dipole Approximation

As an alternative to the point-source approximation (Section 4.1), line-source approximation (Section 4.2), or current-source density description (Section 4.3), the extracellular potential from an MC model can be represented using the so-called *multi-dipole approximation* (Næss et al. 2021). Each axial current in each neural compartment is then treated as an individual dipole, and its contribution to the extracellular potential can be computed with equation (4.50). By summing all individual dipole contributions, the complex spatial distribution of membrane currents is thus taken into account (Figure 4.9). Hence, the multi-dipole approximation is essentially an alternative implementation of the MC+VC scheme.

While the point-source or line-source approximations are the standard choices for computing spikes or LFPs, the multi-dipole approximation is particularly useful in the context of computing EEG signals. The reason is that EEG forward models typically take dipoles as input, and in Section 9.5.1, we use the multi-dipole approximation to validate EEG signal predictions based on the (single-) dipole approximation (equation (4.50)).

We note that the multi-dipole approximation is only valid at distances from a compartment that are large compared to the compartment length. This explains why the multi-dipole predictions and point-source predictions in Figure 4.9 are not in agreement in close proximity to the neuron. This disagreement is, however, not a concern when we

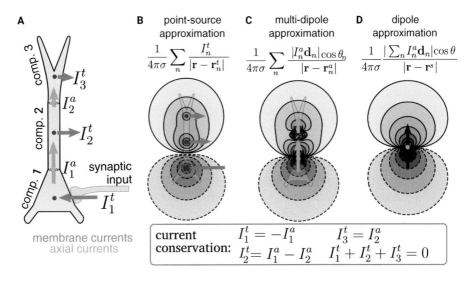

A

comp. 3

comp. 2

comp. 1

I_3^t

I_2^a

I_2^t

I_1^a synaptic input

I_1^t

membrane currents
axial currents

B point-source approximation

$$\frac{1}{4\pi\sigma}\sum_n \frac{I_n^t}{|\mathbf{r}-\mathbf{r}_n^t|}$$

C multi-dipole approximation

$$\frac{1}{4\pi\sigma}\sum_n \frac{|I_n^a\mathbf{d}_n|\cos\theta_n}{|\mathbf{r}-\mathbf{r}_n^a|}$$

D dipole approximation

$$\frac{1}{4\pi\sigma}\frac{|\sum_n I_n^a\mathbf{d}_n|\cos\theta}{|\mathbf{r}-\mathbf{r}^s|}$$

current conservation:
$I_1^t = -I_1^a$ $\qquad I_3^t = I_2^a$
$I_2^t = I_1^a - I_2^a$ $\qquad I_1^t + I_2^t + I_3^t = 0$

Figure 4.9 Different approaches to calculating extracellular potentials. A: Illustration of a cell model with three cellular compartments. A synaptic input initiates membrane currents and axial currents that are related via current conservation. **B**: Extracellular potential computed using the point-source approximation (Section 4.1). **C**: Extracellular potential computed using the numerous dipole moments stemming from the individual axial currents (Næss et al. 2021). This is in essence still the MC+VC scheme, just reformulated so it can be used with (dipole-based) EEG forward models. **D**: Extracellular potential computed using the dipole approximation (Section 4.4).

consider EEG signals, which are far away from the sources. Moreover, in the illustrating example in Figure 4.9, the neuron model had only three compartments. For biophysically detailed MC models with a much higher number of smaller compartments, the MC+VC and multi-dipole approximations will give matching predictions much closer to the neuron (Næss et al. 2021).

4.5 Modeling Recording Electrodes

Earlier, we demonstrated how we can use VC theory to predict the extracellular potential (V_e) evoked by neural activity. However, when measuring V_e experimentally, the measurement equipment will affect what is being measured in several ways, depending both on electrode properties and on the spatial and temporal characteristics of V_e. In this section, we give a short review of different approaches to model such effects.

The exposed contacts of metal electrodes are essentially in direct contact with neural tissue. To accurately measure the potential at the electrode surface, the electrode should be coupled to an amplifier with a very high impedance.[4] In this way,

---

[4] Impedance is essentially resistance, generalized so that it can incorporate capacitive effects and be frequency-dependent.

a negligible amount of current will cross the tissue-electrode interface and cause electrode-polarization effects (Schwan 1992, Moulin et al. 2008, Ishai et al. 2013, Martinsen & Grimnes 2015). The electrode surface itself is typically made of a highly conductive metal like silver or platinum, so the potential directly inside it is in practice homogeneous.

It has been argued that, provided that the proper recording equipment is used, electrode impedance or other electrode properties should not substantially distort recorded potentials (Moulin et al. 2008, Nelson et al. 2008, Nelson & Pouget 2010, Martinsen & Grimnes 2015, Viswam et al. 2019). Assuming that sufficient care has been taken, recorded potentials can therefore be expected to closely reflect the potential in the neural tissue immediately outside the electrode contacts. Here, we will therefore not consider how to model signal distortion introduced by the experimental equipment, except for effects caused by the physical presence of the recording equipment itself. Readers who are interested in more detailed discussions of electrode properties will find in-depth discussions of this elsewhere (Robinson 1968, Franks et al. 2005, Nelson et al. 2008, Nelson & Pouget 2010, Martinsen & Grimnes 2015, Viswam et al. 2019).

We will distinguish between the effects from the electrode contacts (Sections 4.5.1 and 4.5.2) and the electrode shanks (Section 4.5.3)- that is, the non-conducting substrate that the electrodes are embedded in. The motivation for this distinction is that these two effects are essentially independent, and they can be taken into account or ignored independently, depending on the scenario.

We note that although electrode-polarization effects might not be an issue in recording electrodes (since they carry minuscule currents), they may be pronounced in stimulation electrodes, and substantially more-complex electrode models might then be called for (Moulin et al. 2008, Joucla & Yvert 2012, Joucla et al. 2014, Martinsen & Grimnes 2015).

4.5.1   Point-Electrode Contacts

The most straightforward and also most common approach to modeling recording electrodes is to use the *point-electrode approximation*. One then ignores the physical extent of the electrode contact surface and evaluates V_e at a single point in space using VC theory – for example, through equation (4.7). An illustration of this is shown in Figure 4.10, which we will use when comparing the point-electrode approximation with other approaches. Here, two current point sources are positioned in close vicinity to an electrode surface, so that V_e computed with the point-electrode approximation varies substantially over the surface area (Figure 4.10A, B).

For cases when the contact surface is small relative to the spatial variations in V_e, ignoring the spatial extent of the electrode can be expected to be a good approximation (Moffitt & McIntyre 2005, Moulin et al. 2008, Ness et al. 2015, Buccino et al. 2019, Viswam et al. 2019).

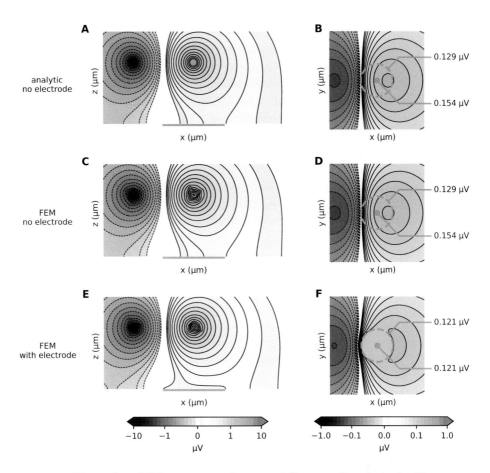

Figure 4.10 Illustration of different approaches to modeling recording electrodes. Two current point sources of opposite polarity ($\pm10\,$nA) at height $z = 10\,\mu$m above a non-conducting plane with a recording electrode centered at the origin with radius $r = 5\,\mu$m. **A**: Cross-section of V_e in the x,z-plane, $y = 0$, calculated analytically using the method of images (see equation (5.32)) without accounting for any effects from the electrode surface (orange). **B**: Same as in panel A but in the same plane as the electrode (x,y-plane, $z{=}0$). If we want to calculate V_e at the location of the hypothetical electrode surface, we can either calculate V_e at a single point (orange dot, value to the lower right) or calculate the average over the electrode surface (orange dashed circle, value to the upper right). **C**, **D**: Same setup as in panels A and B except that V_e is found numerically through the finite element method (FEM), assuming the electrode has the same conductivity as the brain tissue. **E**, **F**: Same as in panels C and D except that the highly conductive electrode ($\sigma \approx \infty$) is included in the FEM model.

4.5.2   Spatially Extended Electrode Contacts

Under some experimental conditions, the electrode surface is large relative to the spatial variations in V_e, either because the electrodes are large (like ECoG electrodes at the cortical surface) or because V_e varies substantially over small distances (like extracellular spikes from nearby axons). Since the conductivity of metal electrodes is much higher than the conductivity of neural tissue, the presence of the electrode may in itself affect V_e, especially in the vicinity of the electrode surface. This effect is not accounted for by the point-electrode approximation, but it can be studied using the finite element method (FEM). This is illustrated in Figure 4.10, where we see that including the highly conductive electrode substantially affects V_e close to the electrode surface (panels E and F versus C and D). Since electrodes are made of highly conductive metal, the potential is practically homogeneous on the electrode-tissue interface (Nelson et al. 2008, Nelson & Pouget 2010, Ness et al. 2015, Vermaas et al. 2020b), as illustrated in Figure 4.10F. In the illustration used here, we found that when using the point-electrode approximation (either analytically or through FEM), we got $V_e = 0.154\,\mu V$ (Figure 4.10B,D), while when including the electrode surface in a FEM simulation (with a realistic conductivity for metals), we got $V_e = 0.121\,\mu V$ (Figure 4.10F), a difference of about 27 percent. Note that this large discrepancy is only present because V_e varies considerably in the vicinity of the electrode contact surface, and this discrepancy would not be present if V_e were close to constant in the vicinity of the electrode contact surface.

A simple approximation to account for the electrode size without comprehensive FEM simulations is through a trivial extension of the point-electrode approximation – that is, by averaging computed point-electrode values of V_e across the surface of the electrode. This can be done by taking the average of the values of V_e computed by the point-electrode approximation at N suitably distributed points across the electrode contact surface:

$$V_{e,\text{disk}} = \frac{1}{s_{\text{disk}}} \iint_{s_{\text{disk}}} V_e(\mathbf{r})\, ds \approx \frac{1}{N} \sum_{n=1}^{N} V_e(\mathbf{r}_n). \qquad (4.58)$$

Here s_{disk} denotes the surface area of the electrode. This approach is referred to as the *disk-electrode approximation* (Camuñas-Mesa & Quiroga 2013, Lindén et al. 2014, Hagen et al. 2015, Ness et al. 2015, Viswam et al. 2019). The disk-electrode approximation accounts for the physical extent of the electrodes and therefore also for the spatial filtering effects introduced by electrodes that are large compared to the spatial variations of the V_e. In Section 7.7.1, we use this approximation to explore the effect of electrode size on recorded extracellular spikes.

The approximation inherent in the disk-electrode approximation is that the potential recorded by the electrode corresponds to the spatial average of the potential that we would have in the tissue surfacing the electrode if the highly conductive electrode was not present. This is illustrated in Figure 4.10, where we found that when we did not explicitly include the electrode in the FEM model (panels C and D) but instead calculated the average of V_e over the hypothetical electrode surface, $V_e = 0.129\,\mu V$ instead of $V_e = 0.121\,\mu V$ (panels E and F), a difference of about 6.5 percent.

Using FEM, Ness et al. (2015) found that the disk-electrode approximation was fairly accurate for estimating measured potentials from a point-current source more than about two electrode radii away from the electrode. The point-electrode approximation was

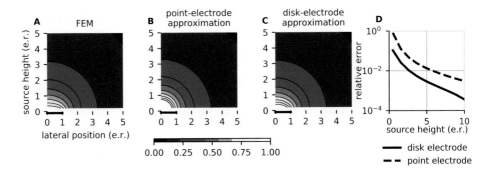

Figure 4.11 Finite element method versus the point- and disk-electrode approximations.
A: Color plot of the value of V_e at the electrode surface $(x, y = 0)$ for different source locations, simulated with the finite element method (FEM). The highly conductive material of the electrode was included in the simulation, see Ness et al. (2015) for full simulation details. The x- and y-axes are shown in units of the electrode radius, which in this example was 10 μm. **B**: Same as panel A but calculated analytically using the point-electrode approximation with the method of images (see equation (5.32)). **C**: Same as panel B but using the disk-electrode approximation. **D**: Relative errors of the point- and disk-electrode approximations (panels B and C) compared to FEM (panel A) as a function of distance from the center of the electrode along the vertical axis.

found to be fairly accurate for distances greater than four electrode radii away from the electrode (Figure 4.11).

In certain cases, like in ECoG measurements from the cortical surface, the recording electrodes are both large ($\approx$ 2–5 mm diameter) and close to the neural sources (the thickness of the human cortex is $\approx$ 3 mm). In such cases, the electrode properties can therefore be expected to have a large effect on measured potentials – see, for example, Rogers et al. (2020) and Vermaas et al. (2020*b*), as well as Chapter 10.

4.5.3 Effect of Electrode Shanks

In addition to the electrode surfaces, the probes used to perform recordings in brain tissue have shanks made of a non-conductive, typically silicon-based, material. In addition to increasing tissue damage (Moffitt & McIntyre 2005) – a topic not covered here – the use of large probes is likely to affect recordings of extracellular potentials since they represent non-conducting volumes (Moffitt & McIntyre 2005, Lempka et al. 2011, Mechler & Victor 2012, Buccino et al. 2019).

It has been demonstrated that shanks of large probes can amplify or dampen recorded potentials from nearby cells by a factor of almost two, depending on whether the cell was in front of or behind the electrode shank (Moffitt & McIntyre 2005, Buccino et al. 2019). This factor of two can be understood by considering the extreme case of the probe being an infinitely large plane: as we will see in Section 5.7.1, this scenario results in exactly a factor of two increase in V_e (Ness et al. 2015). In the other extreme, where the volume of the non-conducting probe goes towards zero, it will not change the extracellular environment and therefore will not affect V_e. A FEM study by Buccino et al. (2019) concluded that small microwires and tetrodes are so small that they hardly affect V_e at all.

4.5.4   Effect of Position of Reference Electrode

When we record extracellular potentials in the brain, we are in principle free to arbitrarily choose the location of the reference electrode. In practice, the optimal placement of the electrode depends on what we want to measure (Sharott 2014). When we simulate extracellular potentials, the reference point for the potential is typically (and often implicitly) placed at "infinity" (see equation (4.5)). However, when trying to mimic experimental data, it can be important to account for the location of the real reference electrode, as this choice can in certain cases substantially affect the measured (and simulated) potentials. To do this, one may simply calculate the potential with respect to "infinity" (for example, with equation (4.7)) both at the location of the measurement electrode and at the location of the reference electrode, then subtract the latter signal from the former.

4.6   Electric and Magnetic Brain Stimulation

This book focuses on the biophysical modeling of electric and magnetic signals generated by neural activity. The modeling of how electric and magnetic brain stimulation affects neural activity is based on the same physical principles. While not a main topic in this book, a brief survey of the modeling of such stimulation is presented here, electric brain stimulation in Section 4.6.1 and magnetic brain stimulation in Section 4.6.2.

4.6.1   Electric Stimulation

Electric brain stimulation already has many clinical applications (Joucla & Yvert 2012) – for example, in cochlear implants to restore hearing for people with hearing impairments (Clark et al. 1977, Zeng et al. 2008), to alleviate movement disorders in patients with Parkinson's disease with deep brain stimulation (DBS) (Benabid et al. 2009), or in electroconvulsive therapy to relieve symptoms of mental health problems. These applications are all based on the same principle: electrodes planted on the scalp, on the cortical surface, or inside brain tissue are used to impose electric currents and voltage gradients through the brain tissue.

Assuming that the tissue is an ohmic volume conductor, the effect of electrode stimulation on the extracellular potential in all points in the tissue can be computed by imposing boundary conditions either on the potential or on the imposed currents at the stimulation electrodes and at the brain surface (or infinity) (Joucla & Yvert 2012). In direct-current (DC) stimulation, the imposed potential or current is a fixed constant;

in alternating-current (AC) stimulation, it is time-dependent.[5] The neural membrane potential is given by the difference between the intracellular potential V_i and the extracellular potential V_e. The effect of the stimulation on the neural dynamics is obtained by letting V_e vary as predicted from the electrode stimulation (Rattay 1999, Fröhlich & McCormick 2010) and imposing this as a boundary condition in the neural simulation (Romeni et al. 2020). This requires the same set of assumptions as those introduced at the start of this chapter. In particular, assumption 1 that the effects of the neurodynamics on V_e does not act back on the neurodynamics is implicit in the assumption that V_e is determined exclusively by the electrode stimulus.

In transcranial electric stimulation (TES), the electrodes are placed on the scalp, and both direct currents (DC) and alternating currents (AC) can be used. The electric field imposed in the brain by this technique is quite diffuse and quite weak, typically on the order of 0.1 V/m (Miranda et al. 2018). This is too weak to induce an action potential by itself (Fröhlich & McCormick 2010), and the effect from TES is thus thought to stem from the modulation of action-potential generation. In DBS, the electric current is delivered by electrodes positioned deep inside the brain and much-more focused electric stimulation can be imposed (Butson & McIntyre 2008).

A complication that is not so relevant when modeling recording electrodes but relevant when modeling stimulus electrodes is that, for a current-carrying metal electrode used in stimulation in an electrolyte-like brain tissue, there can be a sizable drop in potential across the electrode-tissue interface. This effect is referred to as *electrode polarization* (McIntyre & Grill 2001, Joucla & Yvert 2012, Martinsen & Grimnes 2015).

Quantitative modeling of electrode-polarization effects will often require numerically comprehensive approaches, like the finite element method (FEM) (Buitenweg et al. 2003, Moulin et al. 2008, Joucla et al. 2014, Vermaas et al. 2020a) or careful calibration to experimental recordings (Gabriel et al. 1996, Martinsen & Grimnes 2015, Miceli et al. 2017). In the modeling of the electric stimulation of neurons, electrode-polarization effects can be incorporated either by including an interface impedance or by modifying the boundary condition for V_e at the electrode surfaces (Joucla & Yvert 2012).

Further information on the modeling of electric brain stimulations and qualitative insights gathered from studying simplified models are found in Appendix C.

4.6.2   Magnetic Stimulation

Magnetic fields can also stimulate neural activity. Transcranial magnetic stimulation (TMS) (Barker et al. 1985) is a key technique used both in basic research and clinical applications, where a magnetic coil placed immediately outside the head delivers brief pulses of strong magnetic fields into the brain.

The effect of such a pulse stems from one of Maxwell's equations (equation (2.36)), the one that describes how a rapidly changing magnetic field induces an electric field,

---

[5] In the stimulation of neural tissue, charge-balanced AC stimulation is preferred for safety.

and thus electric currents, in brain tissue. To get substantial effects, the applied stimuli are typically brief magnetic impulses (lasting about 0.1 ms) with amplitudes up to teslas (Hallett 2007). The resulting induced currents are large enough to not only modulate the firing of action potentials but also induce action potentials by themselves (Ilmoniemi et al. 2016).

Examples of the modeling of effects from magnetic stimulation on neural dynamics can be found in Nagarajan & Durand (1996), Pashut et al. (2011), and Aberra et al. (2020).

5 Conductivity of Brain Tissue

We have seen that the multicompartment (MC) framework allows us to simulate the distribution of membrane currents in neurons (Chapter 3) and that volume-conductor (VC) theory allows us to compute the extracellular potential resulting from such membrane currents (Chapter 4). Of course, the relationship between the membrane currents and the extracellular potential depends on the medium outside the active neurons. This medium, which we approximate as a continuous medium, is represented through the macroscopic tissue conductivity σ_t.

By definition, σ_t (with unit S/m) is the conductivity for the macroscopic current per cross-section area of the neural tissue. In the literature, the tissue conductivity is often simply denoted as σ, but we have chosen to add a subscript "t" (for tissue) to distinguish it from the two other conductivity measures: the conductivity σ_{ef} of the extracellular fluid and the conductivity σ_e of the extracellular medium. These three conductivity measures are illustrated in Figure 5.1 and defined further in Sections 5.1–5.3, where we also discuss how the three are related.

From Section 5.3 and onwards, we keep our focus entirely on σ_t, which is the relevant conductivity when interpreting extracellular potentials. In many applications of the MC+VC framework, σ_t is assumed to be constant, like we did in Chapter 4. A commonly used value for this constant is $\sigma_t = 0.3$ S/m, which lies within the range of values that have been measured experimentally in grey matter. Other estimates, both experimental and theoretical, are discussed later on in this chapter. Assuming a constant σ_t amounts to assuming that brain tissue is homogeneous, isotropic, and frequency-independent. In the later parts of this chapter (Sections 5.4–5.7), we evaluate these assumptions and discuss ways to relax them when this is deemed necessary.

5.1 Conductivity σ_{ef} of the Extracellular Fluid

The extracellular fluid (often called the interstitial fluid) is the fluid that fills up the small gaps of extracellular space between cellular structures. Its composition is very similar to that of the cerebrospinal fluid (Hladky & Barrand 2014), and its conductivity σ_{ef} is typically measured in a bath of a representative Ringer's solution (Figure 5.1A). Experimentally reported values of σ_{ef} typically range between 1.3 and 1.8 S/m (Okada et al. 1994, Baumann et al. 1997, Logothetis et al. 2007, Martinsen & Grimnes 2015, Miceli et al. 2017, McCann et al. 2019).

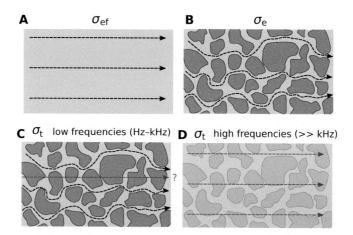

Figure 5.1 Conductivity definitions. A: Currents in a saline solution representing the extracellular fluid, with conductivity σ_{ef}, can travel in straight paths along voltage gradients, facing no obstacles. **B**: Currents assumed to run only through the extracellular part of tissue, with effective (macroscopic) conductivity σ_e. Since extracellular currents (i) are confined to only a fraction α of the total volume and (ii) face cellular obstacles preventing them from following straight paths, $\sigma_e < \sigma_{ef}$. **C**: If real tissue currents to some extent cross membranes and take intracellular pathways (red, dashed arrow), we expect an effective tissue conductivity $\sigma_t > \sigma_e$. **D**: The propensity for crossing membranes will depend on the frequency of the tissue current. High-frequency currents are more prone to pass capacitively through cell membranes, and in the limit of very high frequencies, we may expect these currents "not to notice" the cellular obstacles (Martinsen & Grimnes 2015, Amini et al. 2018). Generally, we therefore expect σ_t to increase with increasing frequencies.

The conductivity generally depends on the availability of free charge carriers, and σ_{ef} is theoretically determined by the ion concentrations in the extracellular space:

$$\sigma_{ef} = \frac{F^2}{RT} \sum_k D_k z_k^2 [k] . \tag{5.1}$$

In Section 2.4.3, we showed how this definition of the conductivity follows from the Nernst-Planck equation. We recall that z_k, D_k, and $[k]$ denote the valency, diffusion coefficient, and concentration, respectively, of ion species k, while $F = 96\,485.3\,\text{C/mol}$ is the Faraday constant, $R = 8.314\,\text{J/(K mol)}$ is the gas constant, and T (K), is the temperature.

Typical concentrations of the most abundant ions in the extracellular fluid were listed in Table 3.1, and their diffusion constants are given in Table 5.1. If we insert the values from Table 3.1 and Table 5.1 into equation (5.1), assuming a body temperature of $T = 310\,\text{K}$, we get a conductivity $\sigma_{ef} = 1.72\,\text{S/m}$.

Note that the extracellular fluid contains many ion species (such as H^+ and $HPO4^{2-}$) that were not included in Table 3.1 and thus do not contribute to the conductivity in our calculation of σ_{ef}. However, concentrations of ions others than those in Table 3.1 are generally quite low, and therefore these ions are not likely to have any major impact on the conductivity. Note also that the extracellular concentrations listed in Table 3.1 were

D_{Na}	1.33×10^{-9} m^2/s
D_K	1.96×10^{-9} m^2/s
D_{Cl}	2.03×10^{-9} m^2/s
D_{Ca}	0.71×10^{-9} m^2/s
D_{Mg}	0.72×10^{-9} m^2/s
D_{HCO3}	1.18×10^{-9} m^2/s

Table 5.1. Diffusion constants for ions in dilute solutions. These values are taken from Bowen & Welfoot (2002) and Lyshevski (2007).

based on reported values in the human cerebrospinal fluid, and saline concentrations may deviate from this in other tissues. The earlier estimated value of σ_{ef} nevertheless seems to be quite representative and lies within the range of experimentally reported values of σ_{ef} listed in the beginning of this section.

5.2 Conductivity σ_e of the Extracellular Medium

Macroscopic currents through brain tissue do not pass through a 3D volume filled entirely by extracellular fluid. As we define it here, the conductivity σ_e of the *extracellular medium* is a theoretical quantity. It represents the conductivity experienced by a hypothetical macroscopic density of (coarse-grained) current that travels exclusively through the extracellular part of brain tissue (Figure 5.1B). This current is (i) restricted to move only through the fraction α of the total medium volume that is extracellular and (ii) must take detours around the cellular obstacles as accounted for by the medium's tortuosity λ (Nicholson & Syková 1998, Nunez & Srinivasan 2006). An estimate of this theoretical quantity can be obtained through the porous-medium approximation.

5.2.1 Porous-Medium Approximation

As illustrated in Figure 5.2A, tissue is densely packed with cellular structures, and the extracellular space occupies only about 20 percent of the total tissue volume. The extracellular space has a highly tortuous geometry, with the average extracellular distance between cellular structures being as narrow as 10–80 nm depending on the kind of structure in question (Syková & Nicholson 2008, Kinney et al. 2013).

At the spatial scale of micrometers and below, the tissue is highly inhomogeneous and anisotropic. How easy it will be for a particle to move in a particular spatial direction will depend on whether there is a local free extracellular pathway in that direction or whether the pathway is blocked by a nearby membrane (Figure 5.2A). However, at a larger spatial scale of, say, a few tens of micrometers, it is reasonable to assume that the micrometer-scale inhomogeneities average out (Figure 5.2B) so that brain tissue can be treated as a continuous, porous medium (Gardner-Medwin 1980, Nicholson & Phillips 1981, Nicholson 2001, Meffin et al. 2014, Gratiy et al. 2017). Such a treatment is further

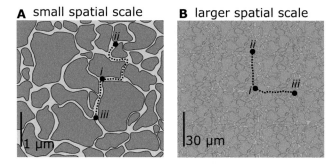

A small spatial scale   **B** larger spatial scale

1 μm   30 μm

Figure 5.2 Illustration of brain tissue. A: Illustration of a tissue cross-section. Extracellular space (light blue) occupies only a fraction of the total tissue volume. On a small spatial scale, the tissue is highly inhomogeneous. The shortest extracellular pathway between two points depends strongly on the microstructure (obstacles faced along the journey). In this example, the travel distance i → ii > i → iii although the euclidean distances are the same in both cases. **B**: On a larger spatial scale, local inhomogeneities average out, and the tissue can be treated as homogeneous.

motivated by the lack of any alternative ways to approach the problem since we, in most experimental settings, do not know the exact microstructure of the brain.

When treated as a continuous porous medium, the structure of brain tissue is defined by two key parameters (Nicholson & Phillips 1981). The first parameter, α, is the extracellular volume fraction, defined as extracellular volume v_e per tissue volume v_t:

$$\alpha = \frac{v_e}{v_t} . \tag{5.2}$$

The second parameter, λ, is the tortuosity of the extracellular medium. A common geometrical definition of tortuosity is that it equals the (average) path distance between two points divided by the euclidean distance between the same two points. However, in studies of neural tissue (and other porous media), it is typically defined as the square root of the ratio between the effective diffusion constant D^* in the tissue compared to its value D in an obstacle-free saline solution (Syková & Nicholson 2008):

$$\lambda = \sqrt{\frac{D}{D^*}} . \tag{5.3}$$

This ratio can generally depend on a variety of different hindrances faced by diffusive agents in the medium, not only geometrical factors causing prolonged travel distances. As reviewed in Syková & Nicholson (2008), the tortuosity of brain tissue may reflect (i) physical obstacles to free diffusion represented by cell bodies, which force diffusive agents to take detours around them, (ii) the transient trapping of diffusive agents in dead-space microdomains, such as cellular cavities, (iii) chemical interaction between diffusive agents and molecules in the extracellular matrix, (iv) unknown electric interactions between diffusive agents and molecules in the extracellular matrix, and (v) physical obstacles to free diffusion represented by molecules in the extracellular matrix.

The quantities α and λ can be estimated experimentally by measuring the diffusion of tracer molecules believed to stay confined within the extracellular part of the tissue

(Nicholson & Phillips 1981, Syková & Nicholson 2008). Estimates of α and λ have been obtained in numerous studies using somewhat different techniques and choices of tracer molecules (see review by Syková & Nicholson (2008)). Typical estimates for brain tissue are $\alpha = 0.2$ and $\lambda = 1.6$, although these values can vary between brain regions and also locally within a brain region, due to brain-state dependent cellular swelling or shrinkage (Syková & Nicholson 2008, Rasmussen et al. 2020).

5.2.2   Estimate of σ_e

When currents are confined to the extracellular space, the extracellular conductivity should theoretically be a factor α/λ^2 lower than the conductivity of the extracellular fluid (Gardner-Medwin 1980, Okada et al. 1994):

$$\sigma_\mathrm{e} = \frac{\alpha}{\lambda^2}\sigma_\mathrm{ef}\,. \tag{5.4}$$

Here, α accounts for the fact that only a fraction of the total volume is extracellular (and conducting), while λ^2 accounts for the hindrances imposed by the tortuous structure of the extracellular medium. We note that the conductivity is reduced by the same factor λ^2 as the effective diffusion constant (equation (5.3)). This follows from the Einstein relation (equation (2.23)), which states that the relationship between the diffusion constant and electric mobility is linear.

With the typical values $\alpha = 0.2$ and $\lambda = 1.6$ for the extracellular space of brain tissue (Nicholson & Phillips 1981, Nicholson & Syková 1998, Syková & Nicholson 2008), σ_e becomes almost a factor 13 lower than σ_ef. With the earlier estimate that $\sigma_\mathrm{ef} = 1.72\,\mathrm{S/m}$, we get an estimated value of $\sigma_\mathrm{e} = 0.134\,\mathrm{S/m}$.

Note that σ_e is defined so that it gives us a (hypothetical) current density $\mathbf{i}_\mathrm{e} = -\sigma_\mathrm{e}\nabla V_\mathrm{e}$ per unit tissue cross-section area, not per extracellular cross-section area, although it by definition only exists in the extracellular part of it. We encountered the same current density previously in equation (4.35).

It is worth taking a sidestep to compare equation (5.4) to the analytically obtained Maxwell-Rayleigh formula for the resistivity of a medium with a homogeneous suspension of identical non-conductive bodies. Expressed in terms of the parameters σ_e, σ_ef, and α, the Maxwell-Rayleigh formula becomes[1]

$$\frac{\sigma_\mathrm{e}}{\sigma_\mathrm{ef}} = \frac{\alpha}{1 + h(1 - \alpha)}\,, \tag{5.5}$$

where the parameter h depends on the structure of the non-conducting bodies so that $h = 1/2$ for spheres, and $h = 1$ for cylinders (assuming that the current direction is perpendicular to the cylinder axes).

By comparing equation (5.5) and equation (5.4), we find that they are in agreement if

$$h = \frac{\lambda^2 - 1}{1 - \alpha}\,. \tag{5.6}$$

---

[1] See e.g. Nunez & Srinivasan (2006, equation 4.1) for the original form of this formula.

With $\lambda = 1.6$ and $\alpha = 0.2$, this gives us $h \approx 2$ for the (average) cellular structures in brain tissue. Hence, the cellular obstacles characteristic for brain tissue reduce the conductivity more strongly than do idealized spherical or cylindrical obstacles.

Although the comparison that gave us equation (5.6) might provide some hint of the tissue-structure complexity, it should be noted that it is inconsistent. The lowest extracellular volume fractions that can be obtained by packing spherical or cylindrical cells are $\alpha = 1 - \pi/(3\sqrt{2}) \approx 0.26$ for spheres (Tao & Nicholson 2004) and $(4 - \pi)/4 \approx 0.22$ for cylinders, thus larger than the observed $\alpha = 0.2$ for brain tissue. Accordingly, a series of studies using other (more packable) cell structures have been conducted (reviewed in Syková & Nicholson (2008)), often with the goal of establishing a relationship between the structure parameters α and λ. Using Monte Carlo simulations of diffusing point particles through a medium packed arbitrarily densely with variously shaped cells, Tao & Nicholson (2004) showed that their results could be fitted with the relation

$$\lambda = \sqrt{(3 - \alpha)/2} \, , \tag{5.7}$$

which gives a theoretical upper limit ($\alpha \to 0$) of $\lambda \approx 1.225$, which is lower than the $\lambda = 1.6$ measured for brain tissue.

The discrepancy between theoretical and experimental estimates of λ can have many explanations. As noted by Syková & Nicholson (2008), a limitation with theoretical models of brain tissue is that they all rely on some simplifying assumption regarding the structure of the cells. Furthermore, the estimates that come out from such models normally only account for the so-called geometrical tortuosity (factor (i) and possibly factor (ii) defined in the discussion following equation (5.3)). Indeed, it has been shown that the theoretical value of λ increases if it is stipulated that only parts of the extracellular space is connected, so that the remaining parts constitute dead-space microdomains (Tao et al. 2005, Syková & Nicholson 2008). However, the mismatches between the theoretically estimated and measured values of λ can also partially be due to other kinds of hindrances (like factors (iii)–(v) following equation (5.3)). Currently, a full understanding of how various factors play together in determining the value of λ is lacking.

Importantly, as we stated in the beginning of this subsection, σ_e is a theoretical conductivity estimate, representing the hypothetical scenario where currents through brain tissue are confined to stay purely extracellular.

5.3   Conductivity σ_t of Brain Tissue

The tissue conductivity σ_t is the macroscopic (coarse-grained) conductivity of the brain tissue, sometimes referred to as the *effective tissue conductivity*. This is the conductivity experienced by a macroscopic (coarse-grained) current traveling through the tissue (Figure 5.1C–D), and it can be estimated experimentally by measuring gradients in V_e resulting from a controlled current injection into the tissue. Such measurements do not require any knowledge of the structure parameters, α and λ. It is σ_t that is the relevant conductivity for use in VC theory.

It has been argued that tissue currents in the frequency range relevant for physiological signals (below a few kilohertz) are predominantly confined to the extracellular part of the medium (Robinson 1968, Foster & Schwan 1989, Nunez & Srinivasan 2006, Martinsen & Grimnes 2015, Amini et al. 2018). In that case, the tissue conductivity σ_t should be similar to σ_e. However, others have argued that a substantial fraction of the tissue currents crosses neural membranes and travels along intracellular pathways (Gardner-Medwin 1980, Okada et al. 1994, Meffin et al. 2014).

If membrane interactions are important, σ_t would be dependent on the cable properties and orientation of cells, which could cause it to differ from σ_e in several ways. Firstly, if tissue currents can follow intracellular pathways in addition to the extracellular ones, we would expect (the resistive component of) σ_t to be greater than σ_e, as we will discuss further in the current section. Secondly, as membrane currents are partly capacitive, this could give σ_t a frequency dependence not present in σ_e, as we discuss in Section 5.4. Thirdly, theoretical studies based on cable theory have also suggested that measurements of σ_t may depend on the distance between the stimulus and recording electrodes, as we discuss   in Section 5.5.

Experimental estimates of the resistive component of σ_t vary between different recordings. Some studies have estimated σ_t in the cortex to be as low as 0.1 S/m or even lower (Gabriel et al. 1996). However, more commonly, the reported values lie between 0.2 and 0.5 S/m (Ranck 1963, Havstad 1976, Hoeltzell & Dykes 1979, Logothetis et al. 2007, Goto et al. 2010, Elbohouty 2013, Wagner et al. 2014, Koessler et al. 2017, Miceli et al. 2017, McCann et al. 2019), and in simulations shown in this book, we have chosen to use 0.3 S/m as a default value.

Gardner-Medwin (1980) compared a theoretically estimated value of $\sigma_e = 0.16$ S/m with values of the resistive component of σ_t that had been reported for various types of brain tissue in earlier experimental papers. The comparatively low σ_e made him conclude that a fair fraction of tissue currents is likely to travel through intracellular pathways. The same conclusion was reached in later experiments by Okada et al. (1994), who determined both σ_e (computed from equation (5.4) after measuring σ_{ef}) and σ_t (measured using DC current injections) in the turtle cerebellum. They found that σ_t was about 50 percent larger than σ_e, thus suggesting that about 1/3 of the tissue currents travel through intracellular pathways. An equivalent comparison of the extracellular conductivity $\sigma_e = 0.134$ S/m estimated in Section 5.2 with the typically reported cortical tissue conductivity $\sigma_t = 0.3$ S/m would suggest that more than half of the tissue currents travel through intracellular pathways.

An exact understanding of how macroscopic currents make their way through brain tissue is lacking. Fortunately, such an understanding is not a prerequisite for using VC theory to predict extracellular potentials. A sufficient criterion for VC theory to work is that we can use the linear ohmic relation $\mathbf{i}_t = -\sigma_t \nabla V_e$ for the tissue current, together with a trusted and experimentally measured value of σ_t. As reviewed in Section 5.4, a number of experimental studies indicate that the macroscopic (coarse-grained) tissue-current density is indeed well approximated by the ohmic relation.

If we accept this ohmic form, we need not worry about details regarding the pathways taken by currents through tissue. We note, however, that the ohmic form suggests that both extracellular and intracellular (if present) tissue currents are linearly dependent on the *extracellular* potential gradient ∇V_e. This may seem counterintuitive, but when it comes to the intracellular components, it can tentatively be explained by a polarization of cellular structures (neurons, glial cells, blood vessels) proportional to the extracellular voltage gradients.

5.4   Frequency Dependence of the Tissue Conductivity

An important question when modeling and interpreting extracellular signals in neuroscience is whether the conductivity of neural tissue (σ_t) has a frequency dependence.[2] If the conductivity of neural tissue is frequency-dependent, extracellular potentials will be filtered by the tissue. In this subsection, we will briefly review some experimental measurements investigating this issue, as well as the theory needed to model such effects, should it be deemed necessary.

If tissue currents were exclusively extracellular, we would expect σ_t to not depend on frequency, since saline is known to be a purely resistive medium (it has no bound charges), at least up to frequencies in the megahertz range (Cooper 1946, Martinsen & Grimnes 2015). A frequency-dependent σ_t would, however, not by itself be surprising. Capacitive neural membranes have a highly frequency-dependent conductivity (Section 3.1.1). One could therefore easily imagine that low-frequency tissue currents are prone to pass resistively through the extracellular part of the tissue, while high-frequency tissue currents – in addition to following resistive extracellular pathways – may pass capacitively through cell membranes (Foster & Schwan 1989, Martinsen & Grimnes 2015, Amini et al. 2018). At the limit of very high (unphysiological) frequencies, we might even expect tissue currents not to experience the capacitive membranes as hindrances at all but instead pass straight through all the components of tissue with the same ease (Martinsen & Grimnes 2015, Amini et al. 2018).

If physiological tissue currents were to pass capacitively through membranes to some degree, it would effectively cause σ_t to *increase* with frequency. As a consequence, the amplitude of extracellular potentials would *decrease* with frequency (see equation (5.19)). If such effects were present, we would for example expect the extracellular signature of neural activity to put a bias on slower neural signals, reflected in the LFP, compared to brief signals such as action potentials, which are reflected in extracellular spikes (Logothetis et al. 2007).

Early investigations indicated that σ_t has little or no frequency dependence over the range of frequencies in endogenous signals (Ranck 1963, Nicholson & Freeman 1975, Pfurtscheller & Cooper 1975). This was later challenged when measurements by

---

[2] Although a frequency-dependent σ_t should be referred to as an "admittance," we will stick with the word "conductivity" here. The frequency dependence is often quantified in terms of an impedance or admittance spectrum.

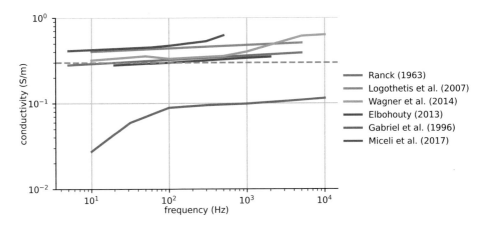

Figure 5.3 **Literature review of reported conductivities in various species and experimental setups.** Most studies seem to indicate a very weak frequency dependence of the tissue conductivity, which would have a negligible effect on measured extracellular potentials (Miceli et al. 2017). The very-low and strongly frequency-dependent values measured by Gabriel et al. (1996) represents an outlier, and although it has received substantial attention, it has to the best of our knowledge not been reproduced by any other study. The plot shows the amplitude $|\hat{\sigma}_t|$ of the possibly complex tissue conductivity $\hat{\sigma}_t$ defined in Section 5.4.1. For details about the data, see Miceli et al. (2017) and references therein (Ranck 1963, Gabriel et al. 1996, Logothetis et al. 2007, Elbohouty 2013, Wagner et al. 2014). Code available via www.cambridge.org/ electricbrainsignals.

Gabriel et al. (1996) implied a substantial frequency dependence of σ_t, in particular below 100 Hz (Figure 5.3). A series of more recent studies again seemed to reconfirm the earlier findings, which reported a σ_t that did not depend, or depended very little, on stimulus frequency (Logothetis et al. 2007, Elbohouty 2013, Wagner et al. 2014, Dowrick et al. 2015, Avery et al. 2017, Miceli et al. 2017, Ranta et al. 2017). Notably, many of these recent studies consistently measured a moderate increase of σ_t by about 20–50 percent when the stimulus frequency was increased from a few hertz to some hundreds or thousands of hertz (Figure 5.3), but two of the studies also measured a similar frequency dependence in (pure) saline, indicating that this frequency dependence at least partially originated in the recording equipment (Logothetis et al. 2007, Miceli et al. 2017).

To our knowledge, the study by Gabriel et al. (1996) is the only experimental study to find a strong frequency dependence for low frequencies. This study also stands out in that it reported a conductivity of only about 0.03 S/m at 10 Hz, about an order of magnitude lower than the commonly reported values. This value seems surprisingly low given that healthy neural tissue contains a substantial fraction of highly conducting saline. The authors were however careful to note that the reported low-frequency values might be distorted by inadequate correction for electrode polarization. In addition, their recordings were done in bovine brains obtained from a slaughterhouse up to a couple of hours post-mortem (Gabriel et al. 1996), which can potentially have resulted in subop-timal tissue preservation and cell swelling, which is known to decrease the conductivity

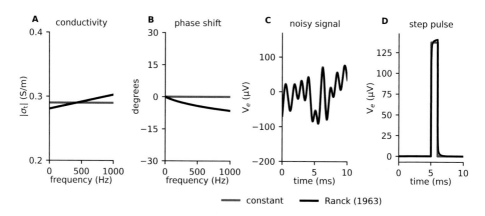

Figure 5.4 Effect of frequency-dependent tissue conductivity on extracellular potentials.
A: Two conductivity profiles, one based on data from Ranck (1963) and one constant, but with the same average value in the frequency range below 1000 Hz. **B**: Phase shifts (see equation (5.15)) required from the conductivity profiles in panel A to preserve causality in accordance with the Kramers-Kronig relations, see Section 5.4.1.3 and Miceli et al. (2017). **C**: Effect of impedance spectrum on the extracellular potential generated by a noisy point-current source with predominately low frequencies (< 1000 Hz), 10 μm away from the recording site. This is meant to represent an arbitrary current source of physiological origin. **D**: Same as in panel C, but the current source is a step pulse, meant to represent a non-physiological external current stimulation. Code available via www.cambridge.org/electricbrainsignals.

of neural tissue post-mortem (Martinsen & Grimnes 2015). Finally, we note that a later study by Wagner et al. (2014), using the same recording equipment, neither observed a strong frequency dependence nor low conductivity values (Figure 5.3).

On balance, experimental evidence thus suggests that neural tissue is predominantly resistive, with a conductivity that increases at most by up to 30 percent with frequency over the frequency range of interest for neurophysiology (below a few thousand hertz). A 30 percent increase might not sound very small, but modeling studies have indicated that frequency dependencies of this order would have a relatively minor effect on measured potentials of physiological origin (Bossetti et al. 2008, Tracey & Williams 2011, Miceli et al. 2017). An example simulation illustrating this point is shown in Figure 5.4.

We note that the notion of a predominantly resistive σ_t has been challenged by theoretical studies suggesting that σ_t depends on spatial scale and should be strongly frequency-dependent when the distances between the source and recording position is sufficiently small (see Section 5.5). In the experimental studies cited earlier, the distances between the stimulus electrodes and recording electrodes were typically several millimeters, but distances down to 100 μm were considered in Miceli et al. (2017) without observing pronounced frequency-dependent effects. Also, we note that the frequency dependence of tissue may become important when considering potentials evoked by extracellular current stimulation using very brief current pulses because of the very-high-frequency content of the pulses (Bossetti et al. 2008).

5.4.1 Complex Conductivity

If an oscillating current is injected into a medium, the resulting electric potential will oscillate with the same frequency as the current. If the conductivity of the medium is frequency dependent, the amplitude of the oscillating potential will depend not only on the amplitude of the injected current, but also on the frequency f.

Up until this point, we have denoted the tissue conductivity simply by σ_t and have mainly focused on its resistive properties. However, if the conductivity of brain tissue is frequency-dependent, it is convenient to use a complex notation and denote it as $\hat{\sigma}_t(f)$.[3] The hat indicates that $\hat{\sigma}_t(f)$ is a complex entity that can contain a real-valued, resistive part $\mathrm{Re}(\hat{\sigma}_t(f))$ (which we, for short, will denote as σ_{Rt}) and an additional imaginary-valued part $\mathrm{Im}(\hat{\sigma}_t(f))$, accounting for non-resistive effects. When investigating the frequency dependence of the conductivity, it is often its amplitude $|\hat{\sigma}_t(f)|$ (see equation (5.16)) that is measured and plotted (as in Figures 5.3 and 5.4).

If the frequency dependence in tissue stems from linear capacitive effects (as supported by Bossetti et al. (2008)), $\hat{\sigma}_t(f)$ represents the (linear) proportionality between the electric field and the current density,

$$\hat{\mathbf{i}}(f,t) = \hat{\sigma}_t(f)\hat{\mathbf{E}}(f,t), \tag{5.8}$$

where both the current density and field are now complex variables. Here, we will derive this relationship, as well as an expression for $\hat{\sigma}_t(f)$.

We start with defining the complex electric field. For simplicity, we assume that it oscillates with a single frequency component f. It can then be expressed in complex notation as

$$\hat{\mathbf{E}}(f,t) = \mathbf{E}_0(f)e^{j2\pi ft}, \tag{5.9}$$

where j is the imaginary unit. The constant vector $\mathbf{E}_0(f)$ contains the (real) amplitude of the field component oscillating with frequency f. Generally, we could have a complex prefactor $\hat{\mathbf{E}}_0(f)$ that also includes the phase of the field relative to, say, a stimulus current. However, we here use the field as the reference and define it to have zero phase.

Next, the total current density in a medium can generally consist of a free current density (of unbound charges) and a displacement current density (of bound charges), as we previously saw in equation (2.39). Assuming that the free current density is purely resistive, the current density associated with the field $\mathbf{E}(f,t)$ can be written in complex notation as

$$\hat{\mathbf{i}}(f,t) = \sigma_{Rt}\hat{\mathbf{E}}(f,t) + \epsilon_t\frac{\partial\hat{\mathbf{E}}(f,t)}{\partial t}, \tag{5.10}$$

where the first term on the right is the resistive current density and the second term is the displacement current density, with ϵ_t being the permittivity of the tissue. If neural tissue possesses a displacement current density, it will presumably be due to capacitive membrane currents.

---

[3] This is, for example, explained in Reitz et al. (1993, section 13.4) for the impedance. The complex conductance (or admittance) is the reciprocal of the impedance.

We can now insert the field defined in equation (5.9) into equation (5.10) to get

$$\hat{\mathbf{i}}(f,t) = \sigma_{Rt}\hat{\mathbf{E}}(f,t) + j2\pi f \epsilon_t \hat{\mathbf{E}}(f,t) . \qquad (5.11)$$

As we seek to express this on the compact form of equation (5.8), we see that the complex conductivity must be given by

$$\hat{\sigma}_t(f) = \sigma_{Rt}\left(1 + \frac{j2\pi f \epsilon_t}{\sigma_{Rt}}\right) . \qquad (5.12)$$

The real and imaginary parts of $\hat{\sigma}_t(f)$ describe the resistive and capacitive properties of the medium, respectively.

Since the electric current density will alternate with the same frequency f as the field, it is possible to simplify equation (5.8) further by removing the time dependency. To do this, we write the electric current density as

$$\hat{\mathbf{i}}(f,t) = \hat{\mathbf{i}}_0 e^{j2\pi f t} , \qquad (5.13)$$

insert it into equation (5.8), and eliminate the factor $e^{j2\pi f t}$ from both sides of the equation. We then arrive at an equation that no longer contains the time variable:

$$\hat{\mathbf{i}}_0(f) = \hat{\sigma}_t(f)\mathbf{E}_0(f) . \qquad (5.14)$$

Equation (5.14) is defined in the frequency domain and relates the frequency and phase of the current density to the frequency and phase of the electric field.

Note that while equation (5.14) applies to a single frequency component, the effects of a frequency-dependent conductivity on the current due to any time-dependent electric field can be computed by the means of Fourier theory. The expression for the electric field given in the time domain is then (i) first decomposed to a sum of sinusoidal components (that is, to the frequency domain). Next, (ii) equation (5.14) is used to find the induced current for each sinusoidal component, and (iii) all sinusoidal current components are summed to get the current in the time domain. More information on how to move between the time domain and the frequency domain through the Fourier transformation is given in Appendix F.

5.4.1.1   Phase Difference

Note that the complex $\hat{\mathbf{i}}_0$ in equation (5.14) contains both the amplitude and phase $\theta(f)$ of the current density. The phase (or, rather, the phase *difference* between the field and current) follows directly from the complex tissue conductivity $\hat{\sigma}_t(f)$. To see this, we express $\hat{\sigma}_t(f)$ in polar form as

$$\hat{\sigma}_t(f) = |\hat{\sigma}_t(f)|e^{j\theta(f)} , \qquad (5.15)$$

with

$$|\hat{\sigma}_t(f)| = \sqrt{\text{Re}\{\hat{\sigma}_t\}^2 + \text{Im}\{\hat{\sigma}_t\}^2} = \sqrt{\sigma_{Rt}^2 + (2\pi f \epsilon_t)^2} \qquad (5.16)$$

and

$$\theta(f) = \tan^{-1}\left(\frac{\text{Im}\{\hat{\sigma}_t\}}{\text{Re}\{\hat{\sigma}_t\}}\right) = \tan^{-1}\left(\frac{2\pi f \epsilon_t}{\sigma_{Rt}}\right). \tag{5.17}$$

Inserting equation (5.15) into equation (5.14) gives us

$$\hat{i}_0(f) = |\hat{\sigma}(f)|\mathbf{E}_0 e^{j\theta(f)}, \tag{5.18}$$

which shows that $\theta(f)$ is the phase difference between the current and the field.

5.4.1.2 Complex Point-Source Equation

If we introduce a complex conductivity into VC theory, the frequency-independent equation (4.6) can be replaced with a frequency-dependent point-source equation (Plonsey & Heppner 1967, Bossetti et al. 2008, Miceli et al. 2017):

$$\hat{V}_e(r, f) = \frac{\hat{I}(f)}{4\pi(\sigma_{Rt} + j2\pi f \epsilon_t)r}. \tag{5.19}$$

Equation (5.19) is valid in frequency space, and $\hat{I}(f)$ represents a singular point source oscillating at a single frequency f. Since the problem is linear, for signals composed of multiple frequency components, one can solve equation (5.19) for each component separately and then sum up the individual solutions (as described after equation (5.14)).

5.4.1.3 Individual Frequency Dependencies of σ_{Rt} and ϵ_t

We focused earlier on the frequency dependence due to capacitive properties in the medium, and we incorporated these into the theory in the form of a complex conductivity. In a more general case, the real parts of the conductivity and the permittivity can have additional "individual" frequency dependencies: $\sigma_{Rt} = \sigma_{Rt}(f)$ and $\epsilon_t = \epsilon_t(f)$. Importantly, these individual frequency dependencies are not independent, as a certain relationship between $\sigma_{Rt}(f)$ and $\epsilon_t(f)$ is required in order for causality not to be violated: the extracellular potential originating from a current source cannot precede the source signal itself. Such a relationship between $\sigma_{Rt}(f)$ and $\epsilon_t(f)$ can be found from the so-called *Kramers-Kronig relations*, according to which $\sigma_{Rt}(f)$ is related to $\epsilon_t(f)$ via

$$\sigma_{Rt}(f) - \sigma_{Rt}(\infty) = 4\int_0^\infty \frac{f'^2 \epsilon_t(f') - f^2 \epsilon_t(f)}{f'^2 - f^2} df'. \tag{5.20}$$

For more on the Kramers-Kronig relations, see Martinsen & Grimnes (2015, section 8.1.6). Equivalent relations can also be found from the *Bode relation* between the magnitude of the conductivity $|\sigma_{Rt}(f)|$ and the induced phase shift $e^{-j\theta(f)}$ (Toll 1956, Foster & Schwan 1989, Bechhoefer 2011, Martinsen & Grimnes 2015, Miceli et al. 2017).

5.4.2   Estimates of the Capacitive Effects in the Brain

From equation (5.12), it is clear that the capacitive contribution (the last term) is negligible compared to the resistive contribution (the first term) when[4]

$$\frac{2\pi f \epsilon}{\sigma_R} \ll 1, \tag{5.21}$$

where σ_R is still the real and resistive component of the conductivity. In equation (5.21), we have not included the subscript "t" for tissue, as we will compute the same ratio for other mediums as well.

An alternative version of this criterion is

$$2\pi f \tau \ll 1, \tag{5.22}$$

which follows from equation (5.21) if we define

$$\tau = \frac{\epsilon}{\sigma_R}. \tag{5.23}$$

Here, τ can be interpreted as the charge-relaxation time in a linear medium (Grodzinsky 2011). To see this, we start by requiring that the total (resistive + displacement) tissue current

$$\mathbf{i}_{tot} = \sigma_R \mathbf{E} + \epsilon \frac{\partial \mathbf{E}}{\partial t} \tag{5.24}$$

should be conserved so that $\nabla \cdot \mathbf{i}_{tot} = 0$. This gives us

$$\sigma_R \nabla \cdot \mathbf{E} + \epsilon \frac{\partial}{\partial t} (\nabla \cdot \mathbf{E}) = 0. \tag{5.25}$$

Next, we use $\nabla \cdot \mathbf{E} = \rho_{free}/\epsilon$ (equation (2.34)) to obtain

$$\frac{\partial \rho_{free}}{\partial t} + \frac{\sigma_R}{\epsilon} \rho_{free} = 0, \tag{5.26}$$

which is a differential equation with the general solution

$$\rho_{free}(\mathbf{r}, t) = \rho_{free}(\mathbf{r}, 0)e^{-\sigma_R t/\epsilon} = \rho_{free}(\mathbf{r}, 0)e^{-t/\tau}. \tag{5.27}$$

Equation (5.27) shows that there is no steady-state free charge in the system considered, and if any nonzero free charge occurs – for example, due to a sudden change in the electric field – it will decay towards zero with the time constant $\tau = \epsilon/\sigma_R$. From this, the interpretation of τ as the charge-relaxation time is clear.

The criterion in equation (5.22) tells us that only for high frequencies, approaching in the order of one oscillation per charge-relaxation time, will capacitive effects be important. For fields of physiological origin, the temporal frequencies are typically less than a few kilohertz. Capacitive effects will therefore mainly be important if the charge-relaxation time is not much smaller than a millisecond. Here, we will explore whether the criterion in equation (5.21) or equation (5.22) is met in three different mediums.

---

[4] Note that with $\mathbf{i}_{free} = \sigma_{Rt}\mathbf{E}$ in equation (2.37), the quasi-magnetostatic approximation of Maxwell's equations (discussed in Section 2.7.2.2) is justified when the criterion in equation (5.21) is fulfilled.

5.4.2.1 Extracellular Fluid

To approximate the charge-relaxation time for the extracellular fluid, we use (i) $\sigma_R = \sigma_{ef} \sim 1.5\,\text{S/m}$ (see Section 5.1) and (ii) that the relative permittivity ($\epsilon_r = \epsilon/\epsilon_0$) of dilute salt-water solutions is approximately the same as that for water. For the low-field frequencies of physiological signals, the relative permittivity of water (ϵ_{rw}) is approximately 80 (Hasted et al. 1948). This gives a value $\tau_{ef} = \epsilon_{rw}\epsilon_0/\sigma_{ef} \sim 80 \times 8.85 \times 10^{-12}\,\text{ns}/1.5 \approx 0.5\,\text{ns}$. A nanosecond-fast charge relaxation implies that the displacement current will mainly be important under conditions when the electric field varies with frequencies in the gigahertz range. Since the relevant fields of physiological origin vary with frequencies that are orders of magnitude lower than this, capacitive effects in saline can safely be neglected. With $\tau_{ef} = 0.5\,\text{ns}$ and $f = 1000\,\text{Hz}$, we get $2\pi f \tau \approx 3 \times 10^{-6}$, and the criterion in equation (5.22) is clearly met.

5.4.2.2 Neural Membrane

The time constant for a neural membrane varies depending on what kinds of ion channels are open, but here we will consider the passive membrane time constant, τ_m, which for mammalian neurons are normally estimated to be in the range from 20 to 100 ms (Koch et al. 1996).

Before we plug τ_m into equation (5.22), let us verify that it rightfully has the interpretation of a charge-relaxation time: $\tau_m = \epsilon_m/\sigma_m$. The permittivity of a parallel plate capacitor is related to specific capacitance through $\epsilon_m = c_m d$, where d is the distance between the conductors. For the neural membrane, the conductors are the intra- and extracellular saline solutions, and d is the membrane thickness. The conductivity (S/m) of the membrane itself is related to its specific conductance g_m (S/m^2) through $\sigma_m = g_m d$.[5] We take g_m to be the passive membrane conductance, which is the inverse of the passive resistance $g_m = 1/r_m$. With this, $\tau_m = \epsilon_m/\sigma_m = (c_m d)/(g_m d) = c_m r_m$, which is the classical definition of the passive membrane time constant (Koch et al. 1996).

As a small sidetrack, we note that we can also calculate ϵ_m and σ_m for the passive membrane. With a membrane thickness $d \approx 10\,\text{nm}$ and a specific membrane capacitance $c_m \approx 1\,\mu\text{F/cm}^2$, we find that $\epsilon_m \approx 10^{-10}\,\text{F/m}$. For a typical membrane time constant $\tau_m \approx 50\,\text{ms}$, we then obtain $\sigma_m = \epsilon_m/\tau_m \approx 2 \times 10^{-9}\,\text{S/m}$. This corresponds to a passive membrane resistance $r_m = d/\sigma_m = 10^{-8}/2 \times 10^{-9}\,\Omega\,\text{m}^2 = 50\,\text{k}\Omega\,\text{cm}^2$. Estimates of r_m based on electrophysiology data vary greatly between cells and experiments, from values below $1\,\text{k}\Omega\,\text{cm}^2$ to about $100\,\text{k}\Omega\,\text{cm}^2$ (Rall 1959, Wilson 1984, Bloomfield et al. 1987, Mainen & Sejnowski 1995, Migliore et al. 1995, Pospischil et al. 2008, Halnes et al. 2011, Hay et al. 2011, Eyal et al. 2016).

Returning to the topic, with a charge-relaxation constant in the typical range (20 to 100 ms) of the membrane time constant and with $f = 1000\,\text{Hz}$, we get $2\pi f \tau_m > 100$,

---

[5] This follows from equating the two (ohmic) expressions, (i) $i = \sigma E = \sigma V_m/d$ (assuming that the field over the membrane is constant), and (ii) $i = g_m V_m$ for the membrane current. Regarding (ii), we have from equation (3.23), that the passive (quasi-ohmic) membrane current is $i_m = g_m(V_m - E_m)$, where E_m is the Nernst-potential, and we may think of it as composed of an ohmic component $i_{m,ohm} = g_m V_m$ and a diffusive component $i_{m,diff} = -g_m E_m$.

and the criterion $2\pi f \tau_m \ll 1$ in equation (5.21) is certainly not fulfilled. This shows (not surprisingly) that capacitive effects clearly cannot be neglected for membrane currents.

5.4.2.3   Neural Tissue

Being a compound medium, neural tissue houses several charge-relaxation processes with different time constants. Because of these various processes, the electric properties of tissue are quite different from those of the cellular membranes and extracellular saline solution individually.

The relaxation processes in biological tissue are often subdivided into three groups, or *dispersion regions*, commonly referred to as α, β, and γ dispersion (illustrated in Figure 5.5). In this context, dispersion refers to the variation of the tissue conductivity (σ_{Rt}) and relative tissue permittivity ϵ_{rt} with frequency. The three dispersion regions are believed to reflect different relaxation mechanisms acting within different frequency ranges (Schwan 1957, Martinsen & Grimnes 2015). Among the three, α dispersion, acting in the low-frequency range (Hz–kHz), is the most relevant for understanding the propagation of signals of physiological origin. Unfortunately, α dispersion is also the least understood among the three, and experimental efforts to understand the dielectric properties of tissue in the low-frequency range have given ambiguous results (Zimmermann & van Rienen 2021).

Several mechanisms have been proposed as possible contributors to α dispersion, including counterion effects near membrane surfaces, dielectric losses, gap junctions, and the polarization of the intracellular sarcotubular system (Pethig 1987, Foster & Schwan 1989, Martinsen & Grimnes 2015, Zimmermann & van Rienen 2021). Contributions from so-called Maxwell-Wagner effects have also been suggested (Monai et al. 2012), although these traditionally are believed to mainly explain the β dispersion. As a further complication, there is evidence that parts of the α dispersion seen in some

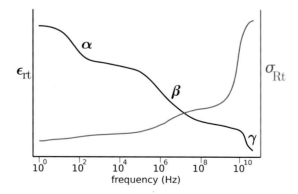

Figure 5.5 Dispersion regions in tissue (illustration). Relative permittivity ϵ_{rt} and conductivity σ_{Rt} in tissue as a function of frequency. The figure is idealized to illustrate regions of α, β, and γ dispersion. Inspired by Martinsen & Grimnes (2015, Fig. 3.14).

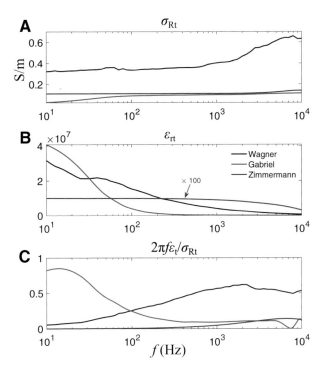

Figure 5.6 Capacitive effects in the brain. A: Frequency dependence of tissue conductivity (σ_{Rt}). **B**: Frequency dependence of relative permittivity (ϵ_{rt}). **C**: The ratio $2\pi f \epsilon_0 \epsilon_{rt}/\sigma_{Rt}$ between the capacitive and resistive effects in brain tissue. Wagner data (black curves) are taken from Wagner et al. (2014, Supplementary material). Gabriel data (red curves) are taken from the online resource https://itis.swiss/virtual-population/tissue-properties/database/dielectric-properties/, containing data based on the compilation in Gabriel (1996). Zimmermann data (blue curves) read off visually from Figure 4 in Zimmermann & van Rienen (2021). Note that the relative permittivity in the Zimmermann data is much lower than in the other data sets and has been multiplied by a factor of 100 to be visible in the plot (panel B).

experiments, such as that of Gabriel et al. (1996), do not reflect passive tissue properties but rather diffusion-polarization effects on the electrode surface (Logothetis et al. 2007, Zimmermann & van Rienen 2021), theoretically described through a Warburg impedance (Bédard & Destexhe 2009). A full mechanistic understanding of α dispersion is currently lacking, and the topic is too complex to go further into here.

As illustrated in Figure 5.6, estimates of the frequency dependence of σ_{Rt} and ϵ_{rt} in the α region vary dramatically between different studies. The figure includes two experimental data sets, from Gabriel (1996) (red curve) and Wagner et al. (2014) (black curve), as well as a third data set from Zimmermann & van Rienen (2021) (blue curve), which can be seen as a theory-based attempt to "correct" the data by Gabriel (1996).

The Gabriel and Wagner data give dramatically different conclusions concerning the relative contribution of capacitive and resistive effects in tissue, as determined by the ratio $2\pi f \epsilon_{rt} \epsilon_0 / \sigma_{Rt}$ (equation (5.21)). If we compute this ratio using the Gabriel data, we arrive at the unexpected conclusion that capacitive effects are most pronounced for the very-low frequencies (around 10 Hz) but decays with f over most of the α region. In contrast, when we use the Wagner data, capacitive effects become gradually more important when f is increased, which seems more plausible. However, the Wagner data also result in the prediction that capacitive effects are quite pronounced already at fairly low frequencies. For example, the ratio between capacitive and resistive effects reaches a value of about 0.25 already at 100 Hz. This is in conflict with other experiments, suggesting that tissue is almost purely resistive up to several kHz (Logothetis et al. 2007, Miceli et al. 2017).

Both the Gabriel and Wagner data were obtained using a two-electrode setup, and it is known that such setups require considerations regarding the polarization effects on the electrode-tissue surface (Ishai et al. 2013). Zimmermann & van Rienen (2021) argued that such effects may only have been partially corrected for in these two data sets, as also originally suggested by Gabriel (1996) for frequencies below about 100 Hz in their own data set. By employing a Kramers-Kronig validity test, making sure that the individual frequency dependencies of σ_{Rt} and ϵ_{rt} are consistent with the principle of causality, they proposed several corrections of the Gabriel data in the form of theoretical dispersion models. In one of these models, assumed electrode effects were removed from the Gabriel data (Figure 5.6, blue curves). The most striking result of this proposed correction was a reduction of ϵ_{rt} by approximately a factor of 100. Moreover, with σ_{Rt} and ϵ_{rt} taken from this corrected model, the ratio between capacitive and resistive effects remains small over the entire α region (Figure 5.6C). This conclusion is in line with measurements based on four-electrode setups, which eliminates electrode polarization effects by design (Logothetis et al. 2007, Miceli et al. 2017).

Perhaps a more direct way to assess whether capacitive effects are present in tissue would be to examine the phase difference θ between recorded potential and the injected current. It follows from equation (5.17) that this phase difference should be

$$\tan \theta = \frac{2\pi f \epsilon_t}{\sigma_{Rt}},\tag{5.28}$$

where we recognize the ratio from equation (5.21) on the right-hand side.

The phase difference between injected current and voltage was recorded by Logothetis et al. (2007) at different cortical depths using a four-electrode setup (see Figure 3 of that work). They observed that the phase difference increased monotonically with stimulus frequency, from about 2 degrees at 10 Hz to about 6 degrees at 1000 Hz. According to equation (5.16), this corresponds to a ratio between capacitive and resistive effects that increases from 0.03 to 0.11 over the same frequency range, which agrees fairly well with the Zimmermann model in Figure 5.6C.

It should be noted that Logothetis et al. (2007) and Miceli et al. (2017) compared the tissue recordings to equivalent recordings in saline and found very similar values for $\theta(f)$ in both cases. Since saline is known to be a purely resistive medium at the low frequencies considered here (Cooper 1946, Martinsen & Grimnes 2015), these findings indicate that electrode effects might play a role even in four-electrode setups, and capacitive effects in tissue (in the $f < 1\,\text{kHz}$ range) might be even smaller than suggested by the earlier estimates. Theoretical studies have, however, suggested that the weak frequency dependence found by these studies may depend on the spatial scale on which the measurements were made. We discuss this further in the next section.

5.5   Spatial Frequency Dependence of the Tissue Conductivity

In a body of theoretical works, Hamish Meffin and coworkers have studied the conductivity of brain tissue using cable theory (Meffin et al. 2012, Tahayori et al. 2012, Meffin et al. 2014, Tahayori et al. 2014, Monfared et al. 2020). In their studies, they predict that, depending on spatial scale, the effective tissue conductivity can have both a temporal and spatial frequency dependence. The mathematical models introduced in these works are quite complex and rest on a number of assumptions and approximations that we will not go into here. We will, however, summarize some of the key model predictions.

A spatial frequency dependence means that the effective conductivity depends on the distance between the current source and the recording position. The proposed explanation of this phenomenon is that currents traveling short distances are prone to stay in the extracellular medium due to the high cost involved in crossing the resistive membranes. In contrast, currents traveling longer distances are prone to cross membranes, since the gain in exploiting an enlarged conductive volume for a long journey is greater than the expense associated with the membrane crossings (Monfared et al. 2020).

One key prediction in the study by Monfared et al. (2020) is that the effective tissue conductivity σ_t increases with the distance between the current source and the recording position, from a predominantly resistive near-field value $\sigma_{t,near}$ to a predominantly resistive far-field value $\sigma_{t,far}$. Another key prediction is that there is a transition regime at intermediated distances (between "near" and "far") where σ_t also has a strong temporal frequency dependence due to tissue currents crossing capacitive membranes.

These theoretical predictions appear to be in conflict with the experimental studies reviewed in Section 5.4, which suggest that σ_t is mostly resistive. A possible explanation to the discrepancy between the theoretical studies and the experiments could be that the spatial scale in the experiments was so large that only $\sigma_{t,far}$ was measured, while $\sigma_{t,near}$ and the transition were never observed. In many of the experiments, the distance between the stimulus electrode and recording electrode were several millimeters. However, Miceli et al. (2017) considered distances down to $100\,\mu m$ without observing any of the spatiotemporal effects predicted by Monfared et al. (2020). The theoretical prediction of a frequency dependence σ_t at small distances is thus so far without experimental support.

5.6   Anisotropic Conductivity

In Chapter 4, we assumed that the tissue conductivity (σ_t) was the same in all the spatial directions (isotropic). However, many brain regions contain tissue with clearly anisotropic geometrical properties, and it is reasonable to expect that such anisotropy should be reflected in the conductivity. For example, the cortex is to a large degree populated with pyramidal neurons that tend to be aligned along the depth direction of the cortex, and indeed, conductivity measurements have found that σ_t is about 1.5 times larger in the depth direction (parallel to the axis of pyramidal cell dendrites) than in the lateral direction (Goto et al. 2010). The cerebellum has an even more pronounced anisotropic geometrical structure, and in the anuran cerebellum, σ_t has been found to be about three times larger in the depth direction than in the lateral direction (Nicholson & Freeman 1975). A pronounced anisotropy in the conductivity has also been observed in white matter because the axons tend to be oriented in similar directions by the formation of fiber bundles (Nicholson 1965, Logothetis et al. 2007, Bangera et al. 2010). This anisotropy can in certain cases cause a 10-fold increase in conductivity along the fiber bundle (Nicholson 1965, Bangera et al. 2010).

The overall effect of anisotropy on extracellular potentials often appears to be quite weak, at least in the cortex (Logothetis et al. 2007, Ness et al. 2015, Miceli et al. 2017). The approximation that σ_t is isotropic therefore often gives good predictions of the potential. An example is given in Figure 5.7, where a 50 percent increase in the conductivity along the axis of the apical dendrite (as suggested by Goto et al. (2010)), only causes a slightly more squeezed shape of the extracellular potential compared to the isotropic case (Ness et al. 2015, Miceli et al. 2017). Since extracellular potentials are rather insensitive to modest anisotropies, it is common to make the modeling approximation that σ_t is isotropic.

When deemed necessary, it is relatively straightforward to expand the VC theory to the case of an anisotropic σ_t. Then, σ_t will no longer be a scalar but instead a tensor with the three components σ_{tx}, σ_{ty}, and σ_{tz}. If we use the point-source approximation (equation (4.8)), the extracellular potential surrounding a set of point-current sources I_k is given by (Nicholson & Freeman 1975, Parasnis 1986)

$$V_e(x, y, z) = \sum_k \frac{I_k}{4\pi\sqrt{\sigma_{ty}\sigma_{tz}(x - x_k)^2 + \sigma_{tx}\sigma_{tz}(y - y_k)^2 + \sigma_{tx}\sigma_{ty}(z - z_k)^2}} . \quad (5.29)$$

A derivation of this equation is found in Appendix D.

If we use the CSD description of the sources (equation (4.22)), the corresponding expression is

$$V_e(x, y, z) = \iiint \frac{C_t(x', y', z')}{4\pi\sqrt{\sigma_{ty}\sigma_{tz}(x' - x_k)^2 + \sigma_{tx}\sigma_{tz}(y' - y_k)^2 + \sigma_{tx}\sigma_{ty}(z' - z_k)^2}} \, dv' .$$
$$(5.30)$$

5.7   Inhomogeneous Conductivity

In Chapter 4, we assumed that the extracellular conductivity σ_t was the same everywhere (homogeneous). As we illustrated in Figure 5.2, this assumption clearly does not hold on the micrometer scale, where neural tissue is highly inhomogeneous (Nicholson &

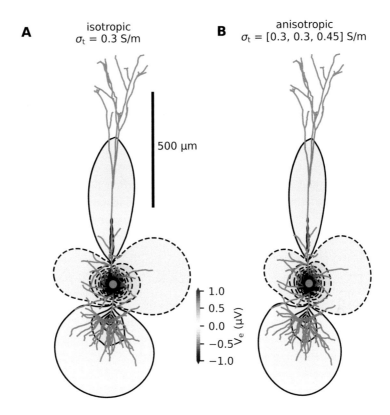

A isotropic
$\sigma_t = 0.3$ S/m

B anisotropic
$\sigma_t = [0.3, 0.3, 0.45]$ S/m

500 μm

1.0
0.5
0.0
−0.5 V_e (μV)
−1.0

Figure 5.7 **Effect of anisotropy on extracellular potential.** **A**: Extracellular potential in isotropic tissue medium. **B**: Extracellular potential in anisotropic tissue medium with a 50 percent higher conductivity along the axis of the apical dendrite. The same simulation was used in both panels, and the extracellular potential is shown at the time when it was at its maximal value, following a single excitatory synaptic input (location marked by blue dot) to the Hay model. In panel B, the extracellular potential was calculated using equation (5.29). Code available via www.cambridge.org/electricbrainsignals.

Syková 1998). Also, it clearly does not hold on the very-large scale since, if we travel far enough through brain tissue, we will pass through boundaries between different tissue types with altogether different conductivities and sooner or later run into the less-conducting skull. However, on a mesoscopic spatial scale, we often assume that the microscale inhomogeneities average out (see Section 5.2.1) and, at the same time, that the boundaries to other tissue types or materials present in experimental settings are too far away to affect the local extracellular potentials notably. In such cases – for example, within a given brain region such as the cortex – a homogeneous conductivity appears to be a reasonable approximation (Nicholson & Freeman 1975, Okada et al. 1994, Logothetis et al. 2007, Goto et al. 2010, Ness et al. 2015).

Sometimes, we may nevertheless need to consider cases when an infinite homogeneous medium is not a reasonable assumption. This may, for example, be relevant if we wish to simulate extracellular potentials near boundaries between neural tissue and other tissue types, or between neural tissue and other types of materials. When the extracellular medium is inhomogeneous, there is no general analytical formula available (like equation (4.22), equation (5.29), or equation (5.30)) that links the extracellular potentials to the underlying current sources. In principle, one can still always solve the forward problem for arbitrarily complex geometries with varying conductivities using numerical methods, like the finite element method (FEM) (Logg et al. 2012). This approach has, for example, been used to model electroencephalography (EEG) signals recorded outside of the head because these potentials are not only affected by the conductivity of brain tissue, but also by the presence of the cerebrospinal fluid (CSF), the skull, and the scalp, which have widely different conductivities. EEG modeling is covered in more detail in Chapter 9. Substantial inhomogeneities can also be introduced into neural tissue through the presence of big electrode shanks, as briefly discussed in Section 4.5.3.

5.7.1   Planar Boundaries: Tissue Interfaces and In Vitro Slice Recordings

As explained above, analytical solutions for extracellular potentials in inhomogeneous volume conductors can generally not be found. However, analytical solutions do exist for idealized cases with planar boundaries (assumed to cover the entire plane) between different tissue types or different materials. In such cases, the *method of images* (MoI) from electrostatics (Jackson 1998) can be used to account for the effect of a planar boundary on the extracellular potential arising from underlying current sources (Gold et al. 2006, Nunez & Srinivasan 2006, Pettersen et al. 2006, Ness et al. 2015, Obien et al. 2019, Viswam et al. 2019). This method is based on introducing virtual current sources to ensure that the necessary boundary conditions on the planar boundary are fulfilled. This works because the solution of Poisson's equation for V_e (equation (2.31)) is unique provided that the boundary conditions are specified and the sources known.

As an example, consider a point-current source near a planar boundary to a non-conducting region. We know that no current can flow into the non-conducting region, and this boundary condition can be fulfilled by ensuring that the potential at the boundary is constant in the direction perpendicular to the planar boundary (since no voltage gradient implies no current). This can be accomplished by mirroring the point-current source across the planar boundary, since the symmetry of such an arrangement removes the potential gradient perpendicular to the boundary at the location of the boundary (Figure 5.8A). Similarly, for a point-current source near a planar boundary to a region of (near) infinite conductivity, we know that the potential at the boundary must be zero, which can be imposed by adding a virtual current source of the same magnitude but opposite sign at the other side of the planar boundary (Figure 5.8B).

In the general case of a planar boundary between two regions of different conductivity, the virtual current sources are always placed on the opposite side of the boundary with the same distance to the boundary as the original point-current sources, with the amplitudes scaled by (Nunez & Srinivasan 2006, Ness et al. 2015)

$$W_{12} = \frac{\sigma_1 - \sigma_2}{\sigma_1 + \sigma_2}, \qquad (5.31)$$

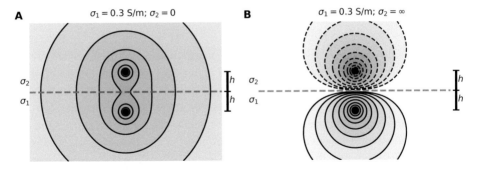

Figure 5.8 Illustration of the method of images. The method of images can be used to impose the boundary conditions of planar boundaries into a homogeneous medium by introducing virtual current sources mirrored across the planar boundary. **A**: For a planar boundary to a non-conducting region, we can mirror the original current source across the boundary. **B**: For a planar boundary to a region of "infinite" conductivity, we can mirror the original current source but invert the amplitude. Code available via www.cambridge.org/electricbrainsignals.

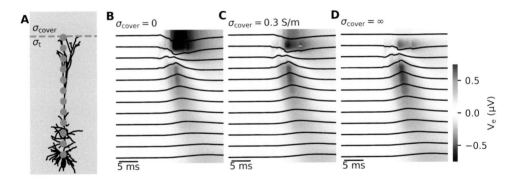

Figure 5.9 Effect on extracellular potential from inhomogeneity at the cortical surface. **A**: Simulation setup. The Hay model receives a wave of excitatory synaptic input to the upper apical dendrite (top 200 µm). The conductivity of cortical tissue is $\sigma_t = 0.3$ S/m, while different conductivities of the region above the cortex are tested. **B**: The cover is an insulator, $\sigma_{cover} = 0$, like a non-conducting mineral oil or air. **C**: The cover has the same conductivity as the tissue, $\sigma_{cover} = 0.3$ S/m, corresponding to an infinite homogeneous volume conductor. **D**: The cover is highly conductive, $\sigma_{cover} = \infty$, like a metal plate. Code available via www.cambridge.org/electricbrainsignals.

where σ_1 is the conductivity of the region containing the real current sources, and σ_2 is the conductivity of the region on the other side of the boundary containing the virtual sources.

The MoI has been used to model effects near the cortical surface during in vivo recordings. In such recordings, the cortical surface must typically be exposed before inserting recording electrodes into the brain, and materials of widely different conductivity can be used to cover the cortical surface during the recordings. Using MoI, the different conductivity of the cortex and the cover material can be accounted for by modifying the point-source equation (4.7) to (Nicholson & Llinas 1971, Pettersen et al. 2006)

$$V_e(\mathbf{r}) = \frac{I_n}{4\pi\sigma_t \|\mathbf{r} - \mathbf{r}_n\|} + \frac{\sigma_t - \sigma_{cover}}{\sigma_t + \sigma_{cover}} \frac{I_n}{4\pi\sigma_t \|\mathbf{r} - \mathbf{r}_n'\|}, \tag{5.32}$$

where $\mathbf{r}_n'$ is the location of the virtual current source, mirrored across the cortical surface. Note that this equation is only valid within the region with the real source, while other formulas apply outside of this region (Nunez & Srinivasan 2006).

The material at the cortical surface can substantially affect potentials measured near the cortical surface, as illustrated in Figure 5.9. In particular, notice from equation (5.32) the special case of a measurement performed at the planar boundary to a non-conducting region, so that $\|\mathbf{r} - \mathbf{r}_n\| = \|\mathbf{r} - \mathbf{r}_n'\|$ and $\sigma_{cover} = 0$. In this case, the measured potential will be exactly a factor of two larger than for an infinite homogeneous medium.

The MoI framework can also be used to model in vitro slice recordings, where a small slice of neural tissue is extracted from a brain and placed on a micro-electrode array (MEA), immersed in a saline bath. In this case, there are two planar boundaries: the lower boundary between the slice and the non-conducting MEA (glass) electrode plate, and the upper boundary between the slice and the saline bath. Having two boundaries instead of one gives rise to an infinite series of virtual current sources (mirrors of mirrors), which for the simplest case of the measurement being performed at the lower boundary to a non-conducting region, can be written as

$$V_e(x, y, 0) = \frac{2I}{4\pi\sigma_t} \Bigg(\psi_{ps}(x, y, z')$$

$$+ \sum_{n=1}^{\infty} W_{ts}^n \Big[\psi_{ps}(x, y, -z' + 2nh) + \psi_{ps}(x, y, -z' - 2nh) \Big] \Bigg), \tag{5.33}$$

with

$$\psi_{ps}(x, y, \tilde{z}) \equiv \left((x - x')^2 + (y - y')^2 + \tilde{z}^2 \right)^{-1/2}. \tag{5.34}$$

Here, h is the thickness of the neural slice, (x', y', z') is the position of the real particle, and (x, y, z) with $z = 0$ is the measurement position. W_{ts} is as defined in equation (5.31), with conductivities for tissue ($t = 1$) and saline ($s = 2$). As the function ψ_{ps} shows, the virtual current sources can have other z-values, denoted by $\tilde{z}$. For further details, see Ness et al. (2015).

In the chapter on extracellular spikes, we show an example using the MoI framework to model spikes in in vitro slices (Figure 7.12). Another application of MoI is given in the ECoG chapter (Chapter 10), where it is used to account for discontinuities in the conductivity at the cortical surface – either due to the CSF or the recording device itself – in the modeling of ECoG signals.

6 Schemes for Computing Extracellular Potentials

As discussed in Sections 2.6.2–2.6.3, the gold standard for simulating extracellular electric signals from single neurons or networks of neurons is the two-step MC+VC scheme introduced in Section 2.6.2, summarized by:

1. Compute the electrical activity of neurons using the multicompartment (MC) framework presented in Chapter 3, assuming that membrane currents are independent of the effect they have on the extracellular potential (V_e).

2. Compute the extracellular potential V_e resulting from the neural membrane currents computed in step 1 using volume-conductor (VC) theory (Chapters 4 and 5).

Biophysically detailed simulations of extracellular potentials may involve large networks containing thousands of neurons. To overcome the computational challenges that come with this, it is quite standard procedure to distribute the simulation task across multiple CPU cores on a single machine or across compute nodes in high-performance computing (HPC) facilities. As we showed in Chapter 4, VC theory provides linear analytical expressions for V_e as a direct function of the neural current sources (step 2), meaning that it is usually the simulations of the neurodynamics (step 1) that is the bottleneck in the simulation (Hagen et al. 2018).

In this chapter, we will describe in detail how constituents of the MC+VC scheme can be implemented and used in different applications for the purpose of predicting extracellular signals in neural simulations. We present computational schemes for the standard MC+VC approach, based on single MC neurons as well as recurrent network models built from multiple MC neurons (Sections 6.1–6.2). We also extend this scheme for related signal types that can be computed from corresponding intracellular (axial) currents and (current-) dipole moments (Section 6.3)[1]. In Section 6.4, we describe different strategies for computing V_e when simulating the network dynamics using computationally efficient spiking point-neuron models or population-rate models. We also describe different simplified methods for predicting V_e and comparing it to "ground-truth" data obtained with the MC+VC framework.

---

[1] For brevity, we will refer to "current dipoles" simply as "dipoles." Charge dipoles will not be encountered in this book.

6.1   Forward-Model Predictions from MC Neuron Models

With the exception of the simplest (idealized) neuron models, models aiming to mimic the dynamics of a particular neuron or neural system are generally too complex to be solved analytically (see Chapter 3). Thus, the complete set of differential equations representing voltage- and concentration-dependent ion channels, synapses, plasticity, neuronal geometry, etc. of the full model will have to be solved numerically with the aid of computers and tailored software.

The numerical solution of morphologically detailed neuron models entails a spatial discretization of the neural geometry into multiple compartments (hence MC), wherein the states of variables and their derivatives are estimated on a temporal grid that may be fixed or irregular (using adaptive step-size methods). While the state variables in such a model could be computed with a general-purpose numerical solver, a more feasible approach is to use previously developed software tools tailored specifically for simulating neurons. Such software greatly simplifies the process of specifying the model and choosing the appropriate numerical schemes for solving the resulting set of partial differential equations (PDEs). Most of these tools are by default based on one key assumption made in standard cable theory – that is, the spatial variation in membrane voltage across the entire morphology can be computed independently of any effect membrane currents may have on the extracellular potential immediately outside the compartments (see Chapter 3 for details).

The NEURON simulation environment[2] (Hines & Carnevale 1997) presently remains a common choice for simulations of MC neuron models as it supports both Windows-, Linux-, and Unix-based operating systems including macOS and can be used with a graphical user interface (GUI) or in a programmatic fashion on laptops, desktop computers, and in parallel on large-scale high-performance computing (HPC) facilities. NEURON is also utilized as a simulator backend for other higher-level software such as PyNN[3] (Davison 2008), NetPyNE[4] (Dura-Bernal et al. 2019), BMTK[5] (Dai et al. 2020), and LFPy[6] (Lindén et al. 2014, Hagen et al. 2018). Alternatives to NEURON with partially overlapping feature sets have been developed in the past in the form of MOOSE[7] (Ray & Bhalla 2008) and GENESIS[8] (Bower & Beeman 1998), while software tailored for point-like neuron models such as Brian[9] (Stimberg et al. 2019) recently received support for biophysically detailed MC neuron models. Another effort named Arbor[10] (Akar et al. 2019) aims to develop MC neuron simulation software from the ground up in order to fully exploit modern high-performance libraries and next-generation hardware in the form of graphical processing units (GPUs) and massively parallel HPC facilities.

---

[2] https://neuron.yale.edu
[3] https://neuralensemble.org/PyNN/
[4] https://www.netpyne.org
[5] https://alleninstitute.github.io/bmtk/
[6] https://lfpy.readthedocs.io
[7] https://moose.ncbs.res.in
[8] http://genesis-sim.org
[9] https://briansimulator.org/
[10] https://docs.arbor-sim.org

6.2   Computing Axial and Membrane Currents

To compute extracellular signals resulting from an MC simulation, we need to first compute either the set of net membrane currents I_m or the set of axial currents I_a across all neural compartments. Each of these sets contains, individually, the complete information about the neural sources, as either one of the two can be computed from the other.

6.2.1   Axial and Membrane Currents in the Continuum Limit

While the point-source and line-source formalisms – as well as current dipoles for scalp EEG and MEG – are typically computed from I_m, the main application of I_a is in formalisms for magnetic fields inside brain tissue. Software tools tailored for simulating neurons typically provide built-in functionality for computing membrane currents, but we here seize the opportunity to go through the required steps and corresponding assumptions.

In the continuum limit, the axial and membrane currents are, respectively, given as a function of membrane potential by

$$I_a(x,t) = -\frac{1}{r_i}\frac{\partial V_m(x,t)}{\partial x} \tag{6.1}$$

and

$$\mathcal{I}_m(x,t) = -\frac{\partial I_a(x,t)}{\partial x} = \frac{1}{r_i}\frac{\partial^2 V_m(x,t)}{\partial x^2}. \tag{6.2}$$

Here, $r_i = 4r_a/\pi d^2$ (with unit $\Omega\,\mathrm{m}^{-1}$) is the intracellular resistance per unit length. As in Chapter 3, r_a (with unit $\Omega\,\mathrm{m}$) and d (with unit m) are the axial resistivity and diameter of the dendritic cable model, respectively. As $I_a(x,t)$ (with unit A) is the total axial current, its spatial derivative $\mathcal{I}_m(x,t)$ (with unit A/m) is the membrane current per unit length. The location x denotes the position along the cable in one dimension (the cable equation is by definition 1D – that is, a function of x, even if the different parts of a neuron may be assigned locations in 3D space).

6.2.2   Axial Currents in Discretized Cables

In this section, we outline two methods for computing axial currents in discretized cables.

6.2.2.1   Via Equivalent Circuits

For an MC model, equations (6.1) and (6.2) must be discretized. For equivalent circuit models corresponding to non-branching cables, such as in Figure 6.1A–C, the discretized equation for I_a for the left- and right-hand side of each compartment becomes

$$I_{a,\text{left}}(x_j,t) = \frac{V_{\text{m}}\left(x_j - \frac{L_j}{2},t\right) - V_{\text{m}}(x_j,t)}{r_{ij}L_j/2} \tag{6.3}$$

and

$$I_{a,\text{right}}(x_j,t) = \frac{V_{\text{m}}(x_j,t) - V_{\text{m}}\left(x_j + \frac{L_j}{2},t\right)}{r_{ij}L_j/2}, \tag{6.4}$$

where L_j is the length of compartment j and $r_{ij} = 4r_a/\pi d_j^2$, with d_j being the compartment diameter.

Note that the axial current terms are expressed as a function of the mid- and endpoint potentials per compartment. These are typically not returned by standard neural simulation software. The NEURON simulator, for example, returns values V_{m} at the center of the compartment (Carnevale & Hines 2009, p. 53). The Arbor simulator solves the cable equation using a finite volume method (FVM) and returns values for V_{m} that may be interpreted as the average membrane potential within each spatial discretization volume (Akar et al. 2019) of the neuron model. We therefore need to outline how the membrane potential can be defined or estimated at the compartment endpoints ($x \in \{x_j - \frac{L_j}{2}, x_j + \frac{L_j}{2}\}$), which may be connected to none (cable terminates), one (non-branching cable), or several (branching cable) endpoints of neighboring compartments (Figure 6.1). Keep in mind that the neural simulation software typically returns the midpoint potential $V_{\text{m}}(x_j,t)$ for every $j \in \{1,\dots,N\}$ compartment. For terminating compartments (i.e. one or both ends have no connecting compartments and sealed ends), we may define

$$V_{\text{m}}\left(x_j - \frac{L_j}{2},t\right) = V_{\text{m}}(x_j,t) \text{ or}$$
$$V_{\text{m}}\left(x_j + \frac{L_j}{2},t\right) = V_{\text{m}}(x_j,t), \tag{6.5}$$

depending on which end(s) have no electric connections. This step effectively uses the forward/backward difference approximation to first derivatives in terminating compartments. The assumption of zero voltage change across the intracellular resistance from the compartment midpoint to the terminating endpoint(s) as illustrated in Figure 6.1B is warranted as the axial current across the terminus must be zero. For a single-compartment model where both ends terminate, the axial current will be zero as the voltage by assumption will be isotropic along the compartment axis.

For a number N_{child} child compartments connected to the endpoint ($x_j + \frac{L_j}{2}$) of the parent compartment (Figure 6.1C, D), the potential at the interconnecting node can be computed from the parent- and child-compartment midpoint potentials as (Hagen et al. 2018)

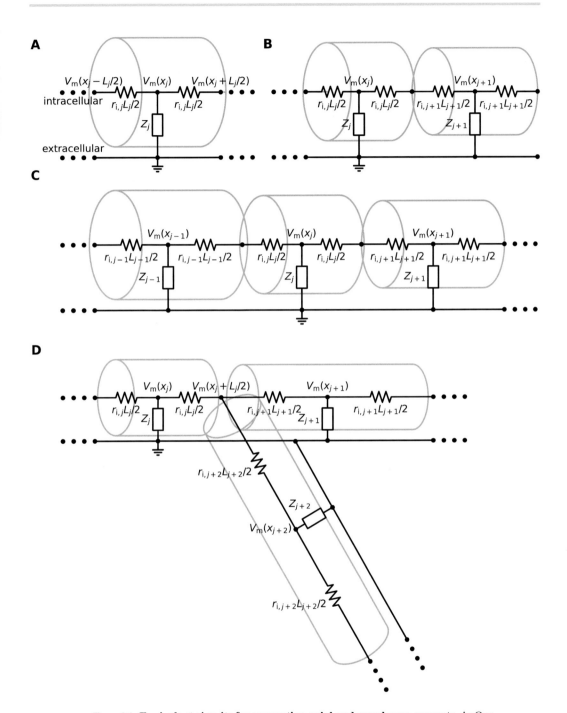

Figure 6.1 Equivalent circuits for computing axial and membrane currents. **A**: One compartment of an MC cable model. **B**: Two compartments of an MC cable model in series, where the second compartment is terminating with a sealed endpoint. **C**: Continuous piece of MC cable model showing a subset of three compartments. **D**: MC cable model with branch point. The symbols $V_{m,j}$, $r_{i,j}$, and L_j denote the membrane potential, the intracellular resistance per unit length, and the length of compartment j, respectively. Rectangular boxes denoted with the symbol Z_j represent various, here unspecified, circuit elements describing the membrane. Code available via www.cambridge.org/electricbrainsignals.

$$V_{\mathrm{m}}\left(x_j + \frac{L_j}{2}, t\right) = \frac{\sum_{n=0}^{N_{\mathrm{child}}} V_{\mathrm{m}}(x_{j+n}, t)/R_{j+n}}{\sum_{n=0}^{N_{\mathrm{child}}} 1/R_{j+n}}. \tag{6.6}$$

For compactness, we introduced the term

$$R_{j+n} = \frac{2r_{\mathrm{a}} L_{j+n}}{\pi d_{j+n}^2}, \tag{6.7}$$

which describes the intracellular resistances between the midpoints of the N compartments to the node where they are connected as a function of model parameters r_{a}, L_{j+n}, and d_{j+n}.

For child compartments connecting to the midpoint of the parent compartment, the corresponding potential at the branch point seen by any number of child segments is per definition the same as the midpoint potential of the parent – that is,

$$V_{\mathrm{m}}\left(x_{j+n} - \frac{L_{j+n}}{2}, t\right) = V_{\mathrm{m}}(x_j, t). \tag{6.8}$$

The same convention is applied to child sections connecting *between* the endpoints of the parent compartment (e.g. spines) (Hagen et al. 2018).

6.2.2.2   Via the Central Difference Approximation

While equations (6.3) and (6.4) are easy to understand from the illustrations in Figure 6.1 – and from the description of MC modeling in Chapter 3 – they are inconvenient in that they rely on redundant information. They associate two axial currents (left and right) with each neural compartment j, while complete information can be contained within one variable per compartment. When computing extracellular signals that depend on axial currents, it can be convenient to compute a single axial current per compartment from midpoint voltages.

For spatially discretized cable models, the (average) axial currents can be estimated per compartment indexed by j via the central difference approximation as

$$I_{\mathrm{a}}(x_j, t) = \frac{\pi d_j^2}{4r_{\mathrm{a}}} \frac{V_{\mathrm{m}}\left(x_j - \frac{L_j}{2}, t\right) - V_{\mathrm{m}}\left(x + \frac{L_j}{2}, t\right)}{L_j}, \tag{6.9}$$

where x_j denotes the compartment midpoint position (Figure 6.1A), and the discretization constant is set equal to the compartment length L_j. Its diameter is assumed to be constant, and the cylindrical compartment is assumed to be straight along its length axis.[11] Via Ohm's law and variable substitution eliminating endpoint potentials, all currents across neighboring parent and child compartments can be computed from midpoint potentials alone using equations (6.5), (6.6), and (6.8).

The central difference approximation allows us to deal with *one* axial current per compartment, which we assume to be constant throughout the compartment, in contrast to computing axial currents per half-compartment. The direction of currents is given by the compartment axis, and we chose the sign convention that a positive axial current

[11] Note that it is possible to swap the central difference approximation with the forward and backward difference approximation estimating axial currents for each half-compartment respectively using equations 6.3–6.4 introduced earlier (see also Hagen et al. (2018)).

corresponds to a positive membrane potential difference from the start point to the endpoint of the compartment (consistent with equation (6.9)).

Thus, up to this point, we have provided the basic "machinery" needed to compute the discrete axial currents in an MC neuron model, independently of any explicit membrane currents present in the MC model. With our application of the central difference approximation to the first derivative of the membrane potential, it is in these derivations implicitly assumed the axial current is constant along the entire length of the compartment, which in turn implies that the membrane voltage varies linearly with position. Note that this assumption is likely to be violated if the spatial discretization chosen for the model neuron(s) is too coarse. Note also that the axial currents calculated using this central difference approximation correspond to the averaged current across *both* resistors within each compartment illustrated in Figure 6.1 (and is thus not in complete correspondence with the circuit diagram depicted in this figure).

6.2.3   Membrane Currents in Discretized Cables

The simulation tools NEURON and Arbor provide native support for computing I_{m} (with unit A) and membrane current density i_{m} per surface membrane area (with unit A/m^2) for each compartment. For completeness, we also show how it can be computed from the membrane potential $V_{\mathrm{m}}(x_j)$.

A starting point is to evaluate the membrane currents on a per-compartment basis by discretizing equation (6.2) using the central difference approximation to the second derivative as

$$\mathcal{I}_{\mathrm{m}}(x_j,t) = \frac{\pi d_j^2}{r_{\mathrm{a}}} \frac{V_{\mathrm{m}}\left(x_j + \frac{L_j}{2},t\right) - 2V_{\mathrm{m}}(x_j,t) + V_{\mathrm{m}}\left(x_j - \frac{L_j}{2},t\right)}{L_j^2}. \tag{6.10}$$

Again, the endpoint potentials can be computed from midpoint potentials using equations (6.5), (6.6), and (6.8).

6.2.4   Matrix Formalism for Computing Currents

In this section, we outline how axial and membrane currents for the whole set of neural compartments can be represented using a matrix formulation. For convenience, we shall assume that the membrane current density, like the axial current, is constant throughout each fixed-diameter cylindrical compartment.

In the formalism outlined in Section 6.2.2, axial currents are *linearly* dependent on the midpoint membrane voltages of the MC model. Thus, the full set of operations needed to compute the axial current magnitudes can be rewritten as a linear combination of the form

$$\begin{bmatrix} I_{\mathrm{a}}(\mathbf{r}_1,t) \\ I_{\mathrm{a}}(\mathbf{r}_2,t) \\ \vdots \\ I_{\mathrm{a}}(\mathbf{r}_N,t) \end{bmatrix} = \mathbf{A}^{-1} \begin{bmatrix} V_{\mathrm{m}}(\mathbf{r}_1,t) \\ V_{\mathrm{m}}(\mathbf{r}_2,t) \\ \vdots \\ V_{\mathrm{m}}(\mathbf{r}_N,t) \end{bmatrix}, \tag{6.11}$$

where $\mathbf{r}_j$ denotes the midpoint location of compartment j while the elements of the matrix $\mathbf{A}$ effectively correspond to intracellular resistances and the specific connectivity between compartments (recall Ohm's law for the current I across a resistor R given a voltage difference ΔV: $I = R^{-1}\Delta V$). The shape of $\mathbf{A}$ is (N, N).

We will next demonstrate that each element of a matrix $\mathbf{G} = \mathbf{A}^{-1}$ can be computed directly. For a two-compartment model connecting the endpoint of the root (parent) compartment with the start point of the child compartment (as in Figure 6.1B), it can be shown with some arithmetic that the axial current in each compartment is

$$I_a(\mathbf{r}_1, t) = I_a(\mathbf{r}_2, t) = \frac{\pi}{4r_a} \frac{d_1^2 d_2^2 \left(V_m(\mathbf{r}_1, t) - V_m(\mathbf{r}_2, t) \right)}{\left(L_1 d_2^2 + L_2 d_1^2 \right)}, \tag{6.12}$$

which can be rewritten as the linear combination

$$\begin{bmatrix} I_a(\mathbf{r}_1, t) \\ I_a(\mathbf{r}_2, t) \end{bmatrix} = \underbrace{\frac{\pi}{4r_a} \frac{d_1^2 d_2^2}{\left(L_1 d_2^2 + L_2 d_1^2 \right)} \begin{bmatrix} 1 & -1 \\ 1 & -1 \end{bmatrix}}_{\mathbf{G}} \begin{bmatrix} V_m(\mathbf{r}_1, t) \\ V_m(\mathbf{r}_2, t) \end{bmatrix}, \tag{6.13}$$

which is mathematically similar to equation (6.11). The same approach holds for other model configurations, such as three-compartment models without and with branching points as illustrated in Figure 6.1C and D, respectively (code available via www .cambridge.org/electricbrainsignals). The approach also holds for an arbitrary number N of compartments. For larger N, the sparsity of $\mathbf{G}$ increases as most compartments are connected to few others.

For some applications, such as computing the magnetic field $\mathbf{B}(\mathbf{R}, t)$ in some location $\mathbf{R}$, the axial currents can be represented as a time-dependent vector,

$$\mathbf{I}_a(t) = \begin{bmatrix} I_a(\mathbf{r}_1, t) \\ I_a(\mathbf{r}_2, t) \\ \vdots \\ I_a(\mathbf{r}_N, t) \end{bmatrix}. \tag{6.14}$$

The path travelled by each axial current is defined by the vector $\mathbf{d}_j = \mathbf{r}_{fj} - \mathbf{r}_{ij}$, where $\mathbf{r}_{ij}$ and $\mathbf{r}_{fj}$ denote the startpoint and endpoint coordinates of compartment j. Consistent with the point-source formalism described previously in Section 4.1, we let the "location" of each axial current be the midpoint $\mathbf{r}_j$ of each cylindrical compartment j. In a similar manner, we can represent the total membrane current per compartment as

$$\mathbf{I}_m(t) = \begin{bmatrix} I_m(\mathbf{r}_1, t) \\ I_m(\mathbf{r}_2, t) \\ \vdots \\ I_m(\mathbf{r}_N, t) \end{bmatrix} = \begin{bmatrix} \mathcal{I}_m(\mathbf{r}_1, t) \|\mathbf{d}_1\| \\ \mathcal{I}_m(\mathbf{r}_2, t) \|\mathbf{d}_2\| \\ \vdots \\ \mathcal{I}_m(\mathbf{r}_N, t) \|\mathbf{d}_N\| \end{bmatrix}, \tag{6.15}$$

under the assumption that the membrane current density $\mathcal{I}_m(x_j, t)$ is constant along the axis of each compartment. Here, $\|\mathbf{d}_j\|$ denotes the norm (length) of $\mathbf{d}_j$. Note that both $\mathbf{I}_a(t)$ (with unit A) and $\mathbf{I}_m(t)$ (with unit A) are time-dependent vectors represented as matrices with shape (N, N_t). Here, N_t is the number of discrete time steps in the simulated output signals.

6.3   Application to Extracellular Potentials and Magnetic Fields

As we just learned, both the axial and membrane currents in neurons are linearly dependent on the spatiotemporal variation in membrane potential. With real-valued tissue conductivities, the resulting electric and magnetic extracellular signals can therefore be computed as a linear function of the distribution of either of the two:

$$
\begin{bmatrix} \psi(\mathbf{R}_1, t) \\ \psi(\mathbf{R}_2, t) \\ \vdots \\ \psi(\mathbf{R}_M, t) \end{bmatrix} = \mathbf{F} \begin{bmatrix} I(\mathbf{r}_1, t) \\ I(\mathbf{r}_2, t) \\ \vdots \\ I(\mathbf{r}_N, t) \end{bmatrix} .
\tag{6.16}
$$

For the sake of compactness, we have here let $I(\mathbf{r}_j, t)$ represent either the axial current or the membrane current in N compartmental sources located at $\mathbf{r}_j$, while $\psi(\mathbf{R}_i, t)$ represents either the electric potential or magnetic field at M different spatial locations $\mathbf{R}_i$. $\mathbf{F}$ is typically a matrix with dimensions (M, N) wherein each element f_{ij} is the chosen forward solution mapping the contribution from each source to the corresponding measurement. The four possible combinations of sources and signals covered by equation (6.16) each require a different $\mathbf{F}$. We note that a forward mapping on the form in equation (6.16) can also be set up for sources expressed as dipoles, which is useful for modeling EEG and MEG signals. This is covered in more detail in Section 6.3.6.

6.3.1   Point Sources

A simple linear forward model that maps membrane currents in MC models to extracellular potentials is the point-source formalism. In an infinite homogeneous, isotropic, and linear volume conductor with conductivity σ_t, the point-source formula (equation (4.6)) results in elements

$$
f_{ij} = \frac{1}{4\pi\sigma_t} \frac{1}{\|\mathbf{R}_i - \mathbf{r}_j\|}
\tag{6.17}
$$

in the matrix formalism in equation (6.16). The full linear system for obtaining the extracellular potential may then be written out in matrix form as

$$
\begin{bmatrix} V_e(\mathbf{R}_1, t) \\ V_e(\mathbf{R}_2, t) \\ \vdots \\ V_e(\mathbf{R}_M, t) \end{bmatrix} = \frac{1}{4\pi\sigma_t} \begin{bmatrix} \frac{1}{\|\mathbf{R}_1 - \mathbf{r}_1\|} & \frac{1}{\|\mathbf{R}_1 - \mathbf{r}_2\|} & \cdots & \frac{1}{\|\mathbf{R}_1 - \mathbf{r}_N\|} \\ \frac{1}{\|\mathbf{R}_2 - \mathbf{r}_1\|} & \frac{1}{\|\mathbf{R}_2 - \mathbf{r}_2\|} & \cdots & \frac{1}{\|\mathbf{R}_2 - \mathbf{r}_N\|} \\ \vdots & \vdots & \ddots & \vdots \\ \frac{1}{\|\mathbf{R}_M - \mathbf{r}_1\|} & \frac{1}{\|\mathbf{R}_M - \mathbf{r}_2\|} & \cdots & \frac{1}{\|\mathbf{R}_M - \mathbf{r}_N\|} \end{bmatrix} \begin{bmatrix} I_m(\mathbf{r}_1, t) \\ I_m(\mathbf{r}_2, t) \\ \vdots \\ I_m(\mathbf{r}_N, t) \end{bmatrix} .
\tag{6.18}
$$

Singular values can occur when the denominator $\|\mathbf{R}_i - \mathbf{r}_j\|$ approaches zero. This can be avoided by setting the minimum allowed distance equal to the segment radius $d_j/2$ (Lindén et al. 2014).

6.3.2   Line Sources

For line sources (equation (4.10)), the elements of $\mathbf{F}$ may be calculated as

$$f_{ij} = \frac{1}{4\pi\sigma_t \Delta s_{ij}} \log \left| \frac{\sqrt{h_{ij}^2 + \rho_{ij}^2} - h_{ij}}{\sqrt{\ell_{ij}^2 + \rho_{ij}^2} - \ell_{ij}} \right|, \tag{6.19}$$

where ρ_{ij} is the distance measured perpendicularly to line source (compartment) j, h_{ij} is the longitudinal distance from the end of the line source, and $\ell_{ij} = \Delta s_{ij} + h_{ij}$ is the longitudinal distance from the start of the line source with length Δs_{ij} to the recording position $\mathbf{R}_i$. Holt (1998) pointed out that this formula as written may produce large roundoff errors due to the subtraction of approximately equal large numbers, and the formula should instead be implemented using the formulas

$$f_{ij} = \frac{1}{4\pi\sigma_t \Delta s_{ij}} \begin{cases} \log \frac{\sqrt{h_{ij}^2 + \rho_{ij}^2} - h_{ij}}{\sqrt{\ell_{ij}^2 + \rho_{ij}^2} - \ell_{ij}} & \text{for } h_{ij} < 0,\ \ell_{ij} < 0 \\ \log \frac{\left(\sqrt{h_{ij}^2 + \rho_{ij}^2} - h_{ij}\right)\left(\ell_{ij} + \sqrt{\ell_{ij}^2 + \rho_{ij}^2}\right)}{\rho_{ij}^2} & \text{for } h_{ij} < 0,\ \ell_{ij} > 0 \\ \log \frac{\ell_{ij} + \sqrt{\ell_{ij}^2 + \rho_{ij}^2}}{\sqrt{h_{ij}^2 + \rho_{ij}^2} + h_{ij}} & \text{for } h_{ij} > 0,\ \ell_{ij} > 0 \end{cases} \tag{6.20}$$

As with the point-source formula, singularities may be avoided by setting the minimum allowed ρ_{ij} equal to $d_j/2$ (Lindén et al. 2014).

6.3.3   Finite-Sized Contacts

In order to predict extracellular signals assuming electrode contacts with finite extents – for instance, by using the disk-electrode approximation – one can incorporate the corresponding forward mapping with the n-point averaging

$$f_{ij} \approx \frac{1}{n} \sum_{i'=1}^{n} f_{i'j}, \tag{6.21}$$

where $f_{i'j} = f_{i'j}(\mathbf{R}_{i'})$ denotes any of the earlier forward solutions for discretely sampled locations $\mathbf{R}_{i'}$ on the exposed surface of each contact indexed by i. For a more in-depth discussion of the effect of finite-sized contacts, we refer to Section 4.5.2.

6.3.4   Ground-Truth Current-Source Density (CSD)

The mapping approach described earlier can also be used to compute the *ground-truth* current-source density (CSD) (Pettersen et al. 2008, Hagen et al. 2016, Hagen et al. 2017). One then assigns the contributions of the membrane currents of different compartments to M different volumetric domains Ω_i as

$$f_{ij} = \frac{\Delta s_{ij \in \Omega_i}}{\Delta s_{ij} V_{\Omega_i}}, \tag{6.22}$$

where $(\Delta s_{ij \in \Omega_i} / \Delta s_{ij}) \in [0,1]$ denotes the fraction of the length of the segment within the domain Ω_i, and V_{Ω_i} is the volume of the domain itself. The geometry of each volume is in principle arbitrary, but they are typically assumed to be cylindrical or cubic. Cylindrical volumes can be useful for computing the ground-truth CSD alongside a laminar probe (Pettersen et al. 2008, Hagen et al. 2016, Hagen et al. 2017), while cubic volumes could be used to compute the ground-truth CSD across evenly distributed voxels filling the entire 3D space occupied by the neuronal sources.

6.3.5 Magnetic Fields in an Infinite Homogeneous Conductor

When computing internal magnetic fields in the brain, it is common to assume an infinite homogeneous head model, meaning that the tissue conductivity is the same everywhere (Section 11.6). Ampère-Laplace's law then takes the form in equation (11.4), and the magnetic field outside neurons depends only on the axial neuronal currents. It can be computed by summing up the contributions from the set of axial currents (Blagoev et al. 2007, Hagen et al. 2018) as

$$\mathbf{B}(\mathbf{R}_i, t) = \frac{\mu}{4\pi} \sum_{j=1}^{N} \frac{I_a(\mathbf{r}_j, t)\mathbf{d}_j \times (\mathbf{R}_i - \mathbf{r}_j)}{\|\mathbf{R}_i - \mathbf{r}_j\|^3}, \tag{6.23}$$

where μ is the magnetic permeability, which for brain tissue is approximately the same as in vacuum (Hämäläinen et al. 1993), so that $\mu \approx \mu_0 = 4\pi \times 10^7$ H/m. Note that this discretized equation is most accurate when the compartments are much shorter than the distance to the recording position – that is, $\|\mathbf{d}_j\| \ll \|\mathbf{R}_i - \mathbf{r}_j\|$.

The matrix formalism introduced earlier may also be used for computing magnetic fields. The set of per-compartment axial current magnitudes can be represented as a time-dependent vector $\mathbf{I}_a(t)$ with shape (N, N_t). The elements of the corresponding mapping matrix $\mathbf{F}_i$ of shape $(3, N)$ between the axial currents and magnetic field in a location $\mathbf{R}_i$ can be constructed via

$$\mathbf{f}_{ij} = \frac{\mu}{4\pi} \frac{\mathbf{d}_j \times (\mathbf{R}_i - \mathbf{r}_j)}{\|\mathbf{R}_i - \mathbf{r}_j\|^3}, \tag{6.24}$$

which defines a vector with three elements for each index pair ij. With $\mathbf{F}_i = [\mathbf{f}_{i1}^\top, \ldots, \mathbf{f}_{iN}^\top]$, one may compute the magnetic field from axial currents in each location $\mathbf{R}_i$ as the product

$$\mathbf{B}(\mathbf{R}_i, t) = \mathbf{F}_i \mathbf{I}_a. \tag{6.25}$$

6.3.6 Forward Models for Dipole Moments

The earlier forward solutions assume axial or membrane currents as signal sources. For forward-model predictions of distal electric and magnetic measures of neural activity, such as EEG and MEG signals, some simplifying steps can be made. As these signals are measured at distances much greater than the typical extent of the neuronal sources, the discrete spatial distribution of membrane currents $\mathbf{I}_m(t)$ may be treated as an equivalent (current-) dipole moment (Sections 4.4 and 9.5.1), defined as the product

$$\mathbf{P}(t) = \begin{bmatrix} \mathbf{r}_1 \\ \mathbf{r}_2 \\ \vdots \\ \mathbf{r}_N \end{bmatrix} \begin{bmatrix} I_m(\mathbf{r}_1, t) \\ I_m(\mathbf{r}_2, t) \\ \vdots \\ I_m(\mathbf{r}_N, t) \end{bmatrix}, \tag{6.26}$$

which is equivalent to equation (4.51). As

$$\mathbf{r}_j = \begin{bmatrix} x_j \\ y_j \\ z_j \end{bmatrix}, \tag{6.27}$$

the resulting shape of the dipole moment array will be $(3, N_t)$.

In the case of an infinite homogeneous volume conductor, the extracellular potential resulting from the dipole moment is

$$V_e(\mathbf{R}, t) = \frac{1}{4\pi\sigma_t} \frac{\mathbf{P}(t) \cdot (\mathbf{R} - \mathbf{r})}{\|\mathbf{R} - \mathbf{r}\|^3}, \tag{6.28}$$

where $\mathbf{R}$ denotes the measurement site and $\mathbf{r}$ the assumed dipole location. For the sake of computing EEG potentials on the scalp, however, an infinite homogeneous volume conductor model is arguably quite a poor approximation of the head. As discussed in more detail in Chapter 9, different forward models for EEG predictions have been developed, representing the head in terms of a series of concentric shells representing different tissue types, such as grey matter, cerebral spinal fluid, and skull (Nunez & Srinivasan 2006, Næss et al. 2017, Næss et al. 2021), or representing the head in even further detail based on anatomical reconstructions such as the New York Head model (Huang et al. 2016). Without going into any mathematical detail here (more on that in Chapter 9), extracellular potential predictions using these head models can generally be written in the form of a linear combination

$$V_e(\mathbf{R}, t) = \mathbf{M}\mathbf{P}(t), \tag{6.29}$$

where the coefficients of the matrix $\mathbf{M} \equiv \mathbf{M}(\mathbf{R}, \mathbf{r})$ are head-model dependent. For the electric potential in an infinite homogeneous volume conductor (equation (6.28)), the corresponding rows of $\mathbf{M}$ for each electrode site $\mathbf{R}_i$ can be computed as

$$m_i = \frac{1}{4\pi\sigma_t} \frac{\mathbf{l}_3 \cdot (\mathbf{R}_i - \mathbf{r})}{\|\mathbf{R}_i - \mathbf{r}\|^3}, \tag{6.30}$$

where $\mathbf{l}_3$ is the shape $(3, 3)$ identity matrix as earlier. The resulting shape of $\mathbf{M}$ is $(M, 3)$.

As the extracellular potential remains linearly dependent on the dipole moment $\mathbf{P}$, which in turn is linearly dependent on the membrane currents, one can with ease estimate a linear predictor of extracellular potentials from membrane currents via the dipole moment by computing the matrices **M** and **F** independently and then estimating the final signal as

$$V_e(\mathbf{R}, t) = \mathbf{M}\mathbf{F}\mathbf{I}_m(t) \, . \tag{6.31}$$

6.3.7 Forward-Model Predictions from Population and Network Models Using the MC+VC Scheme

In the preceding sections, we have defined a formalism for predicting extracellular signals resulting from neuronal current sources. The sources can in principle denote the total compartment count for the populations of neurons (Section 3.5). However, when running simulations on recurrently connected networks of MC neurons, storing the membrane currents from each individual neural compartment will often require too much memory. For the purpose of computing extracellular signals or the dipole moment $\psi \in \{V_e, \mathbf{P}, \mathbf{B}, \ldots\}$ from such simulations, we still have at least two computationally viable options:

1. Compute the signal as the sum of single-neuron contributions

$$\psi(\mathbf{R}, t) = \sum_X \sum_{u=1}^{N_X} \mathbf{F}_u \mathbf{I}_u \, , \tag{6.32}$$

where $\mathbf{F}_u \equiv \mathbf{F}_u(\mathbf{R}, \mathbf{r}_u)$ and $\mathbf{I}_u \equiv \mathbf{I}_u(t)$ denotes the VC forward solution and current sources, respectively, of the number N_X different neurons included in each population X of the network. The outer sum implies that we sum over all contributing network populations X. The index u denotes the neuron index. This formulation requires intermediate storage of single-cell contributions to ψ before summing. If the task at hand involves signal prediction at many locations with high temporal resolution, it may therefore be infeasible for large networks due to the associated use of memory or hard drive space ($\propto N_X M N_t$).

2. Compute the signal as the product

$$\psi(\mathbf{R}, t) = \sum_X \mathbf{F}_X \mathbf{I}_X \, . \tag{6.33}$$

When using this option, the forward solution $\mathbf{F}_X = \mathbf{F}_X(\mathbf{R}, \mathbf{r})$ is precomputed for all neuronal compartments across all N_X neurons in each population X and is multiplied with corresponding currents $\mathbf{I}_X = \mathbf{I}_X(t)$, which represents all contributions from all neurons within the population. This formulation, which avoids explicit for-loops over individual neurons, does not allow for the storing of single-cell contributions to the signal, but as the typical signals of interest are population signals, this is perhaps less of an issue. One can in any case compute single-cell contributions by extracting the corresponding rows and columns of the matrices and multiplying them separately.

For both of these options, it should be noted that the intermediate storage of the current terms $(\mathbf{I}_u(t), \mathbf{I}_X(t))$ for the entire simulation duration is not required if the signal calculations can be done at each time step, or at given intervals, during the simulation. In case the network activity is simulated using parallel computing, one may choose to evaluate the extracellular signal contributions by neurons simulated on each parallel process $\psi_p(\mathbf{R}, t)$ separately and sum over individual contributions as

$$\psi(\mathbf{R}, t) = \sum_p \psi_p(\mathbf{R}, t), \tag{6.34}$$

where p denotes the index of each process. This summing operation is quite trivially implemented in different programming languages – for instance, by calling the *message passing interface* (MPI) "reduce" function in case MPI is used.

6.4   Extracellular Signal Predictions from Network Models

Historically, spike trains (or equivalent population spike rates) and spike train statistics have been the main metric used when comparing model predictions with experimental observations. However, only limited information about ongoing synaptic inputs and dendritic integration may be inferred from each neuron's output spike train alone. This type of information is to a much larger degree provided by the low-frequency parts ($\lesssim 300\,\text{Hz}$) of extracellular signals (i.e. the local field potential (LFP), see Chapter 8). As such, experimental data sets combining, for instance, spike recordings and population signals like the LFP may put stronger constraints on models aiming to explain neural observations.

Whereas recurrent network simulations based on the MC+VC scheme allow the full monty of intra- and extracellular electric measures of neural activity to be compared to experimental data, the less-detailed network models based on point (single-compartment) neurons or population models pose a challenge in terms of extracellular signal predictions (Figure 6.2A). The reason is that point neurons do not provide any sources for extracellular potentials or magnetic fields since all in- and outgoing membrane currents sum to zero (Figure 3.1). Similarly, population models, which typically aim to explain the averaged activity of populations of neurons in terms of fluctuating spike rates or membrane potentials, do not provide any sources for extracellular signals. Hence, in order to relate activity in simplified spike- and rate-based frameworks to extracellularly recorded signals, we require some intermediate modeling steps that can relate either spike trains or firing rates to the signals. In Figure 6.2A, this intermediate step is presented as a "black box" model taking spikes or equivalent population rates as input signals and outputting the signal of interest.

In this section, we venture through different avenues shedding light on the black box itself. The main emphasis shall be put on technical solutions that account for the biophysics of the system that the neural network represents. Illustrated in Figure 6.2B, we start in Section 6.4.1 with the most detailed level of description in the form of MC neuron networks with signal predictions. We then ask which simplified signal-generating model setup may accurately capture the extracellular signals from the biophysically detailed model. Via successive model reduction steps, we first present the

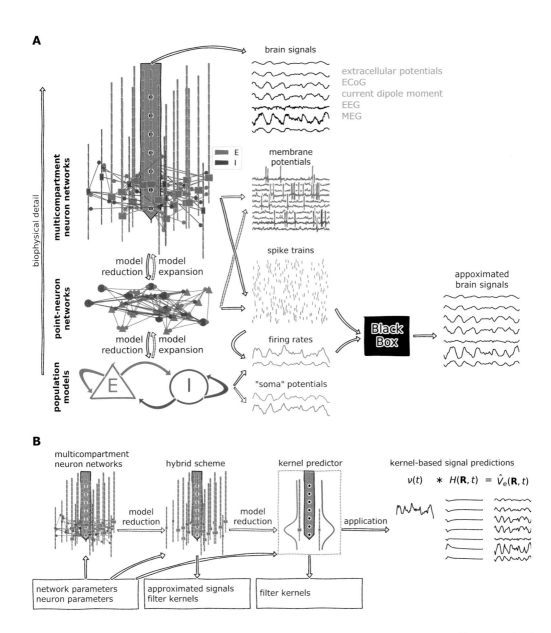

Figure 6.2 Extracellular signal predictions from neuron network models at different levels of detail and the roadmap towards simplified prediction models. **A**: Extension of Figure 3.8, highlighting that extracellular brain signals of detailed network models can be computed using the MC+VC scheme directly. Less-detailed models (e.g. point-neuron networks, rate models) generally require some intermediate model, denoted by the "black box," in order to provide approximated extracellular signals. **B**: Outline of simplifying schemes for computing extracellular signals based on the description of the full model, via the so-called hybrid scheme and kernel-approximating scheme, leading up to signal approximations based on convolving spike rates and appropriate filter kernels. Figure adapted from Hagen et al. (2022) (CC-BY 4.0 International license). Code available via www.cambridge.org/electricbrainsignals.

so-called hybrid scheme (Hagen et al. 2016, Hagen et al. 2022) in Section 6.4.2, and then in Section 6.4.3, we present a biophysics-based scheme for predicting spike-signal impulse response filter coefficients ("kernels") for different extracellular signal types that are applicable with population spike rates (Hagen et al. 2022). Both the hybrid and kernel-prediction schemes account mainly for signal contributions resulting from synaptic activity triggered by presynaptic spikes or corresponding rate changes. Finally, we introduce simple signal approximations ("proxies") that may capture temporal features of extracellular signals in simplified network models (Mazzoni et al. 2015) in Section 6.4.4.

6.4.1   Case Study I: An MC Neuron Network with Extracellular Signal Predictions

We start this section by presenting a generic recurrently connected network of MC neurons. Such a network corresponds to the higher level of biophysical detail in Figure 6.2A, and its extracellular signal predictions are obtained using the MC+VC scheme. Later in this section, this scheme will be used as a basis for simplifying computational schemes that allow us to simulate extracellular signals using networks of more simplified neurons.

As a case study, we consider the network previously described in Hagen et al. (2022) and we refer to the original work for all neuron and network details. This phenomenological network (illustrated in Figure 6.3A) is intentionally kept simple compared to biophysically detailed models aiming to replicate different brain areas in detail (e.g. Markram et al. (2015), Billeh et al. (2020)), but this network shares the main constituents, namely the MC formalism with conductance-based membrane and synapse dynamics.

Population "E" in blue represents excitatory neurons, while population "I" in red represents inhibitory neurons. These populations could represent the main neuron types in a cortical layer – for instance, layer 5. The asymmetric excitatory neurons have prominent apical dendrites pointing upwards (like cortical pyramidal cells), while the smaller inhibitory neurons are symmetric. Recurrent connections between the different populations are set up in such a way that a large fraction of excitatory-to-excitatory synaptic connections are made up along the apical dendrites, while most other recurrent synapses are made closer to the somatic regions (Figure 6.3B, C). The neuron models have active, voltage-gated ion channels in somas and dendrites so that sufficient depolarizing synaptic input from recurrent and extrinsic connections results in somatic APs (Figure 6.3D).

Spike times are recorded from all neurons in this network. An oscillatory activity pattern is seen both in the spike raster plot (Figure 6.3E), the population firing rate (Figure 6.3F), and the corresponding rate power spectra that here have peaks around 55 Hz (Figure 6.3G). The network state shown here resembles the so-called asynchronous irregular (AI) network state (Brunel 2000). The oscillations are clearly visible also in the extracellular potential produced by the network, as shown in Figure 6.3H at virtual electrode contact locations marked by the black markers in panel A. The oscillations are also prominently captured by the vertical z-component of the dipole moment P_z (Figure 6.3I). In this case, the orthogonal components

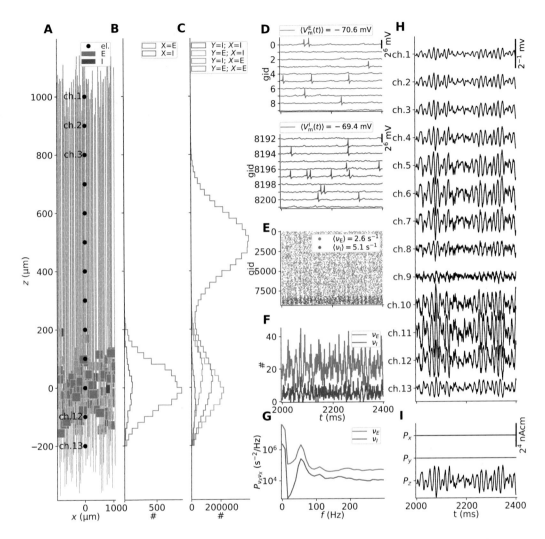

Figure 6.3 Stylized two-population MC neuron network with predictions of extracellular potentials. **A**: Neuronal populations and electrode geometry. The network is constructed of one excitatory ("E") and one inhibitory ("I") population. The neurons are ball-and-sticks neurons, and only a subset of cells is shown from each population. The black point markers along the z-axis denote locations of electrode contact points with separation 50 μm. **B**: Distributions of somatic compartments across depth in bins of 20 μm. **C**: Synapse counts per connection K_{YX} along the vertical z-axis in bins of 20 μm. **D**: Somatic potentials $V_m(t)$ of 10 units in each population, spanning 500 ms of spontaneous activity. The legends provide the time-averaged soma potential computed over 1,024 units in each population. **E**: Network spike raster. The mean population-averaged firing rates are given in the legend. **F**: Per-population spike-count histograms with bin size 1 ms. **G**: Population-firing-rate spectra. **H**: Extracellular potentials $V_e(\mathbf{R}, t)$. **I**: Components of the dipole moment $\mathbf{P}(t)$ along the x, y, and z-axes. Figure taken from Hagen et al. (2022) (CC-BY 4.0 International license). Code available via www.cambridge.org/electricbrainsignals.

(P_x, P_y) cancel as all neuron sections are aligned with the z-axis. Both the extracellular potential $(V_e(\mathbf{R}, t))$ and dipole moment $(\mathbf{P}(t))$ are computed by summing all single-cell contributions as described in Section 6.3.7. The latter can be used for EEG- and MEG-type signal predictions using suitable VC models.

When we set out to predict extracellular signals using networks of more simplified (not MC) neurons, we will use the extracellular signals in Figure 6.3H and I as the ground truth, using either spike trains (panel E) or corresponding spike rates (panel F) as the input signals for the model. This idea was already illustrated in Figure 6.2 by the "black box," representing putative models for extracellular signal predictions from input spike trains or firing rates.

6.4.2   Case Study II: The Hybrid Scheme for Computing Extracellular Signals

The so-called *hybrid scheme* is a physics-based scheme for computing dipole moments or extracellular electric or magnetic signals from neuron networks. The main idea is to divide the computation of network dynamics and corresponding extracellular signals into two steps:

1.   Compute the spike times of neurons using a neuron network.
2.   Compute the extracellular signals using biophysically detailed MC neuron models, letting the spike times computed in step 1 dictate the synapse activation times on the MC neurons.

The advantage of the hybrid scheme is that we can account for anatomical data such as neuronal geometry, cell positions, and the locations of synapses in step 2, even if such details are left out from the network simulation computing the spike times. The scheme is, however, applicable with networks of biophysically detailed neurons (Section 3.5.1) as well as spiking point neurons (Section 3.5.2). This scheme was proposed by Hagen et al. (2016), who later validated it against ground-truth data generated using MC neuron networks as showcased earlier (Hagen et al. 2022).

To compute extracellular signals from a spiking point-neuron network (in step 2), a population representing, say, a set of layer 5 pyramidal cells is represented by a corresponding population of geometrically detailed MC neuron models of this cell type, placed at cortical depths and densities according to the overall network size and available anatomical constraints. Similarly, incoming synaptic connections onto the same point-neuron population from a presynaptic population – for instance, layer 4 inhibitory interneurons – should be positioned onto the layer 5 pyramidal cell geometries according to rules derived from available data on such connections, while also preserving the typical number of incoming synapses (also known as synaptic in-degree).

An illustration of the hybrid scheme is given in Figure 6.4. Panel A shows a simulation of the extracellular potential V_e using the complete MC+VC scheme on a network simulation of recurrently connected MC neurons, which we can use as ground truth. Panel B shows that one can obtain essentially the same V_e from an identical population of neurons without recurrent connections, provided that one preserves the "correct" synapse activation times. The synapse activation times can be computed from spiking data obtained from a recurrent network of MC neurons, as we saw in panel A, but can also be obtained from a computationally efficient point-neuron network, as illustrated in panel C. The hybrid scheme is thus hybrid in the sense that it allows for predicting V_e from spikes in network simulations constructed using arbitrary detailed neurons as in panels A and C with MC neuron simulations as in panel B.

6.4.2.1 Application to a Cortical Microcircuit Model

The first application of the hybrid scheme (Hagen et al. 2016) was to the point-neuron network model by Potjans & Diesmann (2014), representing a generic cortical microcircuit that is illustrated in Figure 6.5A. Several features and parameters of the point-neuron network model were then inherited by the MC+VC modeling step. To preserve the synaptic currents of the point-neuron network, the same current-based synapse type was used (see Section 3.1.4.2), with an exponentially decaying temporal kernel with the same average maximum current amplitude as well as with activation delays drawn from the same conduction-delay distribution as was used in the point-neuron network. Synapse sites on the postsynaptic morphologies were derived from a connectivity data set resolving the numbers of synapses per layer between 16 different excitatory and inhibitory neuron types spanning layers 2/3, 4, 5, and 6. The total numbers of synapses per connection (from each presynaptic to postsynaptic population) were preserved. Furthermore, the membrane time constants of the reconstructed morphologies that utilized purely passive compartments throughout were set equal to those of the point neurons. In Figure 6.5B, the network response in terms of spikes is shown when the cortical populations are targeted by a transient volley of spikes from the external "thalamocortical" (TC) population. The corresponding LFP signal across depth computed using the hybrid scheme (panels C, D) reflects the transient change in cortical spiking activity, resembling responses observed experimentally, for instance, to a brief flash stimulus (Hagen et al. 2016). The response across depth also varies in a non-trivial way.

6.4.2.2 Application to a Network of Biophysically Detailed MC Neurons with Nonlinear Dynamics

Once the pre-recorded spike trains are known, the MC neurons used in step 2 of the hybrid scheme (Figure 6.5B) act like "antennas" for predicting V_e (or magnetic fields) in a feed-forward manner that does not affect network dynamics.

The two-step procedure of the hybrid scheme is beneficial versus networks of biophysically detailed MC neurons that are notoriously difficult to parametrize. For instance, spiking dynamics can be accurately captured by simplified networks that are comparably fast to simulate (e.g. for large-scale parameter scans) and amenable to analytical analyses. The level of biophysical detail in terms of the MC neuron setup in step 2 can be reduced and still provide accurate signal predictions as well as reduce overall computation times.

As investigated thoroughly by Hagen et al. (2022), it is possible to linearize key properties of the MC neurons and their synapses while still retaining key features of the signal predicted in step 2.

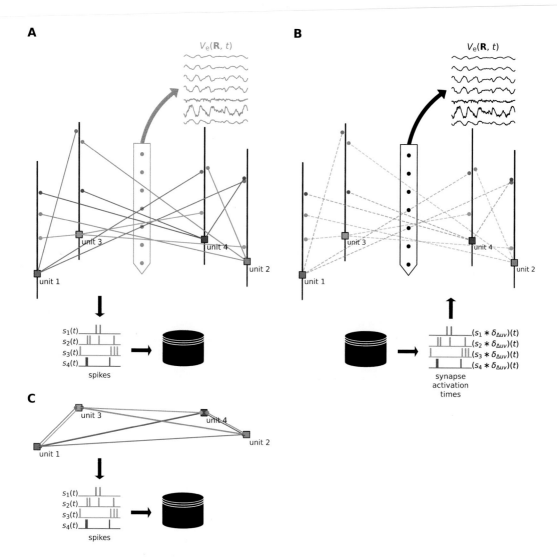

Figure 6.4 Dissociation between simulations of network activity and extracellular potentials in the hybrid scheme. A: Illustration of self-consistent MC+VC simulation of a neural network and the resulting extracellular potential $V_e(\mathbf{R}, t)$. Outgoing colored lines from each ball-and-stick neuron model denote a connection from the soma of presynaptic neurons to synapses distributed on postsynaptic neurons. Single-unit spike trains can be stored in a database or file (bottom). **B**: "Hybrid" simulation where recurrent network connections (dashed lines) have been removed, but the synaptic locations (of panel A) have been preserved. Synaptic activation times are set from pre-recorded spike times from a database (bottom). The synaptic currents and associated membrane currents on the postsynaptic neurons allow for computing $V_e(\mathbf{R}, t)$ separately from the recurrent network dynamics. **C**: "Point-neuron" network simulation that accurately captures the spiking statistics of the original MC network in panel A and can be used to produce input for the simulation in panel B. Code available via www.cambridge.org/electricbrainsignals.

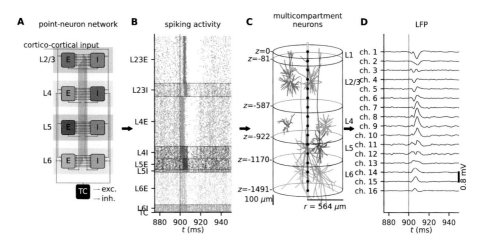

Figure 6.5 Example application of the hybrid scheme to a cortical microcircuit model.
A: Sketch of point-neuron network with eight cortical populations spanning layers 2/3, 4, 5, and
6, with one excitatory (E) and inhibitory (I) population each. The excitatory thalamocortical (TC)
population mainly projects to layers 4 and 6. B: Population spike raster plots. The TC population
is activated at $t = 900$ ms, denoted by the vertical line. C: The hybrid scheme used MC neuron
models, cortical column geometry, and recording electrode geometry used for LFP predictions.
D: Predicted LFP signal resulting from synaptic activations on the postsynaptic MC neuron
models. The figure is taken from Hagen et al. (2016) and reused with permission (CC-BY-NC
4.0 International license). Code available via www.cambridge.org/electricbrainsignals.

Hagen et al. (2022) found that the main simplifying steps that gave well-preserved
signal predictions (in step 2) were approximating (i) the conductance-based synapse
dynamics and (ii) the ion-channel dynamics in MC neurons by equivalent linearized
dynamics. In addition, the leaky properties of the cable models were modified to account
for the (iii) so-called *effective membrane time constant*. For the linearization of dynamics
in points (i) and (ii), dynamics that depend on the membrane potential $V_m(t)$ were
linearized around a chosen constant value $\overline{V}_m$. Hagen et al. (2022) typically chose
this value equal to the median somatic potential in different populations. Then, the
conductance-based synapses (see also Section 3.1.4) can be converted to current-based
synapses as

$$
\begin{aligned}
I_{syn}(t) &= \overline{G}_{syn} f(t) \left(V_m(t) - E_{syn} \right) \\
&\approx \overline{G}_{syn} f(t) \left(\overline{V}_m - E_{syn} \right) \\
&= \overline{I}_{syn} f(t) .
\end{aligned}
\tag{6.35}
$$

Here, $\overline{G}_{syn}$ is the maximum synaptic conductance, $f(t)$ the temporal dynamics of the
synapse, E_{syn} the synapse reversal potential, and $\overline{I}_{syn}$ the magnitude of the equivalent
synaptic current.

In a conceptually similar manner, the voltage-dependent ion-channel dynamics of
some channel(s) (named w) of the form

$$i_w(t) = \overline{g}_w \omega(t, V_m(t)) (V_m(t) - E_w) \tag{6.36}$$

can also be linearized. In contrast to synapse currents, however, the linear approximation is rather applied to the dynamical $\omega(t, V_m(t))$ term (Remme & Rinzel 2011, Ness et al. 2016, Ness et al. 2018, Hagen et al. 2022), effectively reducing the membrane dynamics to linear RLC- or RC-type equivalent circuits. We shall refer to these types as *quasi-active* linearized and *passive-frozen* linearized models, respectively. Here, we shall refrain from repeating the underlying math, as these are amply covered in the cited studies.

For point (iii), the effective membrane time constant is implemented via a change in the membrane leak conductivity $\overline{g}_L$ on a per-compartment basis. The change in $\overline{g}_L$ corresponds to the time-averaged synaptic conductances from all synapses impinging on each compartment; hence, the time-averaged membrane time constants could be better preserved in the linearized neuron models. Implementing this perturbation to $\overline{g}_L$ is necessary for accurate signal predictions as point (i) approximates conductance-based synapses with current-based variants and relies on accurate estimates of synaptic activation rates. We will also here refer to Hagen et al. (2022) for further details.

As a demonstration of signal predictions using the hybrid scheme, Figure 6.6A compares the ground-truth extracellular potentials and the z-component of the dipole

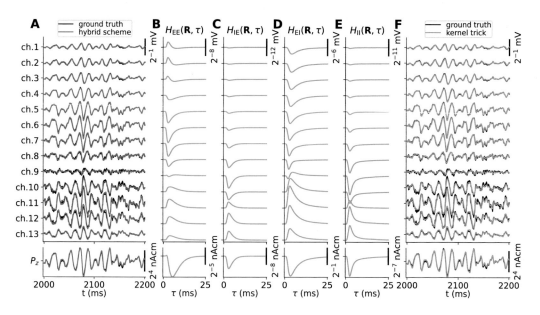

Figure 6.6 Ground-truth signals vs. hybrid-scheme and kernel-based signal predictions.
A: Extracellular potential across depth (top row) and dipole moment (bottom row) predicted from an MC neuron network model (black lines) compared to signal approximations made using the hybrid scheme, with passive-frozen MC neurons accounting for the effective membrane conductance. B–E: Predictions of filter kernels for extracellular potentials and dipole moments for the E-to-E, E-to-I, I-to-E, and I-to-I pathways, respectively. F: Signal approximations using population spike rates and kernels as governed by equation (6.38) compared to ground-truth signals (as in panel A). Figure adapted from Hagen et al. (2022) (CC-BY 4.0 International license). Code available via www.cambridge.org/electricbrainsignals.

moment of the recurrent network in Figure 6.3H and I, respectively, with hybrid scheme predictions encompassing current-based synapse and passive-frozen ion-channel dynamics, while simultaneously accounting for the effective membrane time constant.

The latter signals result from synaptic activations in the receiving populations of neurons, governed by presynaptic spike times in the full recurrent network (Figure 6.3E). In this application of the scheme, the main discrepancies between ground-truth and approximated signals are mainly the lack of higher-frequency jitter in the latter signal. These components are due to neuronal APs in the ground-truth data. As this application of the hybrid scheme implements all linear dynamics, APs cannot occur. The main take-home message is that linearized models accounting for synaptic currents can well explain lower-frequency signals in the brain, such as local field potentials ($\lesssim 300\,\text{Hz}$). In the next section, we exploit this approximately linear relationship between presynaptic spikes and extracellular signals to introduce even-simpler model schemes for extracellular signals.

6.4.3 Case Study III: Predicting Kernels for Computing Extracellular Signals

The hybrid scheme entails including as many MC neurons in the second step of the simulations as there are neurons in a network. It is therefore practically limited to networks on the order of 10^6 neurons (Senk et al. 2018), a number corresponding to the neuron count under $10\,\text{mm}^2$ or so of the cortical surface. The scheme is thus not a satisfactory solution for extracellular signal prediction for network models encompassing very large neuron numbers ($\gtrsim 10^6$ neurons) that are aiming to explain large-scale brain activity (e.g. multi-area and whole-brain models).

In the previous section, we showed by example that a linearized version of the hybrid scheme gave accurate predictions when compared to ground-truth data generated by a nonlinear conductance-based MC neuron network model. By exploiting the approximately linear relationship between presynaptic spike times and postsynaptic responses demonstrated earlier, it is possible to represent the MC neuron part in terms of suitable sets of filters equivalent to spike-signal impulse response functions (IRFs), as known from linear time-invariant (LTI) systems. If such spatiotemporal filters, or kernels, can be computed once via computationally costly MC neuron simulations for later use, the overall reduction in computational demands may still be huge.

6.4.3.1 Via the Hybrid Scheme

As illustrated in Figure 6.7A, the hybrid scheme can be set up for computing such spatiotemporal filters, or kernels $H(\mathbf{R}, \tau)$, directly. Each kernel is associated with pairs of presynaptic and postsynaptic populations. It describes the average extracellular signal resulting from membrane currents of postsynaptic neurons in response to a spike of one of the neurons in the corresponding presynaptic population. Using the hybrid scheme, one derives the kernel by first inserting the same spike time across all neurons in the presynaptic population and next by computing the extracellular signals resulting from

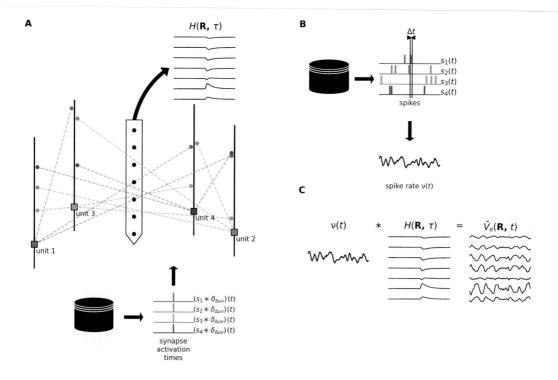

Figure 6.7 Calculation of spike-averaged signal kernel functions and reconstructed signal using the hybrid scheme. A: Computation of population-averaged spike-signal kernels $H(\mathbf{R}, \tau)$. The MC+VC setup is otherwise similar to the "hybrid" scheme illustrated in Figure 6.4B. **B**: Conversion of network-generated spike trains to population spike-count histograms $\nu(t)$ by binning events at some time resolution Δt. **C**: Approximation of the extracellular signal by the convolution $\check{V}_e(\mathbf{R}, t) = (\nu * H)(\mathbf{R}, t)$. Code available via www.cambridge.org/ electricbrainsignals.

the evoked response in the postsynaptic population. The kernel $H(\mathbf{R}, \tau)$ is then simply the computed signal divided by the number of neurons in the presynaptic population (Hagen et al. 2016, Hagen et al. 2022). The value τ denotes the time relative to the presynaptic spike time. By construction, the kernels are causal – that is, $H(\mathbf{R}, \tau) = 0$ for $\tau \leq 0$.

Figure 6.7B illustrates that recorded spike trains can be collapsed to a spike rate $\nu(t)$ or spike-time histogram with a temporal binning operation. Approximate extracellular potentials $\check{V}_e$ (or other signals) may then be computed by the linear convolution of firing rates and predicted kernels (panel C):

$$\check{V}_e(\mathbf{R}, t) = (\nu * H)(\mathbf{R}, t). \tag{6.37}$$

In most network models, however, there are usually multiple populations and connection pathways. Thus, for sets of presynaptic populations X and postsynaptic populations Y, we must compute the corresponding presynaptic firing rates $\nu_X(t)$ and sets of pairwise kernel functions $H_{YX}(\mathbf{R}, \tau)$. For multi-population networks and corresponding signal approximations, we refactor the convolution (applicable with one-population net-works) by the double sum

$$\check{V}_e(\mathbf{R},t) = \sum_X \sum_Y (\nu_X * H_{YX})(\mathbf{R},t). \tag{6.38}$$

6.4.3.2   Via the Direct Kernel Prediction Scheme

In the kernel-based formalism, the challenge is to predict suitable kernels $H_{YX}(\mathbf{R},\tau)$ for different signal types, not firing rates $\nu_X(t)$ as these can be dealt with via simplified networks as illustrated in Figure 6.2A. While the hybrid scheme can be used for this pur-pose, Hagen et al. (2022) provided a framework and corresponding software implemen-tation for computing such kernels by directly relying on only one MC neuron simulation per pathway, corresponding to potentially massive savings in computing requirements. This framework for direct kernel predictions relies on the same linearized synapse and ion-channel dynamics as we introduced for the hybrid scheme in Section 6.4.2.2. The framework accounts for key underlying statistics relevant to the biophysically detailed neuron networks and corresponding populations – that is, the distributions of neurons in space, the distributions of synapses in space, the pairwise connection probabilities and numbers of synapses per connection, the distributions of conduction delays, the typical membrane potentials, the population firing rates, and finally, the choice of VC model for different signal types.

The kernels predicted from the example MC neuron network in Figure 6.3A are shown in Figure 6.6B–E. These kernels were computed while retaining the passive-frozen ion-channel dynamics used for the hybrid scheme predictions shown in panel A. With these sets of kernels, signal approximations are computed using equation (6.38) with firing rates (Figure 6.3F) computed from the spiking activity (Figure 6.3E) of the recurrent MC neuron network. As shown in Figure 6.6F, the resulting signal predictions are quite indistinguishable from the corresponding hybrid scheme signals, and these predictions are able to explain the main oscillatory features of the extracellular potential and dipole moment signals. One main take-home message is that a large fraction of the predicted signals arise from the I-to-E connection pathway in this particular case, as the corresponding amplitudes of the $H_{EI}(\mathbf{R},\tau)$ kernels are $\sim$ 4–16-fold that of the other kernels.

6.4.4   Case Study IV: Proxies for Heuristic Signal Approximations

The work by Hagen et al. (2022) outlined in Section 6.4.3 demonstrated that linearized computational schemes can well explain the extracellular potentials (and by extension, magnetic signals) from recurrently connected networks of MC neurons or spiking point-neuron models. These linear approximations were all derived from the recurrent network specification itself. In order to predict signals from point-neuron networks or rate mod-els, however, various assumptions on the anatomy of the neurons, connections, etc., must be incorporated, which is made difficult by the fact that such anatomical data from experiments are generally lacking.

For the purpose of relating, in particular, point-neuron network activity to extracellular electric or magnetic measures of activity, a number of heuristic signal approximations ("proxies") have been proposed and utilized in the literature. A proxy is, generally speaking, a set of mathematical operations that allows one to compute (approximations of) the extracellular signals from data that be directly extracted from the point-neuron network simulation. The aim is to identify the proxies that give the best predictions of the extracellular signal. These proxies are, however, for the most part, phenomenological and poorly grounded in the biophysics of extracellular electric and magnetic signals. As we shall demonstrate here, some easily implemented proxies can still provide excellent insight into the temporal properties of extracellular signals of neural origin.

Using a computational model setup reminiscent of the hybrid-scheme application outlined in Section 6.4.2, Mazzoni et al. (2015) compared different LFP proxies that could be extracted from two-population spiking point-neuron network models. The method of Mazzoni et al. (2015) was extended to EEG signals by Martínez-Cañada et al. (2021). These proxies were defined by Mazzoni et al. (2015) as the product of linearly separable components as

$$\check{V}_{e,\,\text{proxy}}(\mathbf{R}, t) = f_{\text{proxy}}(\mathbf{R})g_{\text{proxy}}(t) , \tag{6.39}$$

where the spatial function $f_{\text{proxy}}(\mathbf{R})$ defines the overall signal variance and sign, and $g_{\text{proxy}}(t)$ defines a position-independent temporal component. By construction, the temporal component has zero mean and variance of unity (see equation (6.40)).

In the work by Mazzoni et al. (2015), the ground-truth LFP was predicted using the standard MC+VC formalism. Synaptic activation times were determined by the spiking activity of a two-population point-neuron network model taken from a previous study (Cavallari et al. 2014).

The main focus in Mazzoni et al. (2015) was on finding the best choice for the temporal function $g_{\text{proxy}}(t)$. A host of different candidate functions was considered based on dynamical variables available in point-neuron networks with integrate-and-fire type neurons – that is, the membrane potentials and synaptic currents. By computing the LFP signal at different depths, mimicking an experimental recording using a laminar probe inserted down through a cortical column, the performance of the different proxies could be evaluated.

So, which choices for $g_{\text{proxy}}(t)$ can be considered? In spiking point-neuron networks, one can with ease record spike events. Also, one can compute the per-neuron and population-averaged spike rates $\nu_X(t)$ for each population X, the per-neuron and population-averaged membrane voltages, and the summed synaptic currents resolved into excitatory and inhibitory contributions onto different neurons and subpopulations. Then, the comparison can also include linear combinations of the synaptic currents or other measures (which will not be covered here).

Using the nomenclature established throughout this chapter, these temporal components $g_{\text{proxy}}(t)$ of the proxy LFPs can, for current-based synapses, be defined as (Mazzoni et al. 2015):

$$z_{\text{score}}\left(y(t)\right) = \frac{y(t) - \mu_y}{\sigma_y} \tag{6.40}$$

$$g_{\sum v_X}(t) = z_{\text{score}}\left(\sum_X v_X(t)\right) \tag{6.41}$$

$$g_{V_m}(t) = z_{\text{score}}\left(\sum_Y \frac{1}{N_Y}\sum_{v \in Y} V_m^{\langle v\rangle}(t)\right) \tag{6.42}$$

$$g_{\sum I_E}(t) = z_{\text{score}}\left(\sum_{X \in \{E\}}\sum_{Y \in \{E\}} K_{YX}\overline{I}_{\text{syn}YX}\left(v_X * \widetilde{\Delta}_{YX} * f_{\text{syn}YX}\right)(t)\right) \tag{6.43}$$

$$g_{\sum I_I}(t) = z_{\text{score}}\left(\sum_{X \in \{I\}}\sum_{Y \in \{E\}} K_{YX}\overline{I}_{\text{syn}YX}\left(v_X * \widetilde{\Delta}_{YX} * f_{\text{syn}YX}\right)(t)\right) \tag{6.44}$$

$$g_{\sum I}(t) = z_{\text{score}}\left(g_{\sum I_E}(t) + g_{\sum I_I}(t)\right) \tag{6.45}$$

$$g_{\sum |I|}(t) = z_{\text{score}}\left(\left|g_{\sum I_E}(t) + g_{\sum I_I}(t)\right|\right) \tag{6.46}$$

$$g_{\text{WS}}(t) = z_{\text{score}}\left(\left|g_{\sum I_E}(t) + \alpha\left(g_{\sum I_I} * \delta_{\tau_I}\right)(t)\right|\right). \tag{6.47}$$

Here, the function z_{score} ensures that each $g_{\text{proxy}}(t)$ has zero mean and unit variance. As before, $v_X(t)$ denotes the firing rate of population X. $V_m^{\langle v\rangle}(t)$ denotes the membrane potential of a neuron v in population Y. N_Y denotes population size. The terms K_{YX}, $\overline{I}_{\text{syn}YX}$, $f_{\text{syn}YX}$, and $\widetilde{\Delta}_{YX}(t)$ denote the total number of synapses made by population X onto Y, the corresponding current magnitude, the synaptic temporal kernel, and the conduction-delay distributions, respectively. The asterisk ("$*$") denotes a (temporal) convolution operation. For the last temporal component $g_{\text{WS}}(t)$, the parameter α is scalar while the term $\delta_{\tau_I} = \delta(t - \tau_I)$ denotes a temporal delay of duration τ_I. "WS" stands for "weighted sum." In this overview, we only considered the case with current-based synapses impinging onto excitatory neurons. Mazzoni et al. (2015) also compared predictions using conductance-based synapses, which we for the sake of brevity have chosen to ignore.

Snapshots of the different temporal components $g_{\text{proxy}}(t)$ derived from our original MC network are compared in Figure 6.8A. As such, the task at hand is identifying which proxy maximizes the fraction of variance explained (R^2) between the ground-truth LFP signal and the respective proxy predictions in different locations, as summarized in Figure 6.8B.

As expected, the population-firing-rate proxy $g_{\sum v_X}(t)$ performs poorly as it can not account for any spatial or temporal filtering effects captured by actual spike-signal kernel functions $H_{YX}(\mathbf{R}, \tau)$. The membrane potential proxy $g_{V_m}(t)$, in contrast, does account for conduction delays and temporal components of synaptic currents. The resulting performance is indeed better as the membrane voltages reflect membrane currents but do not equate them. That leaves the various linear combinations of excitatory and inhibitory synaptic input currents, which ignore additional linear contributions by intrinsic dendritic filtering and the electrostatic forward model in

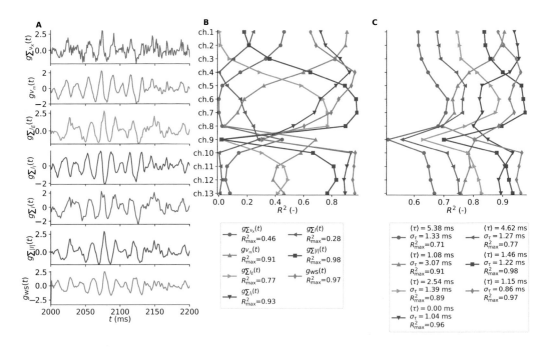

Figure 6.8 Comparison of temporal proxy components $g_{\text{proxy}}(t)$. **A**: Snapshots of normalized temporal components of different candidate proxy signals for mimicking extracellular potentials from the recurrent network model shown in Figure 6.3 for spontaneous, asynchronous irregular (AI) like activity. The different proxy signals are computed using equations (6.40)–(6.47) with input inferred from the network itself. **B**: Ratio of variance explained (R^2) between the different proxies and ground-truth LFP across depth. The ground-truth LFP signals are the same as shown in Figure 6.6A. **C**: Same as panel B but for temporally shifted components with temporal delay values τ maximizing R^2 in each individual channel. The legend shows the corresponding mean and standard deviation of the delays. (CC-BY 4.0 International license) Code available via www.cambridge.org/electricbrainsignals.

question. Here, $g_{\sum I_E}(t)$ displays a poor ability to explain the true signal across depth, while components that include inhibitory synaptic current contributions (from $g_{\sum I_I}(t)$) boost R^2 values across channels. The difference between currents $g_{\sum |I|}(t)$ resulted in the maximum observed $R^2 = 0.98$, closely followed by the summed currents $g_{\sum I}(t)$ and $g_{\text{WS}}(t)$, with maximum observed $R^2 = 0.97$. For the $g_{\text{WS}}(t)$ proxy – which Mazzoni et al. (2015) suggested as the optimal solution – the parameters $\alpha = 5.68$ and $\tau_I = 0.27$ ms, which maximized mean R^2 across all channels, were chosen.

As the R^2 score for different proxies may be affected by systematic phase shifts between the real signal and the different choices of $g_{\text{proxy}}(t)$, Figure 6.8C summarizes the performance of the components if a temporal shift is applied on a per-channel basis in order to maximize R^2. The mean and standard deviations of the temporal shifts are shown in the legend. The time-shifted $g_{\sum v_X}(t)$ here results in about 70 percent explained variance, while the remaining components, which already performed fairly well in terms of R^2 – see overall improved performance and are comparable to predictions using the full hybrid scheme or the kernel-based methods introduced above (Hagen et al. 2022).

Independent of time shift, none of the proxies perform well at the depth of channel 9, which has the lowest overall signal variance due to local cancellation effects. Dependent on location, either the inhibitory synapse current or the absolute sum of synaptic currents maximizes proxy performance. Thus, for the sake of simplicity, inhibitory synapse currents onto excitatory neurons could possibly provide a decent proxy for the temporal properties of extracellular measurements. The $g_{WS}(t)$ proxy also performs at a similar level. However, to reiterate, none of these proxies can accurately account for the spatiotemporal effects of neuronal morphology and the VC model on the signals of interest, which for some applications may be important.

7   Spikes

If a neuroscientist is asked what we have learned about brain networks from electrical recordings, there is a high chance they will first highlight the work done by Hubel and Wiesel in the 1950s and 1960s. In their pioneering studies of neural representations in the cat primary visual cortex, Hubel and Wiesel used sharp extracellular electrodes to measure spikes evoked by visual stimulation. Associating the spikes with action potentials (APs) fired by specific neurons, they found, for example, that many of the cells responded most vigorously to bar-like stimuli oriented in specific directions (Hubel & Wiesel 1959). Later, the same approach has been used throughout the nervous system to map out how different neurons represent information on various sensory stimuli. Since the first spike measurements, which were performed decades before the work of Hubel and Wiesel (see e.g. Adrian (1928), Adrian & Moruzzi (1939)), the spike has arguably been the most important brain signal in neuroscience.

The term *spike* is commonly used interchangeably with the term *action potential* (AP), but in the present chapter, we will let AP refer to a fluctuation in the neural membrane potential and let *spike* refer to the associated fluctuation in the extracellular potential. Hence, the spike is the extracellular signature of the AP.

The goal of spike measurements is normally to infer AP firing times (also called *spike times*) of individual neurons. This is not a trivial problem: since electrodes generally pick up signals from several nearby neurons, it is often difficult to determine which voltage deflection (spike) in the recorded signal belongs to which neuron. In addition, one has to distinguish between voltage deflections that are actual spikes (i.e. reflect actual APs) and voltage deflections caused by ambient noise in the system. To tackle these problems, we need a good understanding of how the size and shape of a spike depends on the properties of the neuron that generated it and on the position of the neuron relative to the recording electrode. Forward modeling, based on the MC+VC framework (see section 2.6.2) can provide good insights in this regard, and in this chapter, we aim to present the most important of these insights.

Throughout most of this chapter, we will consider spikes simulated under the assumption that brain tissue has a homogeneous and isotropic conductivity σ_t, which is expected to be a good approximation when modeling spikes in vivo in cortical tissue.

7.1 Properties of Spikes

The amplitude of the membrane-potential deflection during an AP is about 100 mV. In comparison, the amplitude of its extracellular signature, the spike, is typically less than 1 mV. Also, the shape of the spike is very different from the shape of intracellularly recorded APs (Figure 7.1). The AP is essentially monophasic with a large positive depolarization, though often followed by a weak hyperpolarization. In contrast, spikes are typically biphasic, at least when recorded close to the soma.

When recorded in the vicinity of the neural soma, the spike is typically characterized by an initial sharp negative peak followed by a later positive peak. The initial peak is often referred to as the "sodium peak," as it mainly stems from Na^+ ions flowing into the soma (and axon hillock) during the initial phase of the AP. The later, blunter positive peak is likewise often referred to as the "potassium peak," as it is dominated by K^+ ions flowing out of the soma.

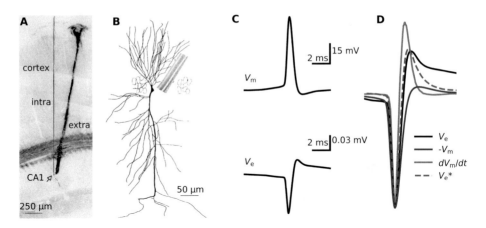

Figure 7.1 Simultaneous recording of intracellular and extracellular APs ("spike") from pyramidal cell in hippocampus. A: Placements of intracellular and extracellular electrode (tetrode device) in hippocampus area CA1. **B**: Illustration of neuron indicating position of extracellular electrode. **C**: Intracellular (top) and extracellular (bottom) recordings of a spike. Depicted potentials are averages over 849 spikes. The extracellular potential corresponds to the average of recordings from the four contacts constituting the multichannel tetrode electrode. The scale bar indicates the different scales for the two recordings. **D**: Comparison of the shape of spike V_e in panel C with the inverted shape of soma membrane potential ($-V_m$), the shape of the derivative of the soma membrane potential dV_m/dt, and the approximate formula (V_e^*) derived in Section 7.3. For the latter, the contribution from each frequency component of the soma membrane potential to the spike is assumed to be amplified by a factor proportional to the square root of the frequency ($f^{1/2}$), intermediate between the assumptions inherent in assuming the spikes to have the waveform of the inverted membrane potential (f^0) and the derivative (f^1). The depicted shapes are normalized to have the same maximal magnitude. Panels A–C are adapted from Henze et al. (2000) with permission from The American Physiological Society. Code available via www.cambridge.org/electricbrainsignals.

When detecting spikes, a main issue is that the spike amplitude must be larger than the ambient noise level. The noise has contributions both from the electrode and recording system, as well as from surrounding physiological processes, and is typically some tens of microvolts. For the detection of spikes, the detailed spike shape is of less importance. However, an extracellular electrode will, in general, measure spikes from several neurons located in the vicinity of the electrode contact. To obtain spike trains from individual neurons, the recorded spikes must thus be sorted in a process referred to as *spike sorting* (Quiroga 2007). In this process, the differences in spike shapes are crucial (Einevoll et al. 2012).

A more ambitious endeavor is to also use the recorded spike shapes to identify the type of neuron the spike originates from (Barthó et al. 2004). The temporal width of the spike is, for example, used to separate putative inhibitory neurons (narrow spikes) from excitatory neurons (broad spikes) (Lemon et al. 2021), and some studies have indicated that a more detailed separation into subgroups is also possible (Buccino et al. 2018, Trainito et al. 2019).

Detailed modeling of spikes is useful for many purposes – for example, (i) to understand the relationship between single-neuron properties and spike shapes (Rall 1962, Rall & Shepherd 1968, Holt & Koch 1999, Gold et al. 2006, Pettersen & Einevoll 2008, Anastassiou et al. 2013), (ii) to investigate which types of neurons are most likely to generate the spikes recorded by electrodes (Pettersen & Einevoll 2008), (iii) to develop and test methods for spike sorting or spike-based neuron identification (Einevoll et al. 2012, Thorbergsson et al. 2012, Camuñas-Mesa & Quiroga 2013, Hagen et al. 2015, Mondragón-González & Burguière 2017, Buccino & Einevoll 2021), (iv) to develop and test methods for estimating positions of neurons (Mechler et al. 2011, Somogyvari et al. 2012, Buccino et al. 2018), and (v) to fit model parameters in multicompartment neuron models (Gold et al. 2007, Buccino, Damart et al. 2022).

As spikes, like APs, have a duration of just a few milliseconds, it is common to study them by concentrating on the high frequencies of the extracellular potential.[1] Accordingly, the signal used for analyzing spiking activity is typically obtained by high-pass filtering the extracellular potential using a lower cut-off set at some hundred hertz. In contrast, the low-frequency part, the local field potential (LFP), which we consider in Chapter 8 – is thought to mostly reflect synaptic inputs onto populations of neurons surrounding the contact (Einevoll, Kayser, et al. 2013).

This does not imply that the spike is built up exclusively of high-frequency components. It is a common misunderstanding that – since a spike typically lasts only a couple of milliseconds, and an oscillation with a period of, say, 2 ms corresponds to a frequency of 500 Hz – frequencies lower than 500 Hz should be absent or at least very weak. This is not generally true for spikes (Pettersen et al. 2008, Ray & Maunsell 2011, Schomburg et al. 2012, Scheffer-Teixeira et al. 2013), and the main reason is that spikes are not single-frequency oscillations. In general, one cannot deduce the frequency content of a

---

[1] A brief introduction to the frequency analysis of signals is given in Appendix F.

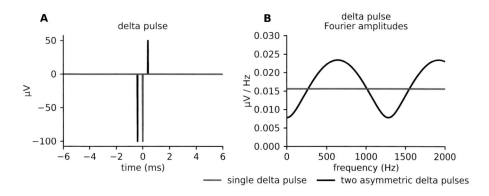

Figure 7.2 Delta pulses and frequency content. A: A single (red) and an asymmetric pair (black) of Dirac delta-pulses. The asymmetric pair caricatures the spike in Figure 7.3B with a negative sodium peak followed by a weaker positive potassium peak. **B**: The Fourier amplitude spectrum of the single Dirac delta-pulse has frequency components with equal amplitudes for all frequencies (red). The asymmetric pair of delta pulses has an oscillating frequency spectrum (black). Its (first) maximum is at 625 Hz, but it also has substantial contributions from frequencies all the way down to zero. Code available via www.cambridge.org/electricbrainsignals.

signal merely from its briefness. This is perhaps best illustrated by the so-called Dirac delta-pulse (or Dirac delta function), which essentially is infinitely sharp and has zero width but is built up of equal amounts of frequency components for all frequencies (Figure 7.2).

Spikes may contain significant power at frequencies at least down to 100 Hz (Pettersen & Einevoll 2008) as illustrated in Figure 7.3 for a spike produced by the Hay neuron model (Section 3.2.6). In addition, an AP may often coincide with slower associated membrane phenomena such as the triggering of calcium currents (Stuart et al. 2007) and afterhyperpolarization currents (Buzsaki et al. 1988). These phenomena may further contribute to increased low-frequency components in the spike (Buzsáki et al. 2012). Anastassiou et al. (2015) found, for example, spike contributions to the LFPs for frequencies as low as 20 Hz for cortical pyramidal neurons, likely stemming from action-potential afterhyperpolarization.

The motivation for applying high-pass filters to extracellular potentials when detecting spikes is not that spikes do not contain frequency components below the cut-off frequency, but rather to isolate the part of the recorded signal specifically reflecting the AP from parts resulting from other slower neural processes. While these other slower processes typically dominate the extracellular potential in terms of overall signal power, they contribute little above a few hundred hertz and can thus be removed by high-pass filtering the recorded signal (Pettersen et al. 2008, Einevoll, Kayser, et al. 2013).

Overviews of the challenges when recording and sorting spikes can be found in Einevoll et al. (2012), Anastassiou et al. (2013), Whittingstall & Logothetis (2013), and Buccino, Garcia et al. (2022).

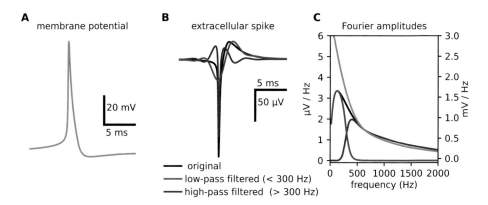

Figure 7.3 Frequency components of spikes. A: Membrane potential during the firing of an AP for the Hay neuron model (see also Figure 7.4). **B**: Corresponding spike measured 20 μm outside the soma: raw signal (black), low-pass filtered signal (red), and high-pass filtered signal (blue). **C**: Frequency content of signals in panel B – that is, amplitudes of frequency components found by Fourier transformation. The grey curve corresponds to the frequency content of the membrane potential in panel A. Code available via www.cambridge.org/electricbrainsignals.

7.2   Modeling Spikes

Wilfrid Rall pioneered the forward modeling of extracellular spikes using stylized ball-and-stick neurons (Rall 1962, Rall & Shepherd 1968). Holt (1998) took a next key step when using the same scheme for biophysically detailed neuron models. In such spike modeling, an MC neuron model with an appropriate set of AP-generating active ion channels must be combined with an appropriate VC model. In this section, we consider three different MC models: (i) the biophysically detailed Hay neuron model, (ii) the two-compartment model – the simplest model generating spikes or other extracellular potentials, and (iii) the ball-and-stick model.

   We here assume that brain tissue is an infinite volume conductor with the same conductivity everywhere and that spikes are recorded with a point electrode. More complicated cases, where we account for effects that real electrodes and recording devices have on the local conductivity, are considered in Section 7.7.

7.2.1   Spikes from Morphologically Detailed Neuron Models

The somatic APs of a given neuron have quite a stereotypical waveform, which they diverge from only in particular cases, such as during burst firing – when the time between two consecutive APs is very short (Hodgkin & Huxley 1952a). Likewise, given that the ambient noise level in the extracellular potential is not too high, the resulting spikes at a given location will typically have a similarly stereotypical waveform unless the interval between spikes is very short (Fee et al. 1996, Hagen et al. 2015).

   The shape and amplitude of a spike will, however, depend strongly on where it is recorded relative to the neuron it originates from. This is illustrated in Figure 7.4, showing computed spikes at various positions around the Hay model neuron during the

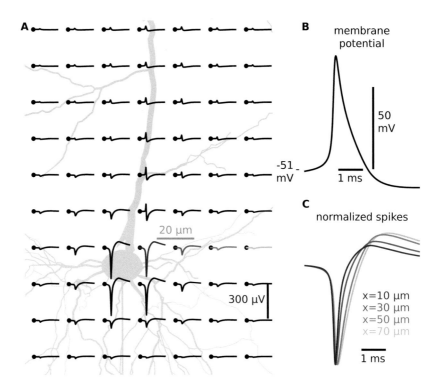

Figure 7.4 Spatial variation of spikes from the Hay model. A: Spikes computed at different positions (marked by black dots at the start of the traces). The line-source approximation (equation (4.11)) is used for the dendritic compartments, and the point-source approximation (equation (4.8)) is used for the soma compartment. **B**: Corresponding membrane potential. **C**: Spike shapes at different lateral positions in panel A as depicted by coloring. Spike shapes are normalized to have the same magnitude of the negative peak. The action potential (AP) resulted from the injection of a constant depolarizing current into the soma compartment of the model neuron. The neuron model has a short stub axon, which only negligibly affects the spike shapes. Code available via www.cambridge.org/electricbrainsignals.

firing of an AP. Two key observations are that (i) the spike's amplitude decays steeply with distance from the soma, while (ii) its shape gets blunter with distance from the soma. Thus, the largest and sharpest spikes are seen for positions close to the soma.[2] The blunting, or low-pass filtering, of the spike with distance (highlighted in panel C) stems from the cable properties of the neuron (as opposed to, say, possible filtering by the extracellular medium itself (Bedárd & Destexhe 2012)) and has been referred to as "intrinsic dendritic filtering" (Lindén et al. 2010, Pettersen et al. 2012). The biophysical origin of this intrinsic dendritic filtering effect was outlined in Section 3.3.3 and is

---

[2] Note that, in reality, the largest spikes may be observed close to the axon initial segment (AIS) (Bakkum et al. 2018, Buccino, Damart et al. 2022), a feature not captured by the Hay neuron model, which only has a stub axon with negligible effects on the shape of the spikes. For a discussion of spikes initiated in the AIS, see Section 7.4.

described in Section 7.3 in the context of spikes generated by ball-and-stick model neurons.

7.2.2   Spikes from Two-Compartment Neuron Model

Figure 7.4 shows how the shape of spikes generated with a biophysically detailed multicompartment model varies with the recording position. Some (but not all) of these observations can be explained using simpler neuron models.

We here consider the simplest neuron model that can generate an extracellular potential: a two-compartment model where each of the two compartments is modeled as a point-current source. For this model, the formula in equation (4.8) for the extracellular potential simplifies to

$$V_e(\mathbf{r}) = \frac{I_1}{4\pi\sigma_t\|\mathbf{r} - \mathbf{r}_1\|} + \frac{I_2}{4\pi\sigma_t\|\mathbf{r} - \mathbf{r}_2\|} = \frac{I_1}{4\pi\sigma_t}\left(\frac{1}{\|\mathbf{r} - \mathbf{r}_1\|} - \frac{1}{\|\mathbf{r} - \mathbf{r}_2\|}\right), \quad (7.1)$$

with compartments 1 and 2 set to represent the soma and dendrite, respectively. In the second equality in this equation, we have used that the membrane current (including the capacitive current) entering the soma compartment must at all times equal the membrane current leaving the dendrite compartment (Section 2.4).

Spikes generated by a two-compartment model are shown in Figure 7.5C. Around the soma, the spikes have a sharp negative peak followed by a slower positive hump, similar to spikes seen experimentally, as exemplified in Figure 7.1. The same spikes with inverted form are observed outside the dendrite compartment. As such inverted spikes are rarely seen in experiments (but see e.g. Gold et al. (2009)), this suggests that while the two-compartment model may account for large spikes recorded next to the soma, it is inadequate for predictions of how the spike shape varies with position around the neuron.

The two-compartment model qualitatively captures the decay of spike amplitudes with distance. However, the two-compartment model differs from the biophysically detailed model in that the (normalized) shape of its spikes (Figure 7.5C, bottom) does not depend on position (except for the complete inversion when crossing the lateral symmetry plane of the neuron). This follows from equation (7.1), where the extracellular potential is given by a product of the membrane current amplitude (I_1) and function of distance (terms inside parenthesis). The only feature that will vary with position is the net amplitude and sign of the spike, and the shape of the spike will thus always be the same.

7.2.3   Spikes from Ball-and-Stick Neuron Model

Unlike the two-compartment neuron model, the so-called ball-and-stick model (Section 3.3.2) qualitatively captures the variation of spike shape with position (Figure 7.5B). In this model, a dendrite cable "stick," here modeled as a chain of small dendrite compartments, is connected to a single-compartment soma. The position dependence is not as pronounced as for the morphologically detailed neuron model

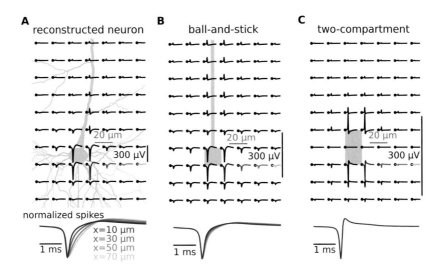

Figure 7.5 Comparison of spikes from different neuron models. A: Spikes computed at different positions around the Hay model neuron (top panel). Spike shapes at different lateral positions are shown in the bottom panel. Spike shapes are normalized to have the same magnitude of the negative peak. **B**: Same as panel A for spikes computed for the ball-and-stick model. **C**: Same as panel A for spikes computed for the two-compartment neuron model. Spikes are generated by imposing the membrane potential from Figure 7.4B in the soma compartments as a boundary condition. In panels A and B, the line-source approximation (equation (4.11)) is used for the dendritic compartments, and the point-source approximation (equation (4.8)) is used for the soma compartment. Code available via www.cambridge.org/electricbrainsignals.

(panel A), but the key qualitative feature remains: the spike gets blunter when moving away from the soma. In particular, the width of the negative sodium peak increases with increased lateral distance. Also, the large positive (inverted) spikes outside the dendrite of the two-compartment model are absent in the ball-and-stick model, in better accordance with predictions from the morphologically detailed model.

7.3   Analysis of Spike Shapes and Sizes

As showcased above, the ball-and-stick model produces spikes with the same key quali-tative features as the morphologically detailed Hay model (see Figure 7.5). Unlike mor-phologically detailed neuron models, the ball-and-stick model is amenable to analytical mathematical studies.

When modeling spikes using the MC+VC formalism, one option is to simulate the neural AP by including active Na^+ and K^+ conductances in the soma compartment. Another option is to impose a voltage waveform corresponding to a somatic AP as a boundary condition in the soma. If this imposed voltage waveform is set so as to be identical to what is found from the first option, the two options will produce an identical pattern of membrane currents returning to the extracellular space through the dendrites

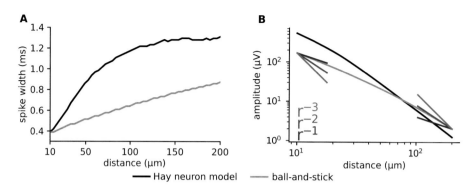

Figure 7.6 Dependence of spike widths and amplitudes on the distance from the soma center.
A: Spike widths as a function of distance from the soma for the Hay neuron model and the
ball-and-stick model in Figure 7.5A and B. The spike width is defined as the width of the spike at
25 percent of maximum negative peak. B: Peak-to-peak spike amplitude for the models in A. The
intracellular AP shown in Figure 7.4B was imposed as a voltage-clamp in the soma. The spike
positions are in the soma plane normal to the stick/primary apical dendrite. In panel B,
guidelines illustrating the power-law decays $1/r$, $1/r^2$, and $1/r^3$ have been added. Code
available via www.cambridge.org/electricbrainsignals.

and the soma, and thus also identical extracellular potentials. In the present section, we
use the second option to investigate the spikes of the ball-and-stick model as it allows
for a clearer and cleaner analysis of the link between the AP and the spike (Pettersen &
Einevoll 2008).

Figure 7.6 shows the distance-dependence of the spike amplitude and spike width,
both for the morphologically detailed Hay model and a ball-and-stick model. The spike
width of the Hay model is larger than for the ball-and-stick model for all distances
considered (Figure 7.6A). The amplitude of the spike is also larger for the Hay model
for the smallest distances. However, this amplitude decays faster with distance than for
the ball-and-stick model, so that beyond 50 μm, the spike amplitude of the ball-and-stick
model is larger (Figure 7.6B).

For the ball-and-stick model, the extracellular potential can be computed with (Pettersen & Einevoll 2008)

$$V_e(z, \rho) = \frac{1}{4\pi\sigma_t} \frac{I_{soma}}{\sqrt{(z - z_{soma})^2 + \rho^2}} + \frac{1}{4\pi\sigma_t} \int_0^L \frac{i_m(z')}{\sqrt{(z - z')^2 + \rho^2}} \, dz', \qquad (7.2)$$

where the first term on the right accounts for the contribution from the somatic
membrane currents, and the second term accounts for the contribution from membrane
currents along the dendritic stick. Here, L is the length of the dendritic stick, which is
pointing in the z-direction with the lower end positioned at $z = 0$, z_{soma} is the depth
position of the center of the soma, and ρ is the lateral distance to the dendritic stick axis.
We recognize the somatic contribution as the point-source equation (equation (4.7)),
while the contribution from the dendritic stick is essentially a continuous version of
equation (4.8) – that is, a continuum of point sources distributed along the stick.

We note that the soma membrane current I_{soma} is not known a priori when a voltage waveform is imposed as a boundary condition for the dendritic stick and must be computed (using equation (7.5) without absolute values). However, it follows from the cable equation that, due to current conservation, the soma membrane current must be the same as the computed axial current entering the dendritic stick at the soma end.

Since the dendritic stick considered here is passive, its response to imposed currents or voltages is linear. It is then convenient to use Fourier theory to decompose the soma AP into a sum of harmonically oscillating components with different frequencies (Appendix F) and consider each frequency component of the AP separately. Each oscillatory component of the soma AP gives rise to an oscillatory axial current entering the stick at the soma end. As we saw in Section 3.3.3, and explore further in Section 7.3.1.2, the spatial pattern of return currents along the stick will depend on the frequency of this oscillation.

A Fourier decomposition of a signal such as the soma AP can be constructed in various ways – for example, as a Fourier integral (Appendix F.2) where a time-dependent signal $x(t)$ can be represented as

$$x(t) = \int_{-\infty}^{\infty} \hat{x}(f)e^{2\pi j f t}\, df \ . \tag{7.3}$$

Here, j is the unit of imaginary numbers, f is frequency, and $\hat{x}(f)$ is the (generally) complex Fourier amplitude. A Fourier decomposition of an example AP (membrane potential) and corresponding spike is illustrated in Figure 7.3C. A key observation is that, unlike for the AP, the largest contributions to the spike do not come from the smallest frequencies but rather from frequencies around 120 Hz.

7.3.1 Approximate Spike Formulas

For a ball-and-stick model with an infinitely long stick, and with an oscillatory voltage imposed in the soma as a boundary condition, equation (7.2) can be solved analytically for the extracellular potential (Pettersen et al. 2008). The resulting formulas are cumbersome and difficult to interpret. However, some useful insights can be obtained in two limiting cases: positions very near to the soma or very far from the soma.

7.3.1.1 Spikes Near the Soma

For positions $\mathbf{r}$ very close to the soma, the contribution from the soma current will dominate over the contributions from the dendrite in the sum giving the extracellular potential in equation (7.2). If we know the amplitude $|\hat{I}_s(f)|$ of the oscillatory soma membrane current with frequency f, the predicted amplitude of the oscillating extracellular potential $|\hat{V}_{\text{e, near}}(f,\mathbf{r})|$ is

$$|\hat{V}_{\text{e, near}}(f,\mathbf{r})| = \frac{|\hat{I}_s(f)|}{4\pi\sigma_t r} \ . \tag{7.4}$$

This oscillating soma membrane current $|\hat{I}_s(f)|$ is identical to the current entering the dendritic stick (see Figure 3.6), and for the relatively high frequencies of most relevance for the spike, it can be shown that the soma current is related to the soma membrane potential $\hat{V}_{m,s}(f)$ through (Pettersen & Einevoll 2008)

$$|\hat{I}_s(f)| = \frac{\pi^{3/2}d^{3/2}}{\sqrt{2}}\sqrt{\frac{f c_m}{r_a}}|\hat{V}_{m,s}(f)|, \tag{7.5}$$

where d is the diameter of the dendritic stick, c_m is the specific membrane capacitance, and r_a is the axial resistivity. The amplitude of the contribution to the spike from this oscillatory component is thus found to be

$$|\hat{V}_{e,\text{near}}(f,\mathbf{r})| = \frac{\sqrt{\pi}}{4\sqrt{2}\sigma_t}\frac{d^{3/2}}{r}\sqrt{\frac{f c_m}{r_a}}|\hat{V}_{m,s}(f)| \propto \frac{d^{3/2}}{r}\sqrt{\frac{f c_m}{r_a}}|\hat{V}_{m,s}(f)|. \tag{7.6}$$

The subscript "near" is added because this expression only applies in the "near-field" limit – that is, close to the soma. $|\hat{V}_{m,s}(f)|$ is the amplitude of the Fourier component of the soma membrane potential at the same frequency.

7.3.1.2   Spikes Far Away from the Soma

For positions far away from the soma, the contributions from both the soma current and membrane return currents must be included in the sum in equation (7.2). The spatial pattern of the return currents will depend on the frequency content of the current entering the dendritic stick or, equivalently, on the frequency content of the oscillating potential imposed in the soma. This frequency dependence of the spatial pattern of return currents will in turn give a frequency dependence of the extracellular potential.

As described in Section 3.3.3, this phenomenon is referred to as intrinsic dendritic filtering. This filtering effect is due to the capacitive membrane: for higher frequencies, the capacitive membrane current will be larger and the membrane effectively more leaky so that the current will travel shorter distances along the stick before returning to the extracellular space. As we will see, the frequency dependence of the return-current patterns implies that the different frequency components will contribute differently to spikes.

The effect of intrinsic dendritic filtering was illustrated in Figure 3.7, and the reader is advised to keep this illustration in mind for the remainder of this section. As described there, the dendritic return currents can, as an approximation, be assumed to leave the dendrite at a single point. The position of this "return-current point" depends on the frequency of the component. With this approximation, each oscillatory component produces a (current-) dipole,[3] where the membrane current entering the soma is accompanied by an oppositely directed current of the same magnitude leaving at a single point on the dendrite. We thus effectively have an individual two-compartment model for each oscillatory component of the spike, where the dipole length in each model is given by the distance between the soma and the return-current position.

---

[3] For brevity, we will refer to "current dipoles" simply as "dipoles." Charge dipoles will not be encountered in this book.

As outlined in Section 3.3.3, a natural choice is to set the dipole length equal to the mean value of the envelope of the sinusoidally varying (normalized) membrane current weighted with the distance from the soma. If the stick is sufficiently long, this frequency-dependent length constant is well approximated by $\lambda_{AC}(f)$ given in equation (3.42):

$$\lambda_{AC}(f) = \lambda \sqrt{\frac{2}{1 + \sqrt{(2\pi f \tau_m)^2 + 1}}}, \tag{7.7}$$

where $\tau_m = r_m c_m$. For high frequencies ($f \gg 1/(2\pi\tau_m)$), after some algebra, this is found to give

$$\lambda_{AC}(f) = \frac{\lambda}{\sqrt{\pi f \tau_m}} = \frac{1}{2\sqrt{\pi}} \sqrt{\frac{d}{f r_m c_m}}, \tag{7.8}$$

where λ is the DC cable length constant from equation (3.32). The extracellular potential contribution from each frequency component can then be approximated by using the dipolar expression in equation (4.50) – that is,

$$|\hat{V}_{e,\,far}(f, \mathbf{r})| = \frac{|p(f)\cos\theta|}{4\pi\sigma_t r^2} = \frac{|\hat{I}_s(f)\lambda_{AC}(f)\cos\theta|}{4\pi\sigma_t r^2}. \tag{7.9}$$

Here, θ is the angle between the position vector and the vertical axis. Thus, for spikes measured far away from the soma, we find

$$|\hat{V}_{e,\,far}(f, \mathbf{r})| = \frac{1}{8\sqrt{2}\sigma_t} \frac{d^2}{r_a} \frac{1}{r^2} |\hat{V}_{m,s}(f)\cos\theta| \propto \frac{d^2}{r_a} \frac{|\cos\theta|}{r^2} |\hat{V}_{m,s}(f)|. \tag{7.10}$$

The suffix "far" is added because this expression only applies in the "far-field" limit – that is, far away from the neuron.

The implications of this formula, as well as the corresponding formula in equation (7.6) for the "near-field" limit, are discussed in Sections 7.3.2–7.3.4.

7.3.2   Distance-Dependence of Spike Amplitudes

A prediction from the near-field equation (7.6) is that the amplitude of each Fourier component decays as $1/r$ when moving away from the soma, as long as the position remains within distances where the "near-field" approximation applies. Since this applies to all frequency components that together constitute the AP, the amplitude of the spike will also decay as $1/r$ in this regime.

Far away, the far-field equation (7.10) predicts that the spike amplitude depends not only on the radial distance r from the neuron, but also the angle θ with the dipole axis, which here corresponds to the direction of the dendritic stick. The amplitude will be largest above and below the neuron where $\theta = 0°$ and $\theta = 180°$, respectively. In the sideways direction, the spike will be smaller, as is characteristic for spatial patterns of potentials around a (current-) dipole. An extreme example is that of the two-compartment neuron in Figure 7.5C, where the extracellular potential in both the near- and far-field limit is identical to zero in the sideways direction when moving from the midpoint between the two compartments. A qualitatively similar dipolar pattern, although not so distinct, is seen for the spike generated by the morphologically detailed neuron in Figure 7.5A.

Another difference between this far-field expression and the near-field expression in equation (7.6) is that the far-field amplitude decays as $1/r^2$, characteristic for potentials around dipolar sources, while the near-field amplitude decays as $1/r$, which is characteristic for potentials around a single source. Such a $1/r^2$ decay is only observed when moving in the radial direction away from the dipole, however. If we instead are moving in the horizontal direction (for example, away from the soma), with a fixed vertical distance from the dipole, a geometric argument implies that the amplitude will instead decay as $1/r^3$ in the far-field limit (see Figure 4.8). Numerical investigations indeed find such a $1/r^3$ dependence for the ball-and-stick spike in Figure 7.6B for larger distances than what is depicted in this panel (see Pettersen & Einevoll (2008, figure 10) for a demonstration of this effect).

A general insight from this analysis is that the spike is quite local, with the amplitude of the spike decaying rapidly with distance from the neuronal soma. For the Hay model considered in Figure 7.6, for example, the spike amplitude decays from $160\,\mu$V at a distance $20\,\mu$m from the soma center to only $6\,\mu$V at a distance $100\,\mu$m away. Qualitatively similar results were found for the neuron models considered in Pettersen & Einevoll (2008). This rapid decay with distance eases the interpretation of recorded spikes since it implies that, in practice, an electrode contact will predominantly pick up spikes from neurons with somas located within a radius of some tens of micrometers.

7.3.3   Spike-Amplitude Dependence on Neuronal Parameters

Both in the near-field and far-field limits, the spike amplitude increases with dendrite diameter d. In the far-field limit, the amplitude is proportional to d^2 and thus to the cross-sectional area of the dendrite (equation (7.10)). In the near-field limit, the spike amplitude increases slightly less with the diameter and is proportional to $d^{3/2}$ (equation (7.6)).

In both limits, the spike amplitude is independent of the membrane resistivity r_m of the dendrite. This reflects that we are considering the high-frequency components ($f \gg 1/(2\pi\tau_m)$) dominating the spike. For the ball-and-stick model, $\tau_m = r_m c_m = 30$ ms so that "high-frequency components" in this case means $f \gg 5$ Hz. In this situation, the ionic membrane current, governed by r_m, is much smaller than the capacitive membrane current, governed by the specific membrane capacitance c_m. Thus, the spatial distribution of the return current along the dendrite will depend only on the capacitive current. This dependence is seen through the presence of c_m in the near-field formula in equation (7.6). Note that in the far-field formula (equation (7.10)), c_m is absent due to cancellation with another factor containing c_m in the mathematical derivation (see Pettersen et al. (2008, equation 23)).

Further, in both limits, the spike amplitude decreases with increasing axial resistivity r_a in the dendrites. The explanation is that, if r_a is increased, the current entering the dendrite will travel a shorter distance before returning to the extracellular space. In effect, this will shorten the distances between the current sinks and sources, giving smaller equivalent dipoles and thus smaller spike amplitudes.

7.3.4 Spike-Shape Dependence on Distance

The near-field and far-field formulas also give qualitative insights regarding the shape of the spike. In the near-field expression (equation (7.6)), $\hat{V}_{e,near}(f,r) \propto \sqrt{f}\,\hat{V}_{m,s}(f)$ so that high-frequency components of the spike are more amplified than the low-frequency components of the soma membrane potential. Close to the soma, the spike is therefore sharper than the AP. In the far-field regime (equation (7.10)), there is no such frequency-dependent amplification: $\hat{V}_{e,far}(f,r) \propto f^0 \hat{V}_{m,s}(f)$ so that $\hat{V}_{e,far}(f,r) \propto \hat{V}_{m,s}(f)$. As a consequence, spikes far away from the soma will have less high-frequency content than those close to the soma. The faraway spikes will therefore be blunter and have larger spike widths, as seen in Figure 7.6A.

The results from this analysis can be qualitatively understood by considering how the distribution of dendritic return currents depends on the frequency of the somatic membrane potential. This was earlier illustrated in Figure 3.7C–D, which showed how the effective dipole length is negatively dependent on the frequency of the injected stimulus. A shorter dipole length in turn implies that the far-field regime, where the dipolar approximation applies, is reached for smaller distances from the neuron. Since the amplitude of the signal contributed by a frequency component decays faster with distance in this far-field regime ($\propto 1/r^2$), the low-frequency components that still remain in the near-field regime (where the signal is $\propto 1/r$) will become increasingly dominant. This implies a low-pass filtering effect and a blunting of the spike.

Very far away from the neuron, all frequency components are in the far-field limit, and there will be no further change in the relative weighting of the different frequency components of the spike. In this regime, the spike will thus have a fixed shape even if the overall amplitude becomes smaller with distance. The exact distance range over which the spike shape changes as the different frequency components change from the near-field to the far-field regimes will depend on the frequency content of the APs as well as the morphology and electrical properties of dendrites – see Pettersen & Einevoll (2008) for examples and discussion.

7.3.5 Proxies for Spike Shapes

The time derivative of the membrane potential has been suggested as a rough proxy for the shape of the spike, essentially assuming that the spike is determined by the capacitive soma current alone (e.g. Henze et al. (2000)). According to equation (7.3), this would correspond to $\hat{V}_e(f) \propto f\hat{V}_{m,s}(f)$ as a time differentiation of a Fourier component $\exp(j2\pi ft)$ is proportional to $f\exp(j2\pi ft)$. This relationship predicts a higher weighting of the high-frequency components, thus a sharper spike shape, than the prediction $\hat{V}_e(f) \propto \sqrt{f}\hat{V}_{m,s}(f)$ for the ball-and-stick model in the near-field limit.

The derivative of the soma membrane potential $V_{m,s}$ indeed predicts a too-narrow spike for the Hay model, unlike the $\sqrt{f}$-relationship, which excellently predicts the

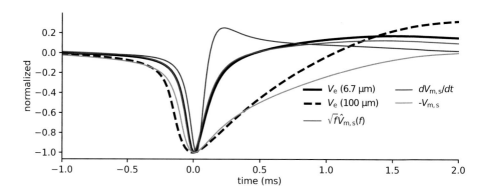

Figure 7.7 Proxies for spike shapes for the Hay model. Comparison of the shapes of spikes V_e computed from the Hay model for two distances from the soma: at the soma surface ($r = 6.7$ µm measured from the soma center) and at $r = 100$ µm, with proxies based on (i) the inverted shape of the soma membrane potential ($-V_{m,s}$), (ii) the shape of the derivative of the membrane potential $dV_{m,s}/dt$, and (iii) the $\sqrt{f}$-relationship found for the ball-and-stick model in the near-field limit ($\hat{V}_e(f) \propto \sqrt{f}\,\hat{V}_{m,s}(f)$, equation (7.6)). For the latter, the contribution from each frequency component of the membrane potential to the spike is assumed to be amplified by a factor proportional to the square root of the frequency ($f^{1/2}$). Code available via www .cambridge.org/electricbrainsignals.

spike shape close to the neuron (Figure 7.7). Far away (100 µm) from the soma, the spike shape is much blunter and is more akin to the inverted shape of the soma membrane potential ($-V_{m,s}$). This again is in accordance with the prediction for the ball-and-stick neuron where, in the far-field regime (equation (7.10)), there is no frequency-dependent amplification of V_e compared to $V_{m,s}$, since $\hat{V}_{e,\,far}(f,r) \propto f^0 \hat{V}_{m,s}(f)$, or, $\hat{V}_{e,\,far}(f,r) \propto \hat{V}_{m,s}(f)$.

The $\sqrt{f}$-relationship also excellently describes the relationship between the intracellular and extracellular signals during an AP for the experimental example in Figure 7.1D. Also, here the derivative formula predicts a too-narrow spike shape compared to the experimental data, while the "inverted-V_m" proxy predicts a too-wide spike shape. This suggests that the spike in this case is recorded in the near-field regime, which agrees with the depicted placement of the extracellular electrode in Figure 7.1B.

7.3.6   Generalization of Findings to Other Neuron Morphologies

While the earlier formulas were derived for a neuron model with a single passive dendritic stick, similar expressions can be derived for more complex neural morphologies where several passive sticks protrude from the soma (Pettersen & Einevoll 2008). The main conclusions earlier also hold for these "ball-and-stick" neuron models – in particular, that spike widths always increase with distance and that the amplitude of a spike is proportional to d^k, where $k \approx 1.5$ close to the soma and $k \approx 2$ or larger further away (figure 9 in Pettersen & Einevoll (2008)). For neurons with many dendrites attached to the soma, the contributions from each dendritic branch superimpose so that

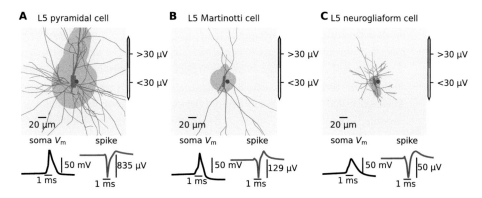

Figure 7.8 Highly variable spike amplitudes and spike detectability for different neuron types. A: AP in a layer 5 pyramidal cell (lower left) and an associated spike immediately outside the soma (lower right). The pink region in the top panel illustrates the region where the peak-to-peak amplitude of the spike is above 30 μV, a typical amplitude required for spike detection in experiments. The red dot denotes the position of the spike. **B**: Same as in panel A but for a layer 5 Martinotti cell. **C**: Same as in panel A but for a layer 5 neurogliaform cell. All neuron models are from Markram et al. (2015). Code available via www.cambridge.org/electricbrainsignals.

the spike amplitudes roughly add up. A simple rule of thumb found from these analytical considerations is that the spike amplitude close to the neuron is roughly proportional to the sum over the diameters of the dendritic branches attached to the soma, each raised to the power $3/2$. Neurons with many thick dendritic branches attached to the soma will thus generate the largest spikes – see Pettersen & Einevoll (2008) for further discussion.

The different amplitudes of spikes generated by three substantially different layer 5 neurons are illustrated by the modeling results in Figure 7.8. Here, the maximum spike amplitudes as well as the regions around neuronal somas where spikes are expected to have amplitudes large enough to be measured are depicted. As observed, the pyramidal neuron has a much larger maximum spike size and likewise a much larger detection region compared to the two other cells. These amplitude differences suggest a bias for recording spikes from pyramidal neurons compared to the other considered cell types when doing in vivo recordings with extracellular electrodes. However, other biases such as differences in the neuron firing rates also come in to play here.

7.4   Spikes from Action Potentials (APs) Initiated in the Axon

In the earlier investigations of spikes generated by a ball-and-stick model, we assumed that the AP is initiated in the soma and that the spike is generated by a set of frequency-dependent (current-) dipoles, reflecting the distribution of return currents in the dendritic stick. However, in most neurons, the AP is initiated in the axon initial segment (AIS) some distance away from the soma (Bender & Trussell 2012, Bakkum et al. 2018, Telenczuk et al. 2018, Goethals & Brette 2020). In the initial phase of the AP, we may then expect there to be a briefly lasting dipole where current enters at the AIS and leaves through the soma before eventually the full soma AP is ignited. Model simulations have indeed confirmed that this is feasible and that a small and narrow positive peak should then be seen prior to the negative sodium peak for positions close to the soma (Telenczuk et al. 2018). Interestingly, detailed experimental studies of the shape and amplitude of spikes recorded simultaneously around the soma and along the axon are now becoming possible by means of high-density micro-electrode arrays (HD-MEAs) (Emmenegger et al. 2019). This allows for the detailed probing of, for example, the position of the AIS (Kumar et al. 2022).

7.5   Spikes from Neurons with Active Dendrites

When we investigated the ball-and-stick model, the assumption of an electrically passive dendritic stick was essential. This assumption made the relationship between intracellular potentials in the soma and extracellular potentials *linear* and independent of the shape of the intracellular waveform, meaning that the Fourier components of the extracellular potential ($\hat{V}_e(f,\mathbf{r})$) for a given frequency depended only on the Fourier components of the soma potential ($\hat{V}_{m,s}(f)$) for the same frequency. Thus, in this case, the analytical insights from Section 7.3.1 apply in principle to all AP waveforms.

However, real neurons have active ion channels in the dendrites also (Stuart et al. 2007), making the dendritic responses and, consequently, the relationship between somatic membrane potentials and extracellular potentials nonlinear. Then, one cannot treat each frequency component independently from the others as for linear dendrites, and one has to resort to numerical investigations using the general formula in equation (3.20), where all active conductances are included explicitly. In our default example of a multicompartment model, the Hay model in Figure 7.4, the effects from active dendritic conductances on spike shape are quite small, as demonstrated in the simulations shown in Figure 7.9. Comparing the active and passive models, we see a slightly larger negativity in V_e along the apical dendrite and a slightly larger positivity in V_e near the soma in the active model (Figure 7.9B). This reflects back-propagation of the AP in active dendrites.

Gold and coworkers (Gold et al. 2006, 2007) performed thorough investigations of the extracellular signatures of spikes from pyramidal neurons in hippocampus CA1, where active dendritic conductances were included in the model. Their results were largely in agreement with many of the observations seen earlier for neurons with passive dendrites: (i) the spike width increased with distance from the soma (figure 5A in Gold et al. (2006)), (ii) the spike amplitude decayed with distance from the soma with a power between 1 and 2 for distances less than 50 μm (figure 14 in Gold et al. (2007)), and (iii) the spike amplitude varied significantly with varying axial resistivity r_a and specific membrane capacitance c_m but not so much with varying membrane resistivity r_m (Gold et al. 2007). They also found that extracellular waveforms provide tight constraints on some of the neuronal model parameters, suggesting (as in Buccino, Damart et al. (2022)) that spikes can be useful for constraining compartmental models.

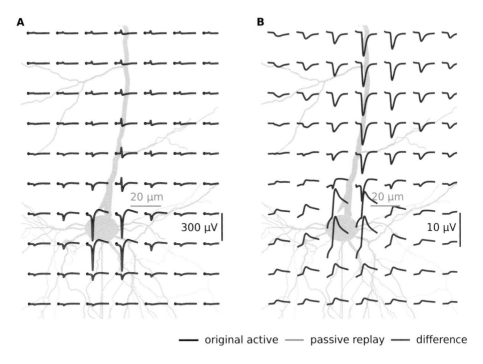

— original active   — passive replay   — difference

Figure 7.9 Effect from active dendritic conductances on spikes. A: Spikes from the original active pyramidal cell model from Hay et al. (2011) (black) and from a passive version of the same model with all active conductances set to zero. Here, the somatic membrane potential from the active model is replayed into the soma of the passive model (red). **B**: Difference between the spikes from the active and passive versions of the model. Code available via www.cambridge .org/electricbrainsignals.

A preliminary conclusion is that active dendritic conductances seem to have modest effects on spike shapes. In Section 8.5.1, in contrast, we find that *subthreshold* active conductances such as the I_h conductance can have substantial effects on the LFP. However, these effects are found for frequencies much smaller than those dominating spikes.

7.6   Axonal Spikes

So far in this chapter, we have focused on somatic spikes, assuming that the dominating membrane currents setting up the spike during an AP are in the soma. However, APs also propagate along the axons, and the associated axonal membrane currents generate so-called *axonal spikes*.

Axons come in two flavors, myelinated and unmyelinated. The myelinated axons are wrapped in a type of glial cells that speed up action-potential propagation (Koch 1999, chapter 6). Figure 7.10 shows example results of spikes recorded outside axons, both unmyelinated and myelinated. A first observation is that the AP propagates much faster

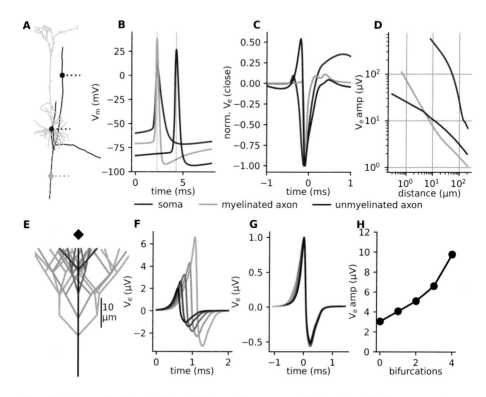

Figure 7.10 **Axonal spikes**. **A**: Pyramidal layer 5 neuron model taken from Hallermann et al. (2012) (available from http://modeldb.yale.edu/144526) with both myelinated axons with nodes of Ranvier (orange) and unmyelinated axons (purple). **B**: Membrane potentials in the unmyelinated axon, in a node of Ranvier for a myelinated axon, and in the soma of the neuron. Colored dots in panel A shows (virtual) recording positions. **C**: Spikes immediately outside the same positions. **D**: Peak-to-peak amplitude of spikes as a function of lateral distance (dotted lines in panel A). **E**: Unmyelinated axon (fixed diameter of 0.3 μm) approaching the cortical surface from below with different degrees of bifurcation. Parameters for the passive and active conductances of the unmyelinated axon model were extracted from the unmyelinated axon sections of the cell model used in panels A–D. **F**: Spikes recorded at the cortical surface (black diamond in panel E) for unmyelinated axons with different numbers of bifurcations (0, 1, 2, 3, 4). **G**: Normalized version of panel F, where spikes also have been temporally aligned. **H**: Peak-to-peak amplitude of spikes in panel F for axons with different numbers of bifurcations. In all panels, the point-electrode approximation (Section 4.5.1) is used. In panels E–H, the electrode is positioned 10 μm above the uppermost axonal tips and has a diameter of 20 μm. In panels F–H, the electrode is assumed to be embedded in a non-conducting substrate at the cortical surface, which is accounted for in the model through the method of images by multiplying the extracellular potential by a factor of two (Section 5.7.1). Panels E–H are adapted from Thunemann et al. (2022). Code available via www.cambridge.org/electricbrainsignals.

in the myelinated axon: the AP recorded at a "node of Ranvier," a gap in the glial-cell wrapping, occurs essentially simultaneously as in the soma (Figure 7.10B). In contrast, the same panel shows that the AP takes several milliseconds to propagate the same distance in the unmyelinated axon.

The axonal spikes are narrower than the spike recorded outside the soma (Figure 7.10C). Further, the spike recorded immediately outside the myelinated axon is much narrower than the spike recorded outside the unmyelinated axon. The axonal spikes also differ in the sense that, outside the unmyelinated axon, they exhibit a sharp positivity prior to the negative sodium peak, which is not seen for the myelinated axon. This stems from the finding that an AP traveling down an unmyelinated axonal cable does not generate a clear dipolar extracellular potential around the axon. Rather, the potential more closely resembles that generated by two opposing dipolar sources around the axon giving a *quadrupolar* contribution (Section 4.4) – see Plonsey & Barr (2007, chapter 8).

Axonal spikes are often assumed to be small, at least much smaller than somatic spikes. A main argument is that the diameters of axons in mammalian brains are typically less than a micrometer. This implies a small surface area compared to the surface area of somas, thus smaller membrane currents and spikes. This is indeed observed in Figure 7.10D, where the amplitude of the axonal spikes recorded 10 μm away from the axon is about a factor of 50 smaller than for the somatic spike (recorded immediately outside the soma). Closer to the axon, the amplitudes of the axonal spikes will be larger but still much smaller than the amplitude of the spike recorded outside the soma.

Axons generally branch out to make many synapses with other neurons, and around such branch points, the amplitude of axonal spikes may become much larger. Such branching and termination of ascending axons have been suggested as giving the largest contributions to spikes recorded at the surface of the somatosensory cortex in mice (Thunemann et al. 2022). Axonal spikes in such terminal regions have also been used as a measure of synaptic inputs onto neuronal populations (McColgan et al. 2017) – for example, in the barn owl auditory brainstem (Kuokkanen et al. 2018). Such an axonal pattern with up to four branching bifurcations is illustrated in Figure 7.10E. The branching substantially increases the recorded spike amplitude recorded by an electrode 10 μm above the uppermost axonal tips (Figure 7.10F, H), even though the spike shape is essentially the same for the different branching patterns (Figure 7.10G).

While spikes from single non-branching axons are small, they can still be measured if the recording electrodes are sufficiently small and sufficiently close to the axons, and the ongoing development of high-density micro-electrode arrays (HD-MEAs) (Emmenegger et al. 2019) allows for more detailed investigations of axonal spikes (Buccino, Yuan et al. 2022). For the special case where neurites grow into tunnels during neural development, the spike amplitudes can be boosted by about two orders of magnitude, as has been demonstrated in cell cultures (Molina-Martínez et al. 2022).

7.7   Effects of Measurement Device on Spike Recordings

In the earlier examples, we assumed that the tissue surrounding the studied neuron was an infinite, homogeneous, and isotropic volume conductor, with the same conductivity everywhere and in all directions. Further, spikes were computed using the point-electrode approximation (Section 4.5.1), assuming that the electrode does not affect the measured signal. However, as discussed in Section 4.5, real recording devices will in general affect the measured potentials in several ways. In the next sections, we discuss some effects of recording devices in the specific context of spike recordings.

7.7.1   Physical Sizes of Contacts and Shafts of the Recording Electrode

Most electrodes presently used for extracellular recordings inside the brain consist of electrode contacts made of a highly conductive material embedded in an electrically insulating electrode shaft. The amplitudes and shapes of recorded spikes are affected by the size of the electrode contacts, and as discussed in Section 4.5.2, this effect can be modeled by use of the disk-electrode approximation. In this approximation, the extracellular potential is computed by averaging the results obtained with the point-electrode approximation across the surface of the electrode contact (equation (4.58)), and it is thus straightforward to implement (see Section 6.3.3). An example of its use is provided by Figure 7.11, showing spikes estimated for circular disk electrodes of different radii. A key observation is that the spike amplitude is reduced with increasing disk sizes (panel B). In the current setup, the spike shape is also affected as the averaging of the potentials across the contact surface reduces the high-frequency components of the spike and makes the spike wider (panel C).

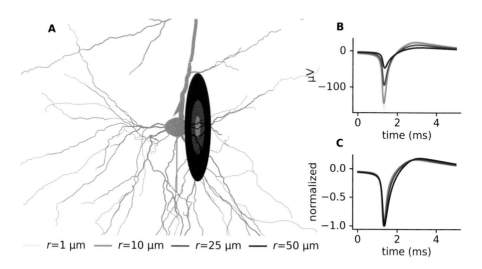

Figure 7.11 Electrode size affects spike shapes. A: Illustration of setup where a spike from a Hay neuron model is recorded by circular electrodes with different radii. The center of the circular electrode is 20 μm from the soma center. **B**: Spike shapes computed with the use of the disk-electrode approximation (equation (4.58)) for the set of different electrode radii. **C**: Spike shapes in B normalized to have the same value of the amplitude for the negative peak. The point-electrode values used to compute the disk-electrode values are computed as in Figure 7.4. Code available via www.cambridge.org/electricbrainsignals.

The insulating electrode shaft in multicontact electrodes may also have a substantial effect on the recorded potentials. Detailed studies of this requires comprehensive numerical investigations with the use of FEM modeling. Such studies have shown that spikes from neurons placed close to the shaft on the contact side may be amplified, while spikes from neurons positioned close to the back side of the shaft may be dampened (Moffitt & McIntyre 2005, Mechler et al. 2011, Mechler & Victor 2012, Buccino et al. 2019).

7.7.2 Spikes Recorded in Micro-electrode Arrays (MEAs)

In in vitro recordings, one either uses neuronal cell cultures that have been grown or small slices of excised brain tissue. In either case, the samples are placed in suitably designed dishes where the physiological properties of cells and networks can be probed in detail for hours. In *micro-electrode arrays (MEAs)* used for such recordings, the bottom of the device is covered by a grid of electrode contacts that record electric signals generated by the neurons above. The brain slice and the MEA are both covered with a liquid, typically artificial cerebrospinal fluid (aCSF), to protect the cells from drying out and to keep them alive for the duration of the experiment.

In MEAs, the electrode contacts are embedded in an insulating plate with very low conductivity, while the covering liquid typically has a higher conductivity than the brain slice it covers. In this setup, the extracellular conductivity around the signal-generating neurons will not be constant as assumed for the in vivo recordings, and this will affect the amplitude and shape of the recorded spikes.

In general, FEM modeling is required to solve the forward modeling for situations such as this where the conductivity σ varies with position. However, if we assume that the MEA substrate, slice, and saline all extend infinitely in the lateral directions so that σ can be assumed to only have planar step-wise discontinuities, formulas analogous to equation (4.6) can be derived by use of the *method of images* from electrostatics (see Section 5.7.1).

An example result is shown in Figure 7.12, where we have compared the cases with the following:

- **Slice in aCSF:** slice embedded between a glass electrode and artificial cerebrospinal fluid (aCSF)
- **Semi-infinite:** slice on glass electrode but with no aCSF (same tissue conductivity everywhere above electrode)
- **Infinite:** infinite homogeneous slice (same tissue conductivity everywhere)

The largest effect on the spike from the measurement device comes from the insulating glass substrate, which roughly doubles the amplitude of the recorded spikes. The effect of the saline is smaller and goes in the opposite direction: it reduces the size of the spike compared to the hypothetical semi-infinite situation where the saline had the same conductivity as the (in reality, less-conductive) brain slice. In Figure 7.12, the neuron is placed closer to the MEA grid than the saline cover. We note that the effects from the saline layer would be larger if the neuron instead were positioned towards the top of slice – that is, close to the saline layer, see Ness et al. (2015, figure 11C).

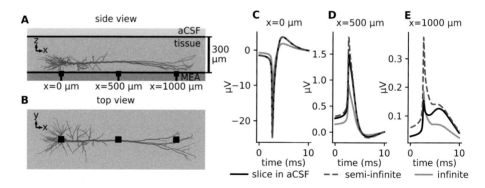

Figure 7.12 Spikes in micro-electrode arrays (MEAs). A: Side-view illustration of the MEA recording where the bottom of the device is covered by a grid of electrode contacts (three contacts in this example) recording signals generated by a Hay model neuron embedded in a brain slice. The brain slice is covered with artificial cerebrospinal fluid (aCSF) with a higher conductivity than the underlying brain slice. **B:** Top view of the measurement in A. **C:** Spike at the electrode contact under the soma ($x = 0\,\mu m$) computed by use of the method of images (MoI) (equation (5.33)) for the cases with (i) a slice with thickness $300\,\mu m$ and $\sigma_t = 0.3\,S/m$ covered by aCSF with $\sigma = 1.5\,S/m$, (ii) a semi-infinite slice above the MEA grid (that is, effectively assuming that σ for the aCSF is the same as for the slice), and (iii) an infinite slice (that is, effectively assuming that σ for the whole MEA device and the aCSF is the same as for the slice). **D, E:** Same as C for electrode contacts positioned at $x = 500\,\mu m$ and $x = 1000\,\mu m$, respectively. Figure is adapted from Ness et al. (2015). Code available via www.cambridge.org/electricbrainsignals.

7.8   Spikes from Many Neurons

In general, a recording contact will pick up spikes from several neurons positioned in its vicinity. If these neurons fire APs roughly simultaneously, their spike waveforms will be superimposed. With a collection of neurons where each neuron i fires a single AP at time t_i, the total extracellular potential is (in the noise-free situation) given by

$$V_e(\mathbf{r},t) = \sum_i V_{ei}(\mathbf{r} - \mathbf{r}_i, t - t_i) . \tag{7.11}$$

Here, $V_{ei}(\mathbf{r} - \mathbf{r}_i, t - t_i)$ is the spike waveform generated by neuron i positioned at $\mathbf{r}_i$.

7.8.1   Synchronous Spikes

If the spikes from the neurons are highly *synchronous*, the individual spike contributions may overlap in time and sum to a broader signal. If so, the extracellular potential of the net signal will contain components at lower frequencies than for the individual spikes. To illustrate this effect, we consider a model where we for simplicity assume that the spike waveforms are identical for all neurons so that $V_{ei}(\mathbf{r},t) = V_{spike}(\mathbf{r},t)$. Then, the sum in equation (7.11) can be written as a convolution (Appendix F.4)

$$V_{\text{e}}(\mathbf{r}, t) = \int_{-\infty}^{\infty} \nu(t - \tau) V_{\text{spike}}(\mathbf{r}, \tau) \, d\tau \, , \tag{7.12}$$

where the spike-density function is given by

$$\nu(t) = \sum_i \delta(t - t_i) \, , \tag{7.13}$$

and $\delta(t)$ is the Dirac delta function.

The advantage of this reformulation is that a convolution in real space corresponds to a multiplication in Fourier space so that

$$\hat{V}_{\text{e}}(\mathbf{r}, f) = \hat{\nu}(f) \hat{V}_{\text{spike}}(\mathbf{r}, f) \, . \tag{7.14}$$

Thus, the frequency content of the extracellular potential will be given as the product of the frequency contents (that is, the Fourier transforms) of the spike density $\hat{\nu}(f)$ and the single-spike waveform $\hat{V}_{\text{spike}}(\mathbf{r}, f)$.

To proceed, we assume a large number of spikes so that we can approximate the spike-density function as a continuous function that is more aptly referred to as a *firing-rate function*. Further, if we assume that the firing-rate function is described with a Gaussian density function

$$\nu(t) = \nu_0 \, e^{-t^2/2\sigma_\nu^2} \, , \tag{7.15}$$

with amplitude ν_0 and spread (standard deviation) σ_ν, the Fourier transform is also a Gaussian (Table F.1):

$$\hat{\nu}(f) \propto e^{-2\pi^2 \sigma_\nu^2 f^2} \, . \tag{7.16}$$

As seen from the two last expressions, the spread of the firing-rate functions in time versus frequency space are inversely related (since σ_ν is in the denominator versus the numerator of the exponents).

The effects of synchronous spiking are illustrated in Figure 7.13. While the power spectral density (PSD) of the single example spike in panel A has its maximum around 150 Hz (panel B), a collection of nearly synchronous spikes may have its maximal frequency contribution at lower frequencies. With spike spreads of $\sigma_\nu = 5$ ms (panels C, D) and $\sigma_\nu = 3$ ms (panels E, F), the maximal frequency contribution instead comes from frequencies less than 10 Hz. For a spread of $\sigma_\nu = 1$ ms, the resulting signal looks essentially as a broad version of a single-neuron spike (panel G), and the largest difference in the shape of the PSD (panel H) from the single-spike counterpart (panel B) is the stronger attenuation at the high-frequency tail. With completely synchronous spikes so that $\sigma_\nu = 0$, one is back to the single-spike situation except that the overall amplitude of the signal is much larger (panels I, J).

The convolution-based formula in equation (7.14) is observed to excellently account for the PSD (grey dashed lines in Figure 7.13). Only some parts of the PSD at the highest frequencies ($\gtrsim 100$ Hz) are not accounted for. In this frequency range, noise stemming from having a finite number of spikes included in the numerical simulation dominates the PSD. This effect of finite sampling is not accounted for in the Gaussian firing-rate function in equation (7.16).

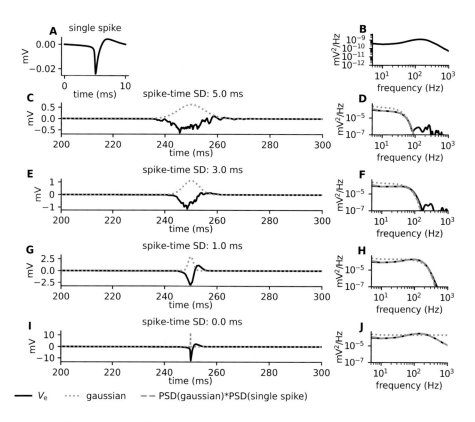

Figure 7.13 **Boosting of signal by synchronous spikes.** A: Spike from single neuron used in present figure – that is, $V_{\text{spike}}(t)$ in text. B: Power spectral density (PSD) of spike in A. C: Extracellular potential (black) from the collection of spiking neurons (equation (7.11)) where the spike-time distribution is Gaussian with a spread of $\sigma_V = 5$ ms (grey). D: PSD of extracellular potential in C (black) and of the PSD of the Gaussian distribution function (grey dotted). The product of the PSDs of a single spike (panel B) and the Gaussian distribution function is shown as grey dashed. E: Same as C with $\sigma_V = 3$ ms. F: Same as D with $\sigma_V = 3$ ms. G: Same as C with $\sigma_V = 1$ ms. H: Same as D with $\sigma_V = 1$ ms. I: Same as C with $\sigma_V = 0$. J: Same as D with $\sigma_V = 0$. In panels C–J, the total number of spikes is in all cases 500. Code available via www.cambridge.org/electricbrainsignals.

The boosting of lower frequencies can be understood from equation (7.14) and the expression for the Fourier-transformed spike density in equation (7.16). The PSD of the total spike signal is given by the product of the PSD of the single spike in Figure 7.13B and the PSD of the firing-rate function – that is, the square of the Fourier amplitude in equation (7.16). As this firing rate PSD is largest for the smallest frequencies, the maximum of the product with the single-spike PSD will shift to a lower frequency compared to the PSD for a single spike.

Schomburg et al. (2012) explored this phenomenon of "bleed-over" to frequencies as low as 100 Hz for synchronized spikes in the context of ripple-wave generation in the hippocampus. We discuss such bleed-over effects further in Section 8.5.2.

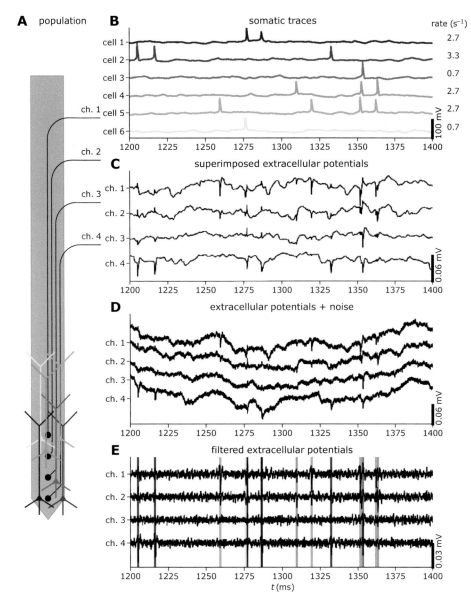

Figure 7.14 Modeling of spikes from multiple neurons recorded with a tetrode. A: Illustration of tetrode geometry with its four contacts surrounded by six cortical pyramidal neuron models driven to action-potential firing by a combination of excitatory and inhibitory synapse activations. **B**: Somatic voltages used to assess ground-truth times for action-potential firing. **C**: Extracellular potential generated by the population of six cortical pyramidal neurons. **D**: Extracellular potentials in panel C with added noise with the same statistical properties as what is seen in real tetrode experiments. **E**: High-pass filtered version of data in panel D, mimicking the typical filtering done on experimental data to produce multi-unit activity (MUA) data. The resulting data in panel E can be used as benchmarking data for the evaluation of spike-sorting methods applied on tetrode recordings as the ground-truth action-potential firing times are known (colored vertical lines in panel E). Figure made based on data from Hagen et al. (2015).

7.8.2   Benchmarking Data for Spike Sorting

Spikes are routinely recorded with multielectrodes – for example, tetrodes with four closely positioned recording contacts (Figure 7.14), linear multielectrodes (polytrodes) with tens of contacts positioned along a straight line, multi-shank polytrodes, or spade-like multielectrodes with many hundreds of tiny contacts arranged in rectangular patterns on an electrode shaft (Jun et al. 2017). On these multielectrodes, the same spike will in general show up on several contacts, and a process known as spike sorting (Quiroga 2007) is required to (i) properly count spikes and (ii) to sort recorded spikes into contributions from individual neurons, as is often the goal.

When developing and validating methods for spike sorting, it is useful to have bench-marking data where the "ground truth" is known. Experimental benchmarking data is hard to come by as they require simultaneous recording of APs and corresponding spikes. Model-based benchmarking data is therefore an attractive alternative (Einevoll et al. 2012), and the generation of such data has been pursued in several projects (Thor-bergsson et al. 2012, Camuñas-Mesa & Quiroga 2013, Hagen et al. 2015, Mondragón-González & Burguière 2017, Buccino & Einevoll 2021). Figure 7.14 shows an example simulation for the generation of model-based benchmarking data.

7.8.3   Population Firing-Rate Estimation from MUA

The high-frequency part of recorded extracellular potentials is often referred to as the *multi-unit activity (MUA)*. As the word "multi" indicates, it is not always possible to reliably identify individual spikes from this signal, partly because individual spikes may be too small to rise above the background noise and partly because the spikes are masked by temporally overlapping spikes from several neurons. The MUA is nevertheless pre-dominantly thought to reflect spiking activity.

The number of spikes that can be individually identified from the MUA depends on several factors. One is the volume density and morphological shapes of active neurons around the contact, a second is the ambient noise level, and a third is the impedance and size of the contact itself. As discussed in Section 7.7.1, large electrode contacts will tend to reduce the spike amplitude through a spatial averaging effect, making it difficult to identify individual spikes from the recorded signal.

In situations where individual spikes cannot be identified, the MUA signal can be used as an alternative to provide useful insight into the combined firing rate of the neurons surrounding the contact (Schroeder et al. 1998, Schroeder et al. 2001, Ulbert et al. 2001). To extract such information, the high-pass filtered extracellular potential recorded by the electrode is rectified, meaning that its absolute value is taken. While this rectified signal cannot give an estimate of the magnitude of the ambient firing activity, a temporally smoothed (low-pass filtered) version of it may provide an estimate of how the firing varies over time in relative terms.

The basic assumption, namely that the rectified MUA signal increases with increasing nearby AP activity, is intuitively justified if only a single neuron contributes with spikes. From the rectified signal, it will then be easy to tell whether a spike has occurred at a

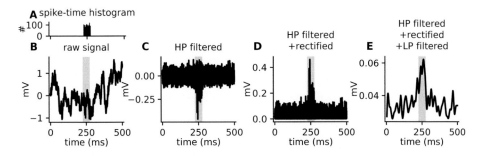

Figure 7.15 Illustration of the MUA rectification approach for the estimation of population firing rates. **A**: Spike-time histogram. **B**: Corresponding raw extracellular potential signal, containing artificially generated noise. To mimic the signal that would be recorded by an electrode surrounded by a population of Hay neurons, the spike waveform for each spike-time is randomly chosen from the Hay-neuron spike waveforms in Figure 7.4A. **C**: High-pass filtered (> 300 Hz) version of the raw signal in panel B. **D**: Rectified version of the high-pass filtered signal in panel C. **E**: Low-pass filtered version of the rectified and high-pass filtered signal in panel D. Code available via www.cambridge.org/electricbrainsignals.

specific instance in time. If this rectified signal then is low-pass filtered, one can get a crude estimate of how the single-neuron firing rate varies over time (see Dayan & Abbott (2001, chapter 1)).

With several neurons contributing with a distribution of spikes of different sizes, the rectified MUA signal will still be expected to be approximately proportional to the number of spikes, as long as the firing is sparse so that there is little overlap in time between the spikes of the contributing neurons. An estimate of the time course of the population firing rate, though not its absolute value, may then be found by low-pass filtering the rectified net spike signal.

The principle behind this "rectification" approach is illustrated in Figure 7.15. Panel B shows the raw extracellular potential from a population of neurons spiking in a narrow time window (panel A). The high-pass filtered signal – that is, the MUA (panel C) – is rectified (panel D) and low-pass filtered (panel E) to give an estimate of the time course of the population firing rate.

When there is a substantial temporal overlap between the spikes, there will be cancellation effects from negative "sodium signals" in one spike overlapping the positive "potassium signal" in another spike. The rectified population signal when several neurons contribute with spikes will then be smaller than the sum of the individual rectified spikes, resulting in an underestimated firing rate.

The rectification approach has been used to estimate the firing rates of cortical populations of neurons based on multielectrode laminar recordings (Schroeder et al. 1998, Schroeder et al. 2001, Ulbert et al. 2001, Einevoll et al. 2007, Blomquist et al. 2009). Its validity was tested in a model study using a simulation that included about a thousand layer 5 cortical pyramidal neurons with somas arranged in a cylindrical disk, mimicking a neuronal population in a rat barrel cortex (Pettersen et al. 2008). The neurons received synaptic inputs resembling those seen experimentally following whisker flicks, and extracellular potentials were computed for a set of contact positions along the central axis of the cylindrical population. The MUA from single trials – that is, single whisker flicks – using a lower cut-off of 750 Hz in the high-pass filtering was found to be too noisy to allow for the estimation of population firing rates. However, accurate estimates for trial-averaged firing rates (commonly referred to as *post-stimulus time histograms* (Gerstein 1960)) could generally be obtained from trial-averaged MUA signals.

A finding in the study was that the MUA-based analysis predicted a too-low maximum firing rate, due to cancellations of signal contributions from temporally overlapping spikes from different neurons. This effect increased with increasing firing rates and could be remedied by taking into account that the MUA signal grows sublinearly with the population firing rate. Further, the accuracy of the population firing-rate estimate could also be improved by using MUA signals from several adjacent recording channels in the estimation. The accuracy of this MUA-based estimation of population firing rates will depend on the specifics of the situation considered. In Pettersen et al. (2008), for example, it was found that the prediction was slightly more accurate when the pyramidal-cell population was driven by excitatory inputs onto the basal dendrites rather than onto the apical dendrites.

8   Local Field Potentials (LFPs)

After the advent of electrical recordings inside the brain, the focus was for a long time on the high-frequency content of the recorded potential since it contained spikes and thus direct information about neural action-potential firing (Chapter 7). Despite containing most of the signal power, the low-frequency part originally received less attention, and its relationship to the underlying neural acivity has taken more time to establish. This part of the signal has been termed the *local field potential* (LFP). The term "local" was presumably chosen because the recording is done close to the neurons generating the signal, at least compared to the EEG signal. The word "field" is somewhat unnecessary and may be confusing as *all* electric potentials are associated with an electric field, and nobody uses the term "membrane field potential" for the membrane potential. However, the term stuck and will also be used here.[1]

While spikes reflect the output of neurons, it was realized that the LFP largely reflects synaptic inputs to neurons. An important breakthrough in extracting useful information from the LFP was the development of current-source density (CSD) analysis (Pitts 1952). This analysis technique allowed for estimation of the depth distribution of membrane currents from LFP recordings spanning layered structures such as the cortex and hippocampus (Nicholson & Freeman 1975, Mitzdorf 1985, Pettersen et al. 2006). The LFP signal was also instrumental in the discovery of long-term potentiation (LTP) in the late 1960s (Bliss & Lømo 1973).

With the development and use of multicontact electrodes for high-density recordings across areas and laminae (Normann et al. 1999, Jun et al. 2017), the interest in LFPs has increased. A host of mathematical techniques for the modeling and analysis of LFPs have been developed and refined, including the biophysical forward-modeling scheme described in this book. It is now recognized that the LFP reflects key integrative synaptic processes that in practice cannot be captured by measuring spiking activity (Buzsáki et al. 2012, Einevoll, Kayser, et al. 2013, Torres et al. 2019).

LFPs have been used to investigate network mechanisms involved in sensory processing, motor planning, and higher cognitive processes including attention, memory, and perception (Buzsáki et al. 2012, Einevoll, Kayser, et al. 2013, Pesaran et al. 2018). The LFP also has potential use in brain-computer interfaces (BCIs) (Andersen et al. 2004, Saha et al. 2021) and for the monitoring of neural activity in humans (Mukamel

---

[1] The label LFP also works as an acronym for "low-frequency potential" (Głąbska et al. 2017), which arguably would be more suitable.

& Fried 2012) because the signal is thought to be more robust and stable than spikes in chronic settings.

Finally, the LFP has an important role in the clinic – for example, for optimizing the location of electrodes used for deep brain stimulation (Telkes et al. 2016). LFPs have also been demonstrated to contain clinically relevant information about the underlying pathology in patients with Parkinson's disease (Lempka & McIntyre 2013, Telkes et al. 2018, Cagnan et al. 2019, Eisinger et al. 2019, Bregman 2021, Vissani et al. 2021).

8.1   Neural Sources of LFPs

In Chapter 7, we focused on spikes, the extracellular signature of action potentials. Spikes are the temporally briefest electric events of interest in extracellular recordings, and their contributions can be isolated from the electrical signals stemming from other neural processes by applying a temporal high-pass filter to the signals.

Rather than defining the LFP as the signal stemming from a particular neural process or set of neural processes, we here take an operational stance and define the LFP as the low-pass filtered version of the extracellular potential. The exact cut-off frequency is a matter of choice and has varied widely, at least between 100 Hz (Perelman & Ginosar 2006) and 500 Hz (Ulbert et al. 2001, Einevoll et al. 2007). In this book, we use 300 Hz.

Several neural processes may contribute to the LFP (Figure 1.2B):

- synaptic currents (and associated return currents)
- "bleed-over" of regular spikes (sodium spikes) into lower frequencies
- slower active currents (e.g. calcium spikes or NMDA spikes)
- contributions from membrane currents in glial cells
- stationary (or approximately so) current loops through neurons, glial cells, and extracellular space
- diffusion potentials stemming from concentration differences of ions in the extracellular space

The synaptic currents are, under most conditions, thought to be the main contributor to the LFP, and the resulting *synaptic LFP* is what we will focus the most on in this chapter. Synaptic effects on LFPs are presented in Sections 8.2–8.4, while other possible contributions are described in Sections 8.5–8.6.

An example of the extraction of the LFP by the low-pass filtering of the raw extracellular signals is shown in Figure 8.1. We have there, again, computed extracellular signals by combining multicompartment (MC) models of neurons with volume-conductor (VC) theory (using the MC+VC scheme defined in Section 2.6.2). Also, we have again used the biophysically detailed Hay model (described in Section 3.2.6) for the neurons. In the Figure 8.1, a population of Hay model neurons (panel A) receives excitatory synaptic inputs that vary over time (panel B). The computed raw extracellular potential at eight equidistant positions at the center axis of the population (panel C) shows sharp spikes superimposed on the more slowly varying signal. These spikes, seen for basal synaptic

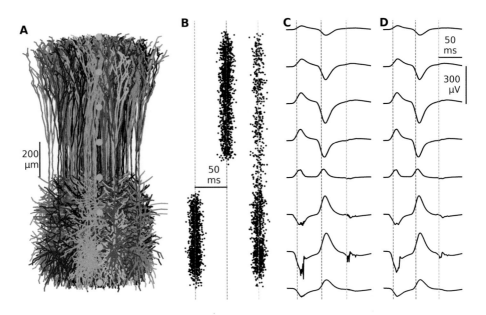

Figure 8.1 Extraction of LFP signal from extracellular potentials. A: Example population with 1,000 Hay model neurons receiving excitatory synaptic inputs. **B**: Depiction of the spatiotemporal distribution of synaptic inputs with three distinct epochs: basal inputs, apical inputs, and uniform inputs. **C**: Extracellular potential computed by MC+VC scheme at positions of electrode contact shown in A. **D**: LFP signal obtained by the low-pass filtering of potentials in C. In this example, the synaptic weights of 5 percent of the neurons have been tuned to make the neurons fire action potentials; the remaining 95 percent do not. Note that the neuronal population in this example lacks recurrent connections so all synaptic inputs are from external sources. Code available via www.cambridge.org/electricbrainsignals.

inputs (first epoch with inputs) and homogeneous synaptic inputs (third epoch with inputs), reflect action-potential firing and are essentially removed by the low-pass filter producing the LFP signal (panel D).

8.2 LFP from Single Postsynaptic Neuron

While spikes from individual neurons are relatively easily extracted from in vivo recordings, the LFP typically stems from thousands of neurons, where each may be receiving synaptic inputs from thousands of other neurons (Einevoll, Kayser, et al. 2013). However, many fundamental features of the LFP can still be illustrated by considering the LFP from a single postsynaptic neuron, which is the focus of the current section.

8.2.1 Single Synaptic Input

The synaptically generated LFP can (as a linear approximation) be thought of as being built up by a sum of numerous small individual LFP contributions, each stemming

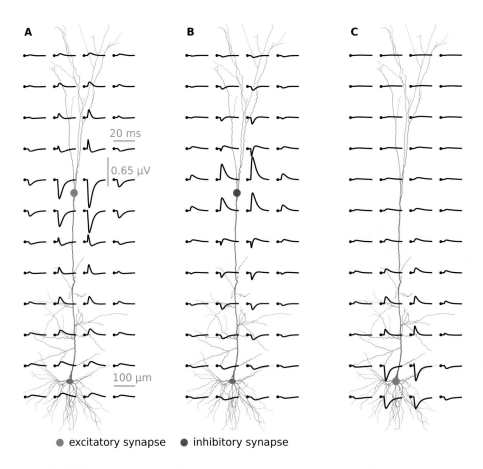

20 ms

0.65 µV

100 µm

● excitatory synapse   ● inhibitory synapse

Figure 8.2 LFP from single synaptic inputs onto the Hay neuron model. The black traces show the extracellular potential in various positions outside the neuron. The positions are marked as black dots at the start of the traces. **A**: Excitatory apical synapse (blue dot). **B**: Inhibitory apical synapse (red dot). **C**: Excitatory synapse on soma (blue dot). The synapse is conductance-based with maximum conductances set below the threshold for inducing action potentials. Code available via www.cambridge.org/electricbrainsignals.

from a single synaptic input onto a single neuron. Examples of such individual LFP contributions are given in Figure 8.2, showing the LFP generated by the Hay model neuron receiving single synaptic inputs.

The spatial pattern of the LFP surrounding the neuron depends on the location of the synaptic input. This pattern is best understood in terms of underlying current sources, or the current-source density (CSD, Section 4.3). When the neuron receives excitatory synaptic input, the compartment containing the synapse becomes a *current sink* (negative CSD), reflecting that the current goes inward from the extracellular space into the neuron. This current sink gives a very large negative contribution in the sum in equation $(4.8)^2$ that dominates the other (positive) contributions from nearby compartments so that the LFP sums to be negative near the excitatory synapse (Figure 8.2A and C). For inhibitory synaptic input, the situation is reversed so that the synapse compartment has a positive CSD and a positive nearby LFP (Figure 8.2B).

As explained in Chapters 2 and 3, current conservation is inherent in the cable equation, and as a consequence, the net membrane current summed across all neural compartments is at all times zero (unless the neuron receives current via intracellular-stimulating electrodes). The current sink at the synapse locations in Figure 8.2A and C must therefore be exactly balanced by the (return-) current sources in the rest of the neuron. Away from the synaptic input sites, these positive return-current sources dominate the sum in equation (4.8) and give rise to positive LFP deflections.

As the Figure 8.2 illustrates, the gross features of the LFP pattern obtained with apical excitatory synaptic input (panel A), with negative deflections near the main bifurcation of the apical dendrite and positive outside the soma, can be inverted in two ways: either by replacing the excitatory synapse with an inhibitory synapse at the same position (panel B) or by moving the excitatory synapse to the soma (panel C).

A common misconception when interpreting LFP signals has been that a negative deflection represents membrane depolarization while a positive deflection represents membrane hyperpolarization. While such an interpretation may be valid for LFPs close to the position of the synapse setting up the signal, it does not hold in general. In fact, the current source implied by the positive LFP around the soma in Figure 8.2A reflects a very different membrane potential than the current source implied by the positive LFP around the position of the inhibitory synapse in Figure 8.2B. In the former case, the membrane potential of the whole neuron, including the soma region, is depolarized by the apical excitatory input, while in the latter case, the whole neuron is hyperpolarized by the inhibitory synaptic input at the apical synapse.

When increasing the distance to the Hay model neuron, the LFP signature does not only become smaller, but also less sharp. This is illustrated through the normalized signal in Figure 8.3A. The same kind of blunting with distance can be captured using a simpler ball-and-stick model for the neuron (Figure 8.3B), although the effect is less pronounced in this case. Such distance-dependent blunting was also seen for the spikes generated by the same neuron models (Figure 7.5). In both of these cases, the biophysical origin of this blunting is the intrinsic dendritic filtering effect described in Section 3.3.3. This effect was described specifically for spikes in Section 7.3 and will be described in detail for the LFP signal in Section 8.4.1. The simpler two-compartment neuron model does not exhibit this blunting effect (Figure 8.3C).

In Figure 8.2, we observed that the LFP pattern stemming from a synaptic input changes completely when moving it from the apical dendrite to the soma. However, since the response to a synaptic input depends on the complex local branching structure

---

^{2} The sum in equation (4.8) is the point-source approximation. In the line-source approximation, it is replaced with the sum in equation (4.11). In the CSD formulation, it is replaced with the integral in equation (4.22).

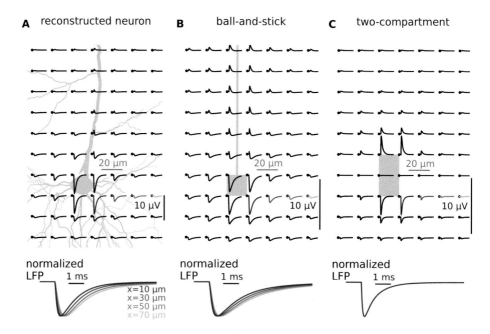

Figure 8.3 **Shape of the LFP signal is dependent on the morphology.** **A**: LFP around the Hay model neuron from a single excitatory synaptic input in the soma. The lower panel shows LFP signal shapes at different lateral positions from the soma, as indicated by the color coding. **B**: Same as panel A for a ball-and-stick neuron model. **C**: Same as panel C for a two-compartment neuron model. Code available via www.cambridge.org/electricbrainsignals.

of the dendrites, smaller shifts of the synaptic position can also give large changes in the resulting LFP pattern. This is illustrated in Figure 8.4A–C, where qualitatively different LFP patterns are found for slightly different positions of a single synapse on the basal dendrite of the same pyramidal cell. Moreover, while the LFP pattern for the synaptic position in Figure 8.4A has a clear dipolar structure, the same does not hold for the patterns seen in panels B and C.

8.2.2   Multiple Synaptic Inputs

While the neural response to a single synaptic input may be sensitive to its exact location, neurons generally have hundreds or thousands of synapses on their dendrites. In in vivo conditions, these neurons will in general receive numerous synaptic inputs at the same time. In this situation, there will expectedly be large cancellation effects so that the net LFP will be less sensitive to exact synapse locations but rather depend on the gross spatial pattern of activated synapses. This is illustrated in Figure 8.4D–F. The joint activation of numerous excitatory synapses on the basal dendrites gives a dipolar pattern, characteristic for a current sink in the basal region accompanied by a current source above (panel D). The joint activation of numerous apical excitatory synapses gives the

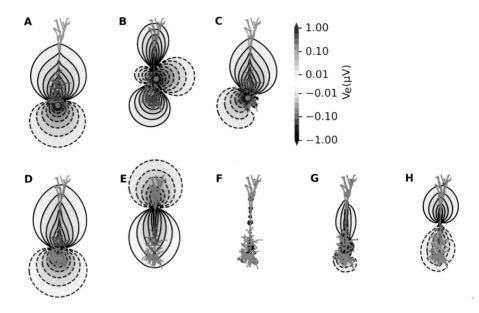

Figure 8.4 LFPs from synaptic inputs onto a pyramidal cell. Snapshots of the LFP at the time of the maximum LFP amplitude around the Hay neuron model for various excitatory synaptic inputs. **A–C**: Single synapse, randomly positioned on basal dendrites. **D**: Many synapses that are randomly positioned on basal dendrites with a uniform probability per membrane area. **E**: Same as panel D but for apical dendrites. **F**: Same as panels D and E but for the whole neuron. **G**: Uniform synaptic input distribution according to length density. **H**: Same as panel F except that the passive membrane conductance is scaled up by a factor of 60 in the apical dendrite. Synapses are conductance-based, and the conductance amplitudes are normalized so that the summed synaptic conductance is the same in all panels. Code available via www.cambridge.org/ electricbrainsignals.

inverted pattern, reflecting an apical current sink accompanied by a basal current source (panel E).

Interestingly, when both basal and apical synapses are activated, the contributions seen in panels D and E will largely cancel. In panel F, where the synaptic input distribution has a uniform density per membrane area, the cancellation is almost perfect, giving a vanishingly small LFP. In panel G, where the synaptic-input distribution instead has a uniform density per unit length of dendrite, the cancellation is less perfect and the LFP slightly larger.

Taken together, these results suggest that the observed LFP will be dominated by neurons receiving simultaneous synaptic inputs of the same type (excitatory or inhibitory) targeting specific sub-domains on the neurons. A characteristic example is pyramidal cells receiving numerous excitatory synaptic inputs on either only the basal or only the apical dendrites as depicted in panels D and E of Figure 8.4, respectively (Lindén et al. 2010, Lindén et al. 2011, Łęski et al. 2013, Hagen et al. 2016, Næss et al. 2021). These situations are commonly referred to as "open-field" configurations, contrasting the "closed-field" configuration in Figure 8.4F.

Note that cells with a more symmetric dendritic morphology, such as so-called stellate cells encountered in layer 4 in the sensory cortex as well as many interneurons, may in principle also give sizable contributions to the LFP if the collection of synaptic input currents is placed non-uniformly – say, only below the soma (Lindén et al. 2010). The small LFP contributions expected from other cells in an in vivo situation rather stem from the assumption that the synaptic inputs are uniformly distributed over the dendrites.

If all neurons in a population receive excitatory synaptic inputs on their distal dendrites, their net LFP contribution will still be small if the distal dendrites point randomly in all different directions. A final factor required for a population of neurons to make a large contribution to the LFP is thus that the dendrites of the neurons are geometrically aligned. This is, for example, the case for populations of pyramidal cells in layered brain structures such as the cortex or hippocampus. As illustrated in the context of pyramidal cell populations in Figure 8.5, we thus identify four different motifs likely dominating the generation of LFP in the cortex – that is, either excitatory or inhibitory synaptic input onto either the apical or basal dendrites. Uniformly distributed synapses will in contrast give small LFPs, even from neurons with elongated dendritic structures such as pyramidal neurons.

Note that non-uniformity of synaptic inputs is not the only type of non-uniformity that may provide large-amplitude LFPs. Non-uniformity of intrinsic electrical properties of the neurons may also contribute. An example is given in Figure 8.4H, where the neuron has the same set of synaptic inputs as in panel F. However, the passive membrane conductance is here (for illustrational purposes) scaled up by a factor 60 in the apical dendrites, making the resulting membrane-current distribution non-uniform, thus generating a sizable LFP signal. Similar effects can also be obtained with non-uniform dendritic distributions of active conductances (see Section 8.5.1).

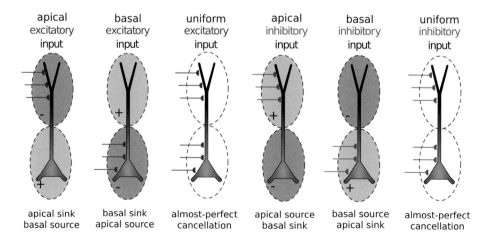

apical excitatory input	basal excitatory input	uniform excitatory input	apical inhibitory input	basal inhibitory input	uniform inhibitory input
apical sink basal source	basal sink apical source	almost-perfect cancellation	apical source basal sink	basal source apical sink	almost-perfect cancellation

Figure 8.5 LFP motifs. Illustration of characteristic qualitative LFP motifs around pyramidal neurons receiving different types of synaptic input. The LFP motifs depend critically on the type of synaptic input (excitatory or inhibitory) and the location (apical, basal, or uniform).

In addition to the placement of synapses, *temporal correlations* in the synaptic inputs also play an important role in determining the LFP. This is discussed in the context of a single neuron in Section 8.2.3 and for populations of neurons in Section 8.3.

8.2.3 Correlations in Synaptic Input

LFP signals are often analyzed in terms of their frequency content – that is, by exploring the properties of the LFP at each frequency separately. One way to do so is to examine the power spectral density (PSD) of the recorded signals (see Appendix F for its definition). Alternative ways that have been pursued have been to explore how tuning properties[3] and information content vary between different frequency bands of the LFP (Liu & Newsome 2006, Belitski et al. 2008, Berens et al. 2008, Mazzoni et al. 2011, Dubey & Ray 2016, Dubey & Ray 2020).

In addition to the spatial organization of synapses, the LFP generated by synaptic inputs also depends on the correlations in time between these inputs. Model simulations suggest that correlations in the synaptic input can increase the PSD of LFP signals by several orders of magnitude compared to the uncorrelated case (Lindén et al. 2011, Łęski et al. 2013). One source of such correlations is so-called *shared-input correlations*, stemming from different neurons receiving synaptic input from the same presynaptic neurons. Another source is *spike-train auto-correlations*, due to correlations within spike trains of individual presynaptic neurons.

The boosting effect from correlations can be illustrated with the toy example in Figure 8.6, where we consider the LFP measured immediately outside the soma of a neuron receiving synaptic inputs. For simplicity, we make the following assumptions:

- The contribution to the LFP signal from each synaptic input is given by a positive-valued exponentially decaying function (Figure 8.6A) resembling that seen near an inhibitory synapse.
- The total LFP signal from a train of incoming synaptic inputs can be found simply by adding the exponentially decaying LFP template every time a presynaptic spike[4] activates a synapse.
- There are several identical synapses on the soma activated by spikes from different presynaptic neurons. The corresponding spike trains can then be tailored to be "uncorrelated" so that each spike train is random and independent from the others or, at the other extreme, completely "correlated" so that all spike trains are identical (Figure 8.6).
- The compound LFP is built up from N_s incoming spike trains,

$$V_e(t) = \sum_{j=1}^{N_s} V_{ej}(t), \tag{8.1}$$

---

[3] The LFP can, like a neuron, be said to be "tuned" in the sense that it responds more strongly to certain "preferred" stimuli.

[4] Unlike in Chapter 7, the term "spike" here refers to a presynaptic event activating the synapse, not the extracellular signature of an action potential.

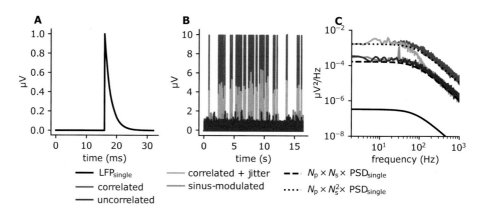

A

μV

1.0
0.8
0.6
0.4
0.2
0.0

0   10   20   30
time (ms)

— LFP$_{single}$
— correlated
— uncorrelated

B

μV

10
8
6
4
2
0

0   5   10   15
time (s)

— correlated + jitter
— sinus-modulated

C

μV2/Hz

10^{-2}
10^{-4}
10^{-6}
10^{-8}

10^1 10^2 10^3
frequency (Hz)

– – $N_p \times N_s \times$ PSD$_{single}$
···· $N_p \times N_s^2 \times$ PSD$_{single}$

Figure 8.6 Toy model illustrating how correlations in synaptic inputs boost LFP. A: LFP contribution, modeled as an exponentially decaying function, from a single synaptic input arriving at $t = 16$ ms. **B:** Compound LFP from $N_s = 10$ spike trains, each containing $N_p = 50$ Poisson-distributed spikes over the 16-second simulation time. Four cases with different correlation cases are considered: uncorrelated ($c_{Ve} = 0$, blue), fully correlated spikes – that is, all $N_s = 10$ spike trains are identical – ($c_{Ve} = 1$, red), fully correlated but with random jitter added to each spike (orange), and uncorrelated but with sinusoidally modulated spike trains (green). **C:** Power spectral densities (PSDs) of LFP signals in panel B. Additionally, curves corresponding to theoretical predictions from combining equation (8.6) and equation (8.8) are shown – that is, $N_p N_s$PSD$_{single}(f)$ ($c_{Ve} = 0$, dashed curve) and $N_p N_s^2$PSD$_{single}(f)$ ($c_{Ve} = 1$, dotted curve). Here, PSD$_{single}(f)$ (solid curve) is the PSD of the signal in panel A. Code available via www.cambridge.org/electricbrainsignals.

where $V_{ej}(t)$ is the LFP due to spike train j. We consider the case where the mean value of the LFP signal from each single spike in spike train j has been pre-subtracted, so that

$$\langle V_{ej}(t) \rangle_t = 0 , \qquad (8.2)$$

where the notation $\langle \cdot \rangle_t$ means averaging over time (equation (E.1)). In this case,

$$\left\langle \sum_{j=1}^{N_s} V_{ej}(t) \right\rangle_t = \sum_{j=1}^{N_s} \langle V_{ej}(t) \rangle_t = 0 . \qquad (8.3)$$

- Each spike train consists of N_p spikes.
- The variance of the LFP contribution (equation (E.3)) from each spike train is the same for all j, so that

$$\langle V_{ej}(t)^2 \rangle_t = V_{es\sigma}^2 \qquad (8.4)$$

for all spike trains j. Here, $V_{es\sigma}$ denotes the standard deviation of the single spike-train LFP, so $V_{es\sigma}^2$ corresponds to the variance.

- The correlation between the LFPs generated by two spike trains j and k is the same for all pairs of spike trains. Then, the (*Pearson*) correlation coefficient (equation (E.8)) for two spike trains are given by

$$c_{Vejk} = \frac{\langle V_{ej}(t)V_{ek}(t)\rangle_t}{\langle V_{ej}(t)^2\rangle_t} = \frac{\langle V_{ej}(t)V_{ek}(t)\rangle_t}{V_{eso}^2} \equiv c_{Ve} \,. \tag{8.5}$$

With this set of assumptions, the variance of the compound signal is given by

$$\begin{aligned}
V_{eo}^2 &= \langle (V_e(t) - \langle V_e(t)\rangle_t)^2\rangle_t = \langle V_e(t)^2\rangle_t \\
&= \sum_{j=1}^{N_s} \left\langle V_{ej}(t)^2\right\rangle_t + \sum_{j=1}^{N_s}\sum_{k\neq j}^{N_s} \langle V_{ej}(t)V_{ek}(t)\rangle_t \\
&= V_{eso}^2 \left(N_s + c_{Ve}N_s(N_s - 1)\right) \\
&= V_{eso}^2 \left((1 - c_{Ve})N_s + c_{Ve}N_s^2\right) \,. \tag{8.6}
\end{aligned}$$

This formula predicts that correlations between the incoming spike trains will result in a strong boost of the LFP. For the uncorrelated case where the correlation coefficient $c_{Ve} = 0$, the formula predicts $V_{eo}^2 = N_s V_{eso}^2$. For the fully correlated case where all N_s spike trains are identical and $c_{Ve} = 1$, the formula predicts $V_{eo}^2 = N_s^2 V_{eso}^2$. In the example in Figure 8.6 where there are ten incoming spike trains ($N_s = 10$), the variance of the compound LFP will thus be ten times larger for the fully correlated case than for the uncorrelated case.

To quantitatively compare this formula with results from numerical simulations, the so-called Parseval's theorem (Appendix F.3) can be used. This theorem relates the variance of a signal to the power spectral density (PSD) via

$$\int_{-\infty}^{\infty} \text{PSD}(f)\,df = V_{eo}^2 \,. \tag{8.7}$$

For the example in Figure 8.6, the corresponding PSDs are given in panel C. A first observation is that while the PSDs for the correlated (red curve) and uncorrelated (blue curve) cases have similar frequency-dependent shapes, the overall magnitude of the PSD is indeed much larger for the correlated case. The magnitude of the correlation boost of the PSD is predicted by the formula in equation (8.6).

The solid black curve in Figure 8.6C shows the PSD of the exponentially decaying LFP induced by a single spike (panel A). Since a Poissonian spike sequence has a flat PSD, the PSD of the resulting LFP will have the same shape as that generated by a single spike. This follows from the convolution theorem, stating that the total PSD will be a product of the PSD from a single-spike LFP and the PSD of the spike sequence (Appendix F.4). However, the magnitude of the PSD is increased by the number N_p of spikes in the spike train. Likewise, the variance V_{eso}^2, which corresponds to the integral of the PSD across all frequencies (equation (8.7)), is for a single spike train given by

$$V_{eso}^2 = N_p V_{epo}^2 \,, \tag{8.8}$$

where V_{epo} is the standard deviation of the LFP for a single spike. Equation (8.6) then implies that, for the case of uncorrelated spike trains, the variance of the compound LFP will be $V_{eo}^2 = N_s N_p V_{epo}^2$. With the simultaneous use of Parseval's theorem, this prediction is indeed confirmed in Figure 8.6C. Here, the PSD for the compound LFP for the uncorrelated case (dashed black line) is seen to be a factor $N_s \times N_p = 10 \times 50 = 500$

larger than for the single-spike PSD. For the case of fully correlated spike trains, the formula instead predicts the PSD to be a factor $N_s^2 \times N_p = 10^2 \times 50 = 5000$ larger than for the single spike PSD, or (again) ten times larger than for the uncorrelated case, and this is also confirmed (dotted black line).

The numerical simulations allow us to also consider cases not covered by the analytical formulas derived here. In Figure 8.6, we have included an intermediate case when the spike trains are correlated but where a small random delay has been introduced to the individual arrival times (panel B, orange). For this case, the PSD values are the same as for the fully correlated signal at low frequencies and the same for the uncorrelated signal at higher frequencies (panel C). The exact transition frequency depends on the magnitude of the introduced random delay.

We have also included a case where the spike trains are sinusoidally modulated but otherwise uncorrelated. A boost is then seen in the PSD for the modulation frequency (30 Hz) but not for other frequencies (Figure 8.6C, green). As we show in Section 8.4.3.1, these frequency dependencies of the PSD can be accounted for by frequency-resolved versions of equation (8.6).

8.2.4   Decay of LFP with Distance

To understand the origin of recorded LFPs, it is important to know how sharply the LFP contribution from a single neuron decays with distance from the neuron. This is difficult to estimate from experiments as the recorded LFPs generally stem from membrane currents from thousands of neurons (Einevoll, Kayser, et al. 2013). However, the question can be readily explored in models (Lindén et al. 2010, Lindén et al. 2011, Łęski et al. 2013).

A modeling example is given in Figure 8.7, showing the spatial LFP decay around the Hay model neuron. The neuron receives numerous uncorrelated synaptic inputs either spread uniformly across the dendrites or only across apical or basal dendrites (Figure 8.7A). A first observation is that the LFP in all cases is smallest for the depths furthest away from any neuronal dendrite ($z = -800$ µm). Where the LFP is largest depends on the distribution of synaptic inputs.

For all input distributions and depths, the LFP is seen to decay as $1/r^2$ with lateral distance far away from the neuron and less steeply closer to the neuron. This $1/r^2$ decay is predicted by the dipole approximation (Section 4.4). Far away from a neuron (i.e. in the far-field limit), the extracellular potential is proportional to the (current-) dipole moment[5] of the neuron and decays as $1/r^2$ (equation (4.50)).

When moving laterally out from the soma, the (near-field) LFP decay approximately follows a $1/r^{1/2}$ law (panel for $z = 0$ µm in Figure 8.7B). At this depth, the transition between this near-field regime ($1/r^{1/2}$ decay) and the far-field dipole regime $1/r^2$ occurs around 60 µm for the situation with uniform and basal synaptic inputs. This cutoff distance r_* is seen in panel A to roughly correspond to the lateral extent of the basal dendrites of the neurons in question. With apical synaptic inputs, a similar transition

---

[5] For brevity, we will refer to "current dipoles" simply as "dipoles." Charge dipoles will not be encountered in this book.

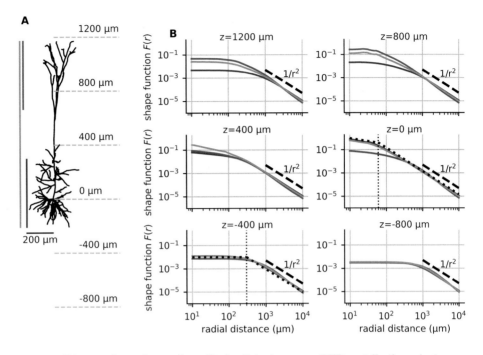

Figure 8.7 Distance-dependence of amplitude of single-neuron LFP contributions. A: A passive version of the Hay neuron (with active conductances removed), receiving uncorrelated synaptic inputs (Poisson sequence). Three spatial regions of synaptic inputs were considered: uniform (grey), apical (red), and basal (blue). **B**: Amplitude of the Hay neuron's LFP contributions as a function of lateral distance at different depths. The LFP is normalized against the value at $r = 10\,\mu\text{m}$ for $z = 0$. The dotted curve in the panel for $z = 0$ shows the fit to the soma-level shape function $F(r)$ in equation (8.9) with $r_* = 60\,\mu\text{m}$. The dotted curve in the panel for $z = -400\,\mu\text{m}$ shows the fit to the alternative shape function $F(r)$ in equation (8.10) for LFPs below the soma level with $r_* = 300\,\mu\text{m}$. The dashed lines indicate a $1/r^2$-dependence added for illustration purposes. The thin vertical dotted lines in the panels for $z = 0$ and $z = -400\,\mu\text{m}$ illustrate the cut-off distances: $r_* = 60\,\mu\text{m}$ and $r_* = 300\,\mu\text{m}$, respectively. The depicted LFPs were computed as the average of the LFP standard deviations over 60 equidistant angles around the neuron. Code available via www.cambridge.org/electricbrainsignals.

between a near-field and a far-field regime is seen to occur at a larger distance, and r_* is about 150 μm or so for this example.

A similar transition between a near-field regime with an (approximate) $1/r^{1/2}$ decay and a far-field regime with a $1/r^2$ decay is observed at a depth about halfway up the apical dendrite ($z = 400\,\mu\text{m}$). However, both below ($z = 800\,\mu\text{m}$, $z = -400\,\mu\text{m}$) and above ($z = 1200\,\mu\text{m}$) the vertical extent of the neuronal dendrites, the near-field decay is almost flat for the smaller lateral distances. This can be understood on geometrical grounds: with a large vertical distance, the absolute distance to the neuron changes less with small changes in the lateral position r, and the LFP may thus be expected to be relatively insensitive to the value of r. Moreover, this geometrical effect also explains why the cut-off distance r_* increases with vertical distance. For $z = -400\,\mu\text{m}$, it is about 300 μm (vertical dotted line in Figure 8.7B), while for $z = -800$ μm, it is about 1000 μm.

The cut-off distance r_* depends on the frequency content of the extracellular potential and is shorter for high-frequency signals than for low-frequency signals (see Figure 8.17). For this reason, the LFP does not decay as steeply with distance as the spikes that we considered in the previous chapter. For the Hay model, the spike amplitude is reduced by about a factor of one hundred when increasing the lateral distance from the soma from 10 μm to 100 μm (Figure 7.6). In comparison, the LFP is reduced by about a factor of ten over the same distance (Figure 8.7B, $z = 0$ μm).

Before moving on, we give a comment on the $1/r^2$-decay indicated in all the panels of Figure 8.7. As we showed analytically in Section 4.4.1, we will always get a $1/r^2$ decay in the far-field limit, except in the special case where we are moving along an axis that is (i) perfectly perpendicular to the dipole orientation and (ii) not radial to the dipole center. In this perfectly perpendicular case, a lateral decay of $1/r^3$ is instead predicted in the far-field limit. Although we in Figure 8.7 are moving laterally away from a somewhat vertically oriented Hay model, we do not see indications of such a $1/r^3$-decay, and the reason is that the dipole moment of the Hay model is not perfectly vertical but contains a lateral component stemming from currents in the many dendritic branches. In the far-field limit, the lateral component will always dominate, resulting in a $1/r^2$-decay. The perfectly perpendicular case giving a $1/r^3$-decay could, for example, be realized if we replaced the Hay model in Figure 8.7 with a vertically oriented ball-and-stick model.

8.2.5   Single-Neuron Shape Function

Later, we will consider the more complex case of LFPs from populations of neurons. It will then be convenient to boil the insights from the study in Section 8.2.4 down to approximate functional descriptions for the spatial decay of single-neuron LFP contributions. For the LFP at the same vertical depth as the soma, we thus assume that this *shape function* $F(r)$ is given by (Łęski et al. 2013, Einevoll, Lindén et al. 2013)

$$F(r) = \begin{cases} F_0 & r < r_\varepsilon \\ F_0 \sqrt{r_\varepsilon/r} & r_\varepsilon \leq r < r_* \\ F_0 \sqrt{r_\varepsilon/r_*} \, (r_*/r)^2 & r \geq r_* , \end{cases} \tag{8.9}$$

where r_* is the cut-off distance between the near-field and the far-field dipole regime, and r_ε is a lower cut-off distance introduced to avoid an unphysical divergence for $r \to 0$ (Figure 8.8A). The distance r_ε may, for example, correspond to the size of the neuronal soma, but we leave it unspecified for now. The distance r_* depends on the synaptic distribution (Figure 8.7B), and for the LFP at the soma level ($z = 0$), the value is between 60 and 200 μm.

For the LFP above or below the vertical extent of the dendrites, we correspondingly assume

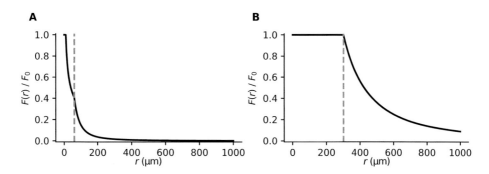

Figure 8.8 Single-neuron shape functions. A: Shape functions $F(r)$ for LFPs at the same depth as the neuronal soma as a function of lateral distance r (equation (8.9)). **B**: $F(r)$ for LFPs at depths below or above the neuron as a function of lateral distance r (equation (8.10)). In A, the parameters used are $r_* = 60\,\mu$m and $r_\varepsilon = 10\,\mu$m, mimicking the situation for $z = 0$ in Figure 8.7B. In this panel, $r_* = 300\,\mu$m is mimicking the situation for $z = -400\,\mu$m in Figure 8.7B. Vertical lines mark r_*. Code available via www.cambridge.org/electricbrainsignals.

$$F(r) = \begin{cases} F_0 & r < r_* \\ F_0\,(r_*/r)^2 & r \geq r_* , \end{cases} \tag{8.10}$$

which can be obtained by setting r_ε equal to r_* in equation (8.9) (Figure 8.8B). As we saw in Figure 8.7, both the values of r_* and F_0 depend strongly on the vertical position. For further discussion on the applicability of equation (8.9) and equation (8.10), see Einevoll, Lindén et al. (2013).

8.3 LFP from Neural Populations

In neuroscience, the term "population" commonly refers to groups of neurons with similar properties, and in the present section, we will investigate LFPs generated by populations of neurons with similar morphological and electrical properties. The compound LFP $V_e(\mathbf{r})$ from a population of N neurons can be found by summing the contributions from the individual neurons:

$$V_e(\mathbf{r}) = \sum_{n}^{N} V_{en}(\mathbf{r}) . \tag{8.11}$$

Here, $V_{en}(\mathbf{r})$ is the contribution from neuron n to the LFP at position $\mathbf{r}$.

To establish a rough understanding of the properties of population LFPs, it helps to consider a simplified scenario where all neurons in a population are (i) identical and (ii) not recurrently connected. The first assumption allows us to use the same model for all neurons in the population. The second assumption means that the LFP generation is driven exclusively by external input, since possible neural APs resulting from it will not lead to additional synaptic activations. This makes the analysis of the generation of LFP signals more transparent (Lindén et al. 2011, Łęski et al. 2013, Hagen et al. 2017).

We saw an example of such a simulation in Figure 8.1, showing the LFPs generated at the center of a population of 1,000 identical Hay model neurons receiving different spatial patterns of external synaptic inputs over time. Since the LFP was computed using equation (8.11) and since there were no recurrent synaptic connections within the population, the population LFP reflects the sum of single-neuron LFPs receiving external synaptic inputs only. The results in Figure 8.1D illustrate how the key role of synapse positions in determining the LFP contributions from single neurons (Figure 8.4) transfers directly to the population LFP.

8.3.1   Spatial Reach of LFPs

The LFP stands for the "local field potential," and an obvious question is then "how local?" This is not a very precise question, and in the following, we will consider two alternative and better-quantified versions of it:

1. If an electrode is surrounded by a large population of neurons, what is the radius R_{reach} of the subpopulation of neurons around an electrode that actually contributes to the measured LFP? This quantity has been referred to as the *spatial reach* of the LFP (Lindén et al. 2011, Łęski et al. 2013).
2. How far does the LFP generated by a population with radius R extend outside the edge of the population? We will refer to this as the *spatial decay* of the LFP (Lindén et al. 2011, Łęski et al. 2013, Hagen et al. 2017).

Both these questions can be addressed by means of computational modeling. In the current subsection, we consider the first of them, while the second will be considered in Section 8.3.2.

Experimental estimates of the spatial reach have been conflicting, suggesting values spanning from a few hundred micrometers (Katzner et al. 2009, Xing et al. 2009) to several millimeters (Kreiman et al. 2006, Hunt et al. 2010) or more (Kajikawa & Schroeder 2011). It turns out that these conflicting experimental observations can be reconciled by considering mathematical population models, where a key insight is that the spatial reach depends strongly on how synchronous the synaptic inputs driving the LFP-generating neurons are (Lindén et al. 2011, Łęski et al. 2013).

The spatial reach of the LFP can, as in Lindén et al. (2011), be examined by considering a population of identical LFP-generating neurons positioned on a disk around a recording electrode (Figure 8.9A). The neurons are synaptically activated, and the LFP can be calculated via equation (8.11). When the radius R of the population is increased, more neurons will contribute to the compound LFP, and the amplitude of the LFP is thus expected to increase. However, the contribution from the ever-more distant neurons will become small (Figure 8.7), and one might intuitively suspect that the LFP amplitude will converge to a fixed value in the limit of an infinite population size R. As we shall see, this turns out to not always be the case. In Sections 8.3.1.1 and 8.3.1.2, we first consider a qualitative model exploring the spatial reach of the LFP and next explore it quantitatively.

8.3.1.1   Qualitative Model

Three factors are a priori expected to be key for determining how the population LFP V_e increases with the population radius R for a disk of neurons as in Figure 8.9A: (i) how sharply the single-neuron LFP contribution decays with distance as given by the shape function $F(r)$ from Section 8.2.5, (ii) the number density $N(r)$ of neurons positioned in a ring of radius r around the electrode, and (iii) the level of correlations between the single-neuron contributions V_{en} that together generate the compound population LFP (Lindén et al. 2011, Einevoll, Lindén et al. 2013). The correlations in the LFP contributions are determined by correlations between the synaptic inputs onto the neurons in the population. As illustrated by the toy model in Section 8.2.3, correlations in inputs to a single neuron may dramatically boost the overall power of the LFP signal, and we expect similar effects of correlations for the population LFP.

Here, we are interested in whether the compound population LFP V_e converges to a finite value or increases without bounds for large values of R. To answer this question, only the form of the shape function $F(r)$ far away from the neuron is of interest, and the far-field dipole approximation $F(r) \propto 1/r^2$ can be used (Figure 8.9C). The number of neurons in each ring will be proportional to the circumference $2\pi r$ of the ring, and $N(r)$ will thus be proportional to r (Figure 8.9D). When it comes to the degree of correlation,

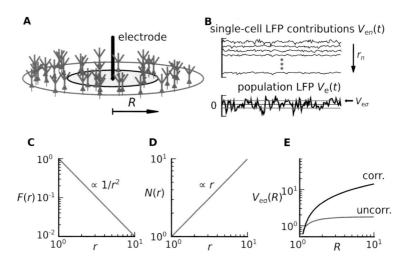

Figure 8.9   Illustration of modeling approach used to explore the question about the "locality" of the LFP. A: Sketch of the model setup where neurons are evenly distributed on a disk of radius R around the electrode tip. **B**: The population LFP $V_e(t)$ is given as a sum over contributions $V_{en}(t)$ from individual cells at distances r_n. **C**: Single-cell LFP shape function $F(r)$ assuming dipolar sources. **D**: Number $N(r)$ of cells on a ring of radius r. **E**: Qualitative illustration of dependence of the compound amplitude of the population LFP defined as the standard deviation of the signal $V_{e\sigma}$ (see also panel B) on population radius R for the cases where the single-neuron LFP contributions are either uncorrelated or fully correlated. The dependence of V_e on the population radius R defines the spatial reach R_{reach} of the electrode (see text). Redrawn based on similar figure in Lindén et al. (2011).

we investigate the two extreme situations where the contributing single-neuron LFPs are either perfectly uncorrelated or perfectly correlated.

Population LFPs from Uncorrelated Single-Neuron Contributions
If the neurons in the population independently receive synaptic inputs at completely random times, their contribution to the LFP will be uncorrelated. Many of the contributions will then cancel each other out so that the amplitude of the compound signal V_e will not be proportional to the number of individual LFP sources. Instead, the variance $V_{e\sigma}^2$ of the population LFP, which is the LFP measure considered here, will be a sum of the variances of the individual single-neuron LFP contributions. The variance is given as the square of the deviation of V_e from the mean value $\langle V_e \rangle$ averaged across time (equation (E.2)).

The proportionality of the variance with the number of neuron sources follows from the argument in Section 8.2.3. According to equation (8.6), the case of uncorrelated sources ($c_{Ve} = 0$) gives a variance proportional to the number of sources:

$$V_{eu\sigma}^2 = \sum_n V_{en\sigma}^2 = \sum_n F(r_n)^2 . \tag{8.12}$$

Here, $V_{en\sigma}^2$ represents the variance of the single-neuron LFP contribution, which is assumed to be proportional to the square of the shape function $F(r_n)$ from Section 8.2.4. The sum is taken over all neurons where $r_n = \|\mathbf{r}_n\|$ is smaller than the population radius R. The subscript "u" indicates that this formula applies to the case of uncorrelated single-neuron LFP contributions.

With many neurons contributing to the population LFP, we can approximate the sum in equation (8.12) with the integral

$$V_{eu\sigma}^2(R) \approx \int_0^R N(r)F(r)^2 \, dr . \tag{8.13}$$

Here, $N(r)dr$ is the number of neurons in a ring with thickness dr at a distance r from the electrode. With $N(r) \propto r$ and the use of the far-field approximation $F(r) \propto 1/r^2$, we can approximate this integral by

$$V_{eu\sigma}^2(R) \approx \int_{R_x}^R N(r) \left(\frac{1}{r^2}\right)^2 dr \propto \int_{R_x}^R \frac{1}{r^3} \, dr \propto \frac{1}{R_x^2} - \frac{1}{R^2} . \tag{8.14}$$

Here, an arbitrary lower integration bound R_x has, for convenience, been introduced to avoid the unphysical singularity that would appear by assuming the far-field relationship $F(r) \propto 1/r^2$ for distances r approaching zero.

The key insight from equation (8.14) is that the variance of the population LFP will converge to a fixed finite value when $R \to \infty$ (since $1/R^2 \to 0$ in this limit). Thus, the LFP can be said to have a finite spatial reach R_{reach}, which can be defined by the population radius for which the population LFP has reached a specific fraction of the infinite-population value. In Lindén et al. (2011), the criterion

$$V_{e\sigma}(R_{reach}) = \alpha \, V_{e\sigma}(R \to \infty) , \tag{8.15}$$

with $\alpha = 0.95$, was chosen to define R_{reach}, where $V_{\text{e}\sigma}(R)$ is the standard deviation of the population LFP.

Population LFPs from Correlated Single-Neuron Contributions

If all the neurons in the populations receive correlated spike trains, the situation becomes different. We here consider the idealized case where (i) all neurons in the populations are identical, (ii) all neurons have identically positioned synapses, and (iii) all incoming spike trains are identical (fully temporally correlated). In this case, which we refer to as the fully correlated case, the single-neuron contributions to the population LFP will overlap in time and sum up to a large population LFP.

In this fully correlated case, the variance of the population LFP is proportional to the square of the number of sources. This follows from inserting $c_{Ve} = 1$ into equation (8.6). We can approximate the resulting sum over fully correlated sources with the integral

$$V_{\text{ec}\sigma}(R)^2 \approx \left(\int_0^R N(r)F(r)\, dr \right)^2 , \tag{8.16}$$

or equivalently

$$V_{\text{ec}\sigma}(R) \approx \int_0^R N(r)F(r)\, dr , \tag{8.17}$$

where the subscript "c" has been added to denote fully correlated.

When inserting $N(r) \propto r$ and $F(r) \propto 1/r^2$ in this integral, we find that the LFP standard deviation from contributions from the neurons positioned outside a radial distance R_x from the population center is given by

$$V_{\text{ec}\sigma}(R) \propto \int_{R_x}^R r \frac{1}{r^2}\, dr = \int_{R_x}^R \frac{1}{r}\, dr = \ln \frac{R}{R_x} . \tag{8.18}$$

From equation (8.18), it follows that $V_{\text{ec}\sigma}(R) \to \infty$ when $R \to \infty$. In contrast to the uncorrelated case, the LFP for the correlated case increases without bounds as the population size R increases and can thus be said to have an "infinite" spatial reach.

8.3.1.2 Quantitative Model

If we expand this conceptual model with the single-neuron shape functions in equation (8.9) and equation (8.10), we obtain a more quantitative analytical model, which we present in this section. Its predictions for how single-neuron LFPs sum to population LFPs have been validated elsewhere against numerical simulations of populations of up to 10,000 neurons (Lindén et al. 2011, Łęski et al. 2013).

Population LFPs from Uncorrelated Single-Neuron Contributions

According to the conceptual model (equation (8.13)), the standard deviation $V_{\text{e}\sigma}$ of the population LFP from uncorrelated sources can be approximated by the integral

$$V_{\text{eu}\sigma}(R) = \left(\int_0^R N(r)F(r)^2\, dr \right)^{1/2} = \left(2\pi\rho \int_0^R r F(r)^2\, dr \right)^{1/2} . \tag{8.19}$$

Here, ρ is the area density of neurons with somas positioned on the disk.

Insertion of the soma-level shape function $F(r)$ from equation (8.9) into the integral in equation (8.19) gives, after some algebra, the following expression for $V_{\text{eu}\sigma}$ as a function of population radius R,

$$
V_{\text{eu}\sigma}(R) = \begin{cases}
F_0 \sqrt{\rho\,\pi}\,R & R < r_\varepsilon\,, \\
F_0 \sqrt{\rho\,\pi}\sqrt{r_\varepsilon(2R - r_\varepsilon)} & r_\varepsilon \le R < r_*\,, \\
F_0 \sqrt{\rho\,\pi}\sqrt{r_\varepsilon(3r_* - r_\varepsilon - r_*^3/R^2)} & R \ge r_*\,.
\end{cases}
\tag{8.20}
$$

The function is illustrated in Figure 8.10A, and a first observation is that $V_{\text{eu}\sigma}(R)$ approaches a finite value as $R \to \infty$. According to equation (8.20), this value is given by

$$
V_{\text{eu}\sigma}(R \to \infty) = F_0 \sqrt{\rho\,\pi}\sqrt{r_\varepsilon(3r_* - r_\varepsilon)}\,.
\tag{8.21}
$$

This formula demonstrates that the large-R magnitude of the population LFP not only increases with the prefactor F_0 of the single-neuron shape function, but also with the cut-off distance r_* describing how far away from the neuron the "slow-decay" regime where $F(r) \propto 1/\sqrt{r}$ extends. For the numerical example in Figure 8.10A, a value of the standard deviation of the compound LFP of about $50\,\mu\text{V}$ is found.

The analytical formulas in equation (8.20) further suggest a new definition of the spatial reach: R_{reach} can be set to be the distance from the population center at which the intermediate-R ($r_\varepsilon \le R < r_*$) and the infinite-$R$ (equation (8.21)) formulas intersect (Einevoll, Lindén et al. 2013) (see Figure 8.10A). For this particular value of R, we have $2R_{\text{reach}} - r_\varepsilon = 3r_* - r_\varepsilon$ – that is, $R_{\text{reach}} = 1.5r_*$. Thus, a simple rule of thumb is that, in the case of uncorrelated inputs, the spatial reach is 50 percent larger than the single-neuron cut-off value r_*.

Insertion of $R = R_{\text{reach}} = 1.5r_*$ into equation (8.20) results in

$$
V_{\text{eu}\sigma}(R_{\text{reach}}) = \sqrt{23/27}\, V_{\text{eu}\sigma}(R \to \infty) \approx 0.92 V_{\text{eu}\sigma}(R \to \infty)\,,
\tag{8.22}
$$

independently of the value of r_* (as long as $r_\varepsilon \ll r_*$ as in the example in Figure 8.10A). Thus, with uncorrelated neuronal sources, the model implies that 92 percent of the LFP recorded by an electrode in the center of a disk-like population stems from neurons closer than $R_{\text{reach}} = 1.5r_*$. For the numerical values used in Figure 8.10A where $r_* = 60\,\mu\text{m}$, this corresponds to $R_{\text{reach}} = 90\,\mu\text{m}$. This value of r_* was chosen by inspection of the single-neuron shape functions in Figure 8.7B computed for the Hay model neuron. Inspection of Figure 8.7A shows that $90\,\mu\text{m}$ is about the same as the lateral extent of the basal dendritic branches of this neuron.

We can now ask the question of how many neurons contribute to the LFP at the soma level. To get a rough idea, we can estimate the number of neurons with somas positioned within a sphere of radius R_{reach} around the electrode. With a neuron volume density of $\rho_{\text{v}} = 50{,}000$ neurons per mm^3, which is typical for the cortex (Beaulieu 1993), we find that this sphere will contain $N_{\text{neuron}} = 4\pi R_{\text{reach}}^3 \rho_{\text{v}}/3 = 150$ neurons.

This result pertains only to the LFP at the same depth level as the neuronal somas. However, the same reasoning and modeling can be used for positions above or below the vertical extent of the dendrites (Einevoll, Lindén et al. 2013). In this case, the

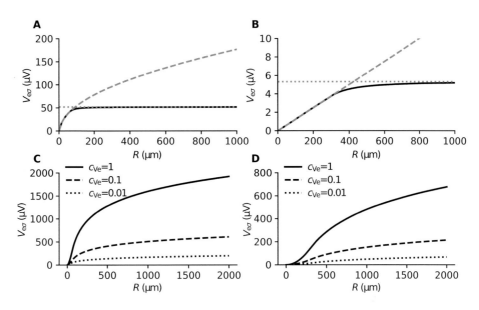

Figure 8.10 Population LFP at the center of neuron population. A: Solid line: standard deviation of the LFP at soma level for uncorrelated single-neuron LFP contributions, equation (8.20). Dashed line: intermediate-R expression in equation (8.20) ($r_\varepsilon \leq R \leq r_*$) assumed for all values of R. Dotted line: infinite-R limit ($R \to \infty$), equation (8.21). Parameters for shape function set to mimic situation for basal or uniform input for $z = 0\,\mu$m in Figure 8.7B (i.e. $r_* = 60\,\mu$m, $r_\varepsilon = 10\,\mu$m, $F_0 = 10\,\mu$V). Note that the choice of F_0 depends on the number of synaptic inputs onto each neuron in the population (see Figure 8.7). The resulting value of the population LFP is proportional to F_0, and the present choice is arbitrarily set to give an experimentally plausible value of the population LFP. **B:** Solid line: as in panel A, but for LFP below soma level, equation (8.23). Dashed line: small-R expression in equation (8.23) ($R \leq r_*$) assumed for all values of $R > r_\varepsilon/2$. Dotted line: infinite-R limit ($R \to \infty$) of equation (8.23). Parameters for shape function set to mimic situation for basal or uniform input for $z = -400\,\mu$m in Figure 8.7B – that is, $r_* = 300\,\mu$m. $F_0 = 0.1\,\mu$V, which is one-hundredth of the value used in panel A, mimicking the ratio of the small-R LFP amplitude at $z = -400\,\mu$m compared to $z = 0\,\mu$m in Figure 8.7. **C:** As in panel A, but for correlated neurons for different values of correlation coefficient c_{Ve}, equation (8.26). **D:** As in panel C, but for LFP below soma level, equation (8.27). Parameters for shape function $F(r)$ as in panel B. For all panels, the area density is $\rho = 5.000\,$mm^2, corresponding to a volume density of $\rho_V = 50.000\,$mm^3 and a population "thickness" of $100\,\mu$m. Code available via www.cambridge.org/electricbrainsignals.

single-neuron shape function in equation (8.10) is inserted into the integral in equation (8.19), and the expression for the LFP standard deviation is found to be

$$V_{\text{eu}\sigma}(R) = \begin{cases} F_0\,\sqrt{\rho\,\pi}\,R & R < r_* , \\ F_0\,\sqrt{\rho\,\pi}r_*\sqrt{2 - r_*^2/R^2} & R \geq r_* \end{cases} \tag{8.23}$$

(see Figure 8.10B for an illustration). From simple analytical investigations of equation (8.23), it follows that $V_{\text{eu}\sigma}(R \to \infty) = F_0\sqrt{2\pi\rho}\,r_*$. Thus, here $V_{\text{eu}\sigma}$ also converges to a finite value in the large-R limit, and the crossing of the small-R and the large-R regimes occurs for $R = \sqrt{2}r_* \approx 1.4r_*$. For this value of R, which can be used as a measure of spatial reach, $V_{eu\sigma}(R)$ is found to be $\sqrt{3}/2 \approx 0.87$ of $V_{eu\sigma}(R \to \infty)$.

For the example results in Figure 8.10, we observe that the soma-level population LFP has a larger magnitude than the population LFP found 400 μm below the soma. This reflects the much larger value of the parameter F_0 in the shape function at the soma level. However, the soma-level spatial reach R_{reach} is smaller, reflecting that the value of the cut-off distance r_* is much smaller there.

Population LFPs from Correlated Single-Neuron Contributions
As shown in Appendix G.1, the standard deviation of the population LFP in the general case with an arbitrary level of correlations between the single-neuron LFPs is

$$V_{e\sigma}(R, c_{Ve}) = \sqrt{(1 - c_{Ve})V_{eu\sigma}(R)^2 + c_{Ve}V_{ec\sigma}(R)^2} \,. \tag{8.24}$$

Note that this formula is analogous to the formula in equation (8.6) for the toy model in Section 8.2.3, with $V_{es\sigma}^2 N_s$ replaced by $V_{eu\sigma}(R)^2$ and $V_{es\sigma}^2 N_s^2$ replaced by $V_{ec\sigma}(R)^2$.

The correlation coefficient c_{Ve} measures how correlated the single-neuron LFP contributions are, where $c_{Ve} = 0$ corresponds to the completely uncorrelated case and $c_{Ve} = 1$ to the fully correlated case. Further, $V_{eu\sigma}(R)$ is the standard deviation of the population LFP in the uncorrelated case given in equation (8.19), while $V_{ec\sigma}(R)$ is the standard deviation of the population LFP in the fully correlated case given by

$$V_{ec\sigma}(R) = \int_0^R N(r)F(r)dr = 2\pi\rho \int_0^R r F(r) \, dr \,. \tag{8.25}$$

The correlation between the single-neuron LFPs can be induced in several ways (Section 8.2.3). In Lindén et al. (2011), it was induced by randomly drawing the pre-synaptic spike trains driving the neurons in the population from a shared pool of uncorrelated Poisson spike trains. With a finite number of spike trains in this pool, some of the spike trains will by chance be drawn multiple times, thus inducing what is known as shared-input correlations (Renart et al. 2010, Tetzlaff et al. 2012). The smaller the pool, the larger the correlations will be. While the correlation coefficient c_{Ve} will monotonically decrease with pool size, the exact value of c_{Ve} must be found for each realization separately by numerical simulations; for details, see Lindén et al. (2011).

For the case with fully correlated single-neuron LFPs ($c_{Ve} = 1$), insertion of the soma-level shape function $F(r)$ from equation (8.9) into the integral in equation (8.25) gives, after some algebra, the following expression for the LFP:

$$V_{ec\sigma}(R) = \begin{cases} F_0\rho\,\pi R^2 & R < r_\varepsilon \,, \\ F_0\rho\frac{1}{3}\pi \left(4r_\varepsilon^{1/2}R^{3/2} - r_\varepsilon^2\right) & r_\varepsilon \leq R < r_* \,, \\ F_0\rho\frac{1}{3}\pi \left((6\ln(R/r_*) + 4)\,r_\varepsilon^{1/2}r_*^{3/2} - r_\varepsilon^2\right) & R \geq r_* \,. \end{cases} \tag{8.26}$$

Here, $V_{ec\sigma}(R \to \infty) \propto \ln(R/r_*)$ does not converge to a finite value as $R \to \infty$. This is again in accordance with the results from the earlier conceptual model (equation (8.18)).

For the LFP above or below the vertical extent of the dendrites, insertion of the shape function $F(r)$ from equation (8.10) into the integral in equation (8.25) instead gives

$$V_{ec\sigma}(R) = \begin{cases} F_0 \, \rho \, \pi R^2 & R < r_*, \\ F_0 \, \rho \pi r_*^2 \, (2 \ln (R/r_*) + 1) & R \geq r_* \end{cases} \qquad (8.27)$$

(see Figure 8.10D for an illustration). The qualitative point regarding the logarithmic divergence of the population LFP in the large-R limit still holds.

As for the uncorrelated example results, we observe also for the correlated example in Figure 8.10 that the soma-level population LFP (panel C) has a larger magnitude than the population LFP below the soma level (panel D). Again, this reflects the much larger value of the prefactor F_0 in the shape function at the soma level.

Population LFPs from Partially Correlated Single-Neuron Contributions
The formulas for $V_{eu\sigma}(R)$ and $V_{ec\sigma}(R)$ represent the two extreme cases where the single-neuron LFPs are completely uncorrelated ($c_{Ve} = 0$) or completely correlated ($c_{Ve} = 1$). In real situations, one would expect the situation to be somewhere in between, with partially correlated ($0 < c_{Ve} < 1$) single-neuron LFPs. Results for the population LFP in this intermediate case can readily be found by plugging the results for the uncorrelated and fully correlated LFPs into equation (8.24), which is valid for any value c_{Ve}.

Figure 8.10C and D show results for three different values of c_{Ve}: 1, 0.1, and 0.01. Relative to the uncorrelated case (panel A), the weakest ($c_{Ve} = 0.01$) and strongest ($c_{Ve} = 1$) correlation levels boost the LFP for maximum population size ($R = 2\,\mathrm{mm}$) by a factor 4 and 40, respectively. The fact that increasing c_{Ve} by a factor of 100 increases the LFP boosting by a factor of 10 is predicted from equation (8.24) for a situation where the correlated contributions $V_{ec\sigma}(R)$ dominate: the LFP standard deviation should then be proportional to $\sqrt{c_{Ve}}$. A similar boosting of non-soma-level LFP signals by correlations is seen by comparing results in Figure 8.10B and D.

Equation (8.24) allows for an assessment of the relative importance of the uncorrelated ($V_{eu\sigma}(R)$) and correlated ($V_{ec\sigma}(R)$) contributions to the LFP. In addition to the correlation coefficient c_{Ve} itself, the number density ρ of neurons is a key parameter here. While $V_{ec\sigma}(R)$ is proportional to ρ (equation (8.26) and equation (8.27)), $V_{eu\sigma}(R)$ is proportional to $\sqrt{\rho}$ (equation (8.20) and equation (8.23)). Thus, with all else equal, a higher-number density ρ will give larger weight to the correlated contributions.

Importantly, the value of c_{Ve} does not only depend on the level of correlations in the synaptic input. While correlations in a spatially non-uniform synaptic input onto a pyramidal cell like the Hay neuron may give a large value of c_{Ve}, the same input uniformly spread over the entire dendritic tree will not (Lindén et al. 2011, figure 4G). This point was previously illustrated in Figure 8.4, where large input correlations only resulted in large LFPs when the synapses were positioned non-uniformly on the neuron (Figure 8.4D, E).

For LFP signals dominated by correlated sources, the quantitative analytic model predicts that the spatial reach in practice will be set by the size of the population of neurons receiving correlated synaptic inputs surrounding the electrode (see Lindén et al.

(2011, figure 5)). This in turn will depend on the particular recording conditions – for example, brain state – and may explain why experimental estimates of the spatial reach have varied so widely, from a few hundred micrometers to centimeters (Kreiman et al. 2006, Liu & Newsome 2006, Berens et al. 2008, Katzner et al. 2009, Xing et al. 2009, Hunt et al. 2010, Kajikawa & Schroeder 2011).

Correlation Levels in Real Networks
Values for the correlation coefficient c_{Ve} for single-neuron LFP contributions in real neural networks are difficult to determine from experiments, and we do not here make any attempt to give estimates of c_{Ve} for any particular biological scenario. The value will obviously vary, not only from network to network, but also dynamically between different states of a given network.

Values for c_{Ve} can be computed from network models, though. For the Potjans-Diesmann model network (Potjans & Diesmann 2014) that included about 80,000 neurons mimicking a $1\,mm^2$ patch of cat visual cortex, Lindén et al. (2011) found values of c_{Ve} in the range 0.001–0.01 for the considered pyramidal-cell populations (Lindén et al. 2011, figure 6). This corresponded to being in the correlation-dominated regime where the second term under the square root in equation (8.24) is much larger than the first term. However, how representative this result is for cortical networks in general is unclear.

Finally, it should be noted that all analyses presented in the current section pertain to the standard deviation of the population LFP. Thus, a DC LFP contribution set up by a static (current-) dipole – for example, due to sustained activation of basal dendrites on pyramidal neurons – is not described. For a discussion on such "slow potentials," see Section 8.6.

8.3.2   Spatial Decay of LFPs

The spatial reach represents an "electrode-centric" perspective on the spatial extent of the LFP. An alternative is to take a "population-centric" perspective and study how sharply the LFP signal decays outside an LFP-generating population. The analytical model used to investigate the spatial reach can readily be generalized to explore this question.

Away from the population center, the formula in equation (8.24) for how uncorrelated and correlated contributions add up in the LFP standard deviation becomes

$$V_{e\sigma}(R,r;c_{Ve}) = \sqrt{(1-c_{Ve})V_{eu\sigma}(R,r)^2 + c_{Ve}V_{ec\sigma}(R,r)^2}\,. \tag{8.28}$$

Here, R still describes the population radius and r the lateral distance from the population center. The special case $r = R$ thus corresponds to the edge of the population, and $r = 0$ corresponds to the center of the population.

In equation (8.28),

$$V_{\text{eu}\sigma}(R,r) = \left(\iint_{\|\mathbf{r}'\| \leq R} F(\|\mathbf{r}' - \mathbf{r}\|)^2 \, ds' \right)^{1/2} \tag{8.29}$$

is the LFP standard deviation for the uncorrelated case ($c_{\text{Ve}} = 0$), where the integral goes over the disk area of the population. Using cylindrical coordinates and considering recording positions along the x-axis, the area integral in equation (8.29) can be written as

$$V_{\text{eu}\sigma}(R,r) = \left(\rho \int_0^{2\pi} \int_0^R r' F\left(\sqrt{(r - r'\cos\theta)^2 + (r'\sin\theta)^2} \right)^2 \, dr' \, d\theta' \right)^{1/2}, \tag{8.30}$$

for which solutions in general must be found numerically.

The LFP standard deviation for the fully correlated case ($c_{\text{Ve}} = 1$) is generally given by

$$V_{\text{ec}\sigma}(R,r) = \rho \iint_{\|\mathbf{r}'\| \leq R} F(\|\mathbf{r}' - \mathbf{r}\|) \, ds', \tag{8.31}$$

or in cylindrical coordinates as

$$V_{\text{ec}\sigma}(R,r) = \rho \int_0^{2\pi} \int_0^R r' F\left(\sqrt{(r - r'\cos\theta)^2 + (r'\sin\theta)^2} \right) \, dr' \, d\theta'. \tag{8.32}$$

Note that for the special case $r = 0$, the integrals in equation (8.30) and equation (8.32) reduce to equation (8.19) and equation (8.25), respectively, as they should.

Results for how the LFP in the soma layer of a cylindrical population of neurons of fixed size ($R = 500$ μm) varies with lateral position r from the center of the population are shown in Figure 8.11.

The dominant role played by synaptic-input correlations for the case of non-uniform input is highlighted in Figure 8.11A, where the synaptic input is delivered apically to a population with radius of 500 μm. The LFP is highly amplified for $c_{\text{Ve}} = 1$ and $c_{\text{Ve}} = 0.1$ compared to the uncorrelated case ($c_{\text{Ve}} = 0$) and thus extends much further outside the edge of the population. In the most correlated case ($c_{\text{Ve}} = 1$), the LFP 1000 μm outside the population ($r = 1500$ μm) is of similar magnitude as the LFP in the center of the population with uncorrelated input ($c_{\text{Ve}} = 0$).

Relative to the population radius R, the decay is less sharp for small populations than for large populations (Figure 8.11B). This reflects that the spatial extension of the LFP decay zone outside the population in absolute terms is almost independent of the size of the population. In relative units r/R, the decay zone therefore decreases with population size.

How the LFP decays outside a population is determined by the value of r_*. This is best illustrated if the LFP standard deviation is normalized to have a maximal value of one. Normalized LFP standard deviations for the fully uncorrelated and correlated cases are shown in Figure 8.11C and Figure 8.11D, respectively. In both cases, the decay length is much larger when $r_* = 200$ μm, mimicking a situation with apical input (red curve), than when $r_* = 60$ μm, mimicking a situation with basal or uniform input (black curve).

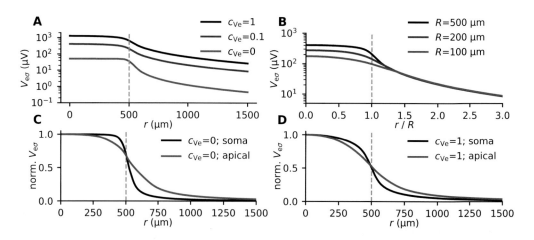

Figure 8.11 Spatial decay of population LFP. A: Dependence of the standard deviation of the population LFP at the soma level on the horizontal distance r from the population center for different input correlations: $c_{Ve} = 1$ (black), $c_{Ve} = 0.1$ (blue), $c_{Ve} = 0$ (red). Here, $R = 500\,\mu m$, $r_* = 60\,\mu m$, and $r_\varepsilon = 10\,\mu m$, mimicking the situation with uniform or basal input for single-neuron LFPs at the soma level ($z = 0\,\mu m$ in Figure 8.7) and $F_0 = 10\,\mu V$. **B**: Dependence of LFP standard deviation on population sizes: $R = 500\,\mu m$ (black), $R = 200\,\mu m$ (blue), and $R = 100\,\mu m$ (red) at correlation level $c_{Ve} = 0.1$. Values of r_*, r_ε, and F_0 as in panel A. The x-axis depicts distances r/R relative to the population size. **C**: LFP standard deviation outside the center of population for uncorrelated single-neuron LFPs ($c_{Ve} = 0$). Black curve: $R = 500\,\mu m$, $r^* = 60\,\mu m$, and $r_\varepsilon = 10\,\mu m$, mimicking the situation with uniform or basal input for single-neuron LFPs at the soma level. Red curve: $R = 500\,\mu m$, $r^* = 200\,\mu m$, and $r_\varepsilon = 10\,\mu m$, mimicking the same situation for apical input ($z = 0\,\mu m$ in Figure 8.7). **D**: Same as panel C for $c_{Ve} = 1$. Dashed vertical lines mark the population edge. Code available via www.cambridge.org/electricbrainsignals.

A comparison of Figure 8.11C and D shows that, relative to the LFP in the population center, $V_{e\sigma}(R, R)$ at the population edge is larger for uncorrelated cases than the correlated cases. This can be qualitatively understood by first considering $V_{e\sigma}(R, R)$ at the edge of an infinitely large population. The limit $R \to \infty$ corresponds to being at the edge of an infinitely large half-plane. In this situation, the LFP signal at the center of the population will correspond to being in the middle of an infinite plane.

For uncorrelated input ($c_{Ve} = 0$), the variance of the LFP signal will correspond to the sum of the LFP signal variances from two infinitely large half-planes. We thus have $[V_{e\sigma}(R, 0)]^2 = 2[V_{e\sigma}(R, R)]^2$ so that $V_{e\sigma}(R, R) = V_{e\sigma}(R, 0)/\sqrt{2} \approx 0.7 V_{e\sigma}(R, 0)$. For the fully correlated case ($c_{Ve} = 1$), we find by the same argument that $V_{e\sigma}(R, R) = V_{e\sigma}(R, 0)/2$ so that the LFP at the edge should be about half of the LFP in the population center for large population sizes R. For the case with $R = 500$ μm, we indeed observe that $V_{e\sigma}(R, R)/V_{e\sigma}(R, 0) \approx 0.7$ in Figure 8.11C and $V_{e\sigma}(R, R)/V_{e\sigma}(R, 0) \approx 0.5$ in Figure 8.11D for the uncorrelated and fully correlated cases, respectively. This good agreement is seen even though R/r_* (which is the relevant measure for how large R is) is only 2.5 for the example curves in the figures with $r_* = 200$ μm.

At very large distances ($r \gg R$), it can be derived from the integral expressions in equation (8.30) and equation (8.32) that the population LFP will inherit the far-field behavior $F(r) \propto 1/r^2$ of the single-neuron shape function in equation (8.9) (Einevoll, Lindén et al. 2013). This $1/r^2$ decay can be seen in the large-r tails of the population-LFP curves in Figure 8.11. It can likewise be derived that, in this large-r limit, the population LFP will be independent of population radius R when plotted as a function of r/R, as observed in Figure 8.11B.

In Lindén et al. (2011), predictions obtained using equation (8.28) were compared with results from numerical simulations using the direct summation formula in equation (8.11). Generally, a very good agreement was found, except in the large-r limit for correlated ($c_{Ve} \neq 0$) apical or basal inputs, where the simulations predicted a $1/r^3$-decay, while equation (8.28) predicts a $1/r^2$-decay (Lindén et al. 2011, figure 7). This deviation reflects a subtle point, which has to do with the cancellation effects of lateral dipole-components being stronger in the numerical simulations than in the analytical model. For discussion of this point, see Lindén et al. (2011).

A finding from this model study is that a recorded cortical LFP does not necessarily stem from the neurons closest to the electrode contact. It might, for example, be the case that the synaptic inputs onto these closest neurons are weak or uncorrelated, and the electrode mainly picks up LFP signals from a distant population of neurons receiving non-uniform correlated synaptic inputs. The LFP from such a population may be very strong and extend far outside the population itself and overpower any LFP from a population close by receiving uncorrelated inputs. For example, in Figure 8.11A, it is seen that the LFP 1 mm outside an LFP-generating population of radius 0.5 mm receiving fully correlated input ($c_{Ve} = 1$) is similar in magnitude to the LFP at the center of the same population receiving uncorrelated input ($c_{Ve} = 0$).

8.3.3 LFP from Single Presynaptic Neurons: Unitary LFPs

In Section 8.2, we studied the LFP generated by membrane currents from a single neuron receiving synaptic inputs. This "postsynaptic" single-neuron LFP is difficult to measure in experiments since, in practice, the recorded LFP will always have contributions from membrane currents in several neurons. An alternative to this postsynaptic perspective is to instead take a "presynaptic" perspective and consider the LFP generated by a single neuron sending action potentials to its afferent targets. This LFP signal will in general have contributions both from the low-frequency components of the spike of the single presynaptic neuron and the synaptic and associated return currents of the typically numerous postsynaptic neurons. Such presynaptic single-neuron LFPs can be measured (Swadlow et al. 2002, Bazelot et al. 2010), and the LFP generated by a single action potential is sometimes referred to as the monosynaptic, or *unitary*, LFP.

Swadlow et al. (2002) measured the unitary spike-time averaged LFP in the sensory cortex following action potentials in single thalamic neurons projecting to the cortex (Figure 8.12). An example measurement is shown in panel C. Here, even the LFP due to the membrane currents associated with the arriving action potential on the presynaptic side of the synapse can be seen (arrow in the figure). After a time lag of less than a

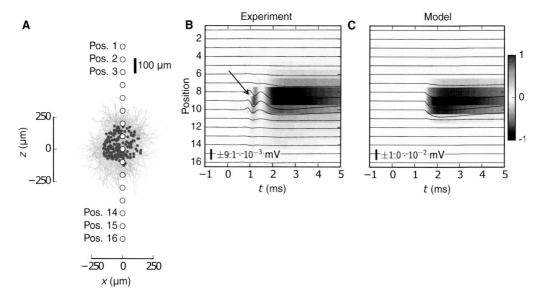

Figure 8.12 Comparison of experimental and modeled unitary LFPs. A: Qualitative illustration of postsynaptic model populations in cortex showing a small subset of the 4,000 reconstructed regular spiking (RS) cells and 1,000 reconstructed fast-spiking (FS) cells used in the forward modeling of unitary LFPs. Locations of synaptic inputs from the thalamocortical (TC) cell are denoted by red dots, all within a radius of 165 µm from the center of the synaptic projection pattern from the TC cell. The locations of micro-electrode-contact points penetrating the population are denoted by circles labeled Pos. 1–16. **B**: LFP responses due to a single synaptic input from a single TC recorded in a rabbit somatosensory cortex, from Swadlow et al. (2002, figures 1B–N1). The arrow points to the short presynaptic contribution to the LFP. **C**: Corresponding LFP responses from model (see Hagen et al. (2017) for details). Figure modified from Hagen et al. (2017) (CC BY 4.0 license).

millisecond when the LFP signal is close to zero, reflecting the synaptic transmission delay, a prominent LFP due to the direct synaptic activation of the population of some hundreds of postsynaptic target neurons is seen. In a modeling study by Hagen et al. (2017), the latter (postsynaptic) part of this measured unitary LFP was found to be well accounted for (Figure 8.12D) by a model based on known properties of the population of postsynaptic target neurons.

While the unitary LFPs investigated by Swadlow et al. (2002) stemmed from excitatory neurons in the thalamus, similar recordings in the hippocampus could only detect unitary LFPs from inhibitory interneurons (Bazelot et al. 2010). This is in agreement with other studies that also have suggested that LFPs recorded in the cortex are dominated by synaptic inputs from interneurons (Hagen et al. 2016, Teleńczuk et al. 2017, Teleńczuk et al. 2020*b*). This is not at odds with the general notion suggested earlier that cortical LFPs largely stem from excitatory pyramidal cells. Rather it reflects the difference between the postsynaptic and presynaptic perspectives on LFP generation. While it is expected that the LFP measured in layered structures such as the cortex and hippocampus typically is dominated by membrane currents in populations of pyramidal cells (Einevoll, Kayser, et al. 2013), the synaptic inputs driving these currents may very well be dominated by contributions from inhibitory neurons.

In contrast to the unitary LFP, the single-neuron LFP – that is, the LFP generated by the membrane currents of a single neuron – is very difficult to measure in vivo since it will be masked by the contributions from other neurons. In the general case of recurrent networks, it is also difficult to use single-neuron LFPs as a template for computing population LFPs since the single-neuron contribution will vary over time depending on the spike dynamics of the network. The unitary LFP, however, is well suited as a basic template for computing LFPs originating from synaptic inputs. If we knew both the times for action-potential firing as well as the unitary LFPs for all neurons in the brain, we could – assuming linearity of postsynaptic responses to multiple inputs – sum up all the individual contributions to get the total LFP. This idea is the starting point for the development of so-called kernel methods for modeling LFPs (Hagen et al. 2016, Skaar et al. 2020, Teleńczuk et al. 2020a, Hagen et al. 2022), which we presented in Chapter 6.

8.4   Frequency Content of LFPs

Frequency spectra of LFPs will in general be determined by a combination of different biophysical phenomena, including intrinsic dendritic filtering (Section 3.3.3), temporal filtering reflecting the shape of synaptic currents, the frequency content of the presynaptic firing rates, and the level of correlations in the synaptic inputs. These filter mechanisms will operate in concert to determine the overall frequency content of the LFP (Figure 8.13).

With current-based synapses (Section 3.1.4.2) and passive or "quasi-active" linearized cable models (Section 6.4.2), the LFP contributions from numerous synaptic inputs sum linearly. Under this linearity assumption, the total effect of several such filtering mechanisms on the LFP is in frequency space given as a product of the individual filtering contributions as given by the convolution theorem (Appendix F.4). The Fourier transform of the LFP can thus be written as

$$\hat{V}_e(f) \propto \hat{h}_{dend}(f)\hat{h}_{syn}(f)\hat{v}(f),\tag{8.33}$$

where $\hat{h}_{dend}(f)$, $\hat{h}_{syn}(f)$, and $\hat{v}(f)$ denote the Fourier transforms of the intrinsic dendritic filtering, synaptic current, and presynaptic firing rate, respectively. Note that the same approach of separating between different filter contributions in the spatial and temporal domain rather than the frequency domain was already introduced for accurate kernel-based signal predictions from network models in Section 6.4.3 (described fully in Hagen et al. (2022)).

8.4.1   Intrinsic Dendritic Filtering

As outlined in Section 3.3.3, the intrinsic dendritic filtering effect causes a low-pass filtering of extracellular potentials, which depends on the distance between the recording site and the neural sources. The effect was observed in Figure 8.3A–B, where we saw

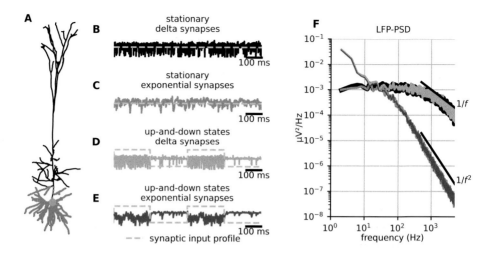

Figure 8.13 Frequency content of LFP signals. A: Hay neuron model receiving current-based synaptic inputs to the basal dendrites (blue). **B**: LFP signal immediately outside the soma for delta-function synaptic currents following a stationary Poisson distribution. **C**: Same as B but with exponentially decaying synaptic currents. **D**: Same as B but with up-and-down modulated synaptic input – that is, square-wave firing rate. **E**: Same as C but with up-and-down modulated firing rates. **F**: LFP PSDs for the examples in B–E. Code available via www.cambridge.org/electricbrainsignals.

that the LFP becomes blunter with distance from the neuron it originates from. This occurs because high-frequency components are attenuated more than low-frequency components.

The physical origin of the intrinsic dendritic filtering effect is the frequency dependence of capacitive membrane currents. In the context of synaptically generated LFPs, we can study the effect by considering each frequency component of the synaptic current separately. In Figure 8.14, we do this by injecting sinusoidal synaptic-current components into one end of a compartmentalized passive dendritic stick. A stimulation frequency of 1 Hz gives rise to a large dipolar LFP pattern around the dendritic stick, reflecting a spatially extended distribution of the membrane currents (Figure 8.14A). For the same current amplitude but with a frequency of 200 Hz, the resulting LFP is more spatially confined (Figure 8.14B).

The explanation of the LFP patterns in Figure 8.14 is two-fold:

1.   We know from earlier that the frequency-dependent length constant λ_{AC} of the dendrite decreases with f (Figure 3.7). This implies that the low-frequency components of I_{syn} give larger average distances between the sink and returning source currents. Since we know that the LFP (at least when recorded at some distance from the neuron) increases with the separation distance between sink and source, this, in turn, explains why the low-frequency input gives higher LFP amplitudes.

2.   In the far-field regime where the dipole approximation applies, the intrinsic dendritic low-pass filtering effect is stronger than it is close to the neuron

(Figure 8.14C). This far-field regime is reached at distances substantially larger than the dipole length λ_{AC}. Since λ_{AC} decreases with frequency, the far-field regime is reached for smaller distances from the stick for the high-frequency components of the stimulus (small λ_{AC}) than for the low-frequency components (large λ_{AC}). For this reason, the LFP contribution from the high-frequency components will not only be smaller but also fall off faster with distance compared to the low-frequency contributions. This explains why not only the amplitude, but also the shape of the LFP signal varies with distance.

Both these effects are seen for the example simulations in Figure 8.14C. Very close to the input site where the LFP signal is dominated by the contribution from the synaptic input current, the LFP amplitude is very similar for 1 Hz and 200 Hz inputs. The LFP amplitude decays faster with distance for the 200 Hz input, however.

An actual synaptic current is not a single sinusoid, but will contain many frequency components. As illustrated in Figure 8.14, the high-frequency components in the LFP will be more attenuated with distance than the low-frequency components. This is why the intrinsic dendritic filtering effect will cause a blunting of the LFP signal (Lindén et al. 2010). Note, however, that very far away from the synaptic current source, all frequency components contained in the signal are in the far-field limit where the dipole approximation is valid, and these components will have the same $\propto 1/r^2$ decay with

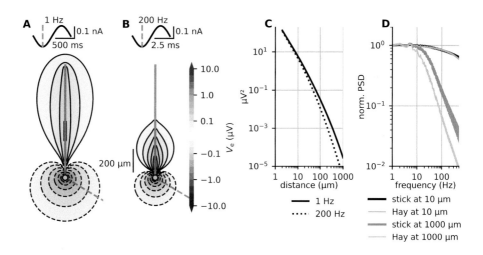

Figure 8.14 Intrinsic dendritic filtering of LFP. A: Illustration of the LFP at a snapshot in time in response to a 1 Hz sinusoidal synaptic current injected into one end of a passive dendritic stick with a diameter of 1 μm. **B**: Same as in panel A but with a frequency of 200 Hz. **C**: Amplitude of sinusoidally oscillating LFP for 1 Hz and 200 Hz inputs as a function of distance from the input site along an axis that is -30° with the horizontal plane (dashed line in panels A and C). **D**: LFP PSDs 10 and 1000 μm (along the same axis as in panel C) from a passive dendritic stick and the Hay model neuron receiving white-noise current inputs at the end of the stick and in the soma, respectively. Code available via www.cambridge.org/electricbrainsignals.

distance. In this regime, the LFP signal will thus have a fixed shape even if the overall amplitude still decays with distance.

A transparent way of illustrating how intrinsic dendritic filtering affects the frequency content of LFP signals is to inject a white-noise synaptic current into a neuron. A white-noise current has, by definition, equal (Fourier) amplitude at all frequencies (Table F.1). The power spectral density (PSD), which is given by the square of these Fourier amplitudes (equation (F.4)), thus also has the same magnitude for all frequencies and is said to be flat. The PSD of the resulting LFP signal is then a direct measure of the strength of the intrinsic dendritic filtering effect. From the general expression in equation (8.33), we then find that

$$\mathrm{PSD_e}(f) = |\hat{V}_\mathrm{e}(f)|^2 \propto |\hat{h}_\mathrm{dend}(f)|^2 \ . \tag{8.34}$$

Figure 8.14D shows the results from such an analysis both for a dendritic stick and the morphologically detailed Hay model. For LFPs close by (at a distance of 10 μm), notable effects of intrinsic dendritic filtering are only seen for frequencies larger than 100 Hz or so for the two models. For LFPs recorded 1000 μm away, the effect kicks in for lower frequencies, at around 10 Hz for the Hay model and around 20 Hz for the ball-and-stick model. The numerical values of these low-pass "cut-off" frequencies will depend on the morphology and electrical parameters of the neuron.

8.4.2   LFP Filtering and Power Laws

Observations of power-law relationships in empirical data have for a long time intrigued scientists in many fields. By power-law behavior, one typically means that an observed quantity or probability distribution $y(x)$ satisfies (Stumpf & Porter 2012)

$$y(x) \propto \frac{1}{x^\alpha} \ \text{ for } x > x_0 \ . \tag{8.35}$$

Here, α is the *power-law exponent*, and the power-law behavior occurs in the tail of the distribution – that is, for $x > x_0$.

In neuroscience, the frequency content of certain brain signals has been suggested to follow power laws. In particular, power-law behavior in the PSD,

$$\mathrm{PSD}(f) \propto \frac{1}{f^\alpha} \ , \tag{8.36}$$

has been suggested, even if the requirement regarding power-law behavior also in the high-frequency tail (equation (8.35)) has not always been fulfilled.

Power laws have been suggested in the high-frequency tails of LFP signals (Bédard et al. 2006, Martinez et al. 2009, Milstein et al. 2009, Baranauskas et al. 2012) and also in PSDs of EEG signals (Freeman et al. 2003, Fransson et al. 2012), ECoG signals (Freeman et al. 2000, Freeman 2009, Miller et al. 2009, Zempel et al. 2012), as well as the soma membrane potential and soma membrane current (Diba et al. 2004, Jacobson et al. 2005, Rudolph et al. 2005, Bédard et al. 2006, Bédard & Destexhe 2008, Yaron-Jakoubovitch et al. 2008). Power laws have also been suggested in recordings of so-called neural avalanches (Beggs & Plenz 2003, Milton 2012), which was taken as

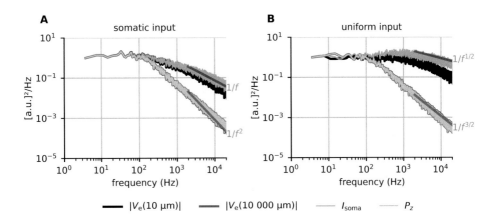

Figure 8.15 Power laws from intrinsic dendritic filtering. A: PSD of the LFP outside a ball-and-stick model receiving Poisson-distributed, current-based delta-function synaptic inputs to the soma. **B**: Same as in panel A but with input uniformly distributed over the membrane. The LFP amplitude is measured in a direction of -30° from a rightward horizontal axis passing through the soma. I_{soma} is the soma membrane current, and p_z is the dipole moment in the stick direction. Code available via www.cambridge.org/electricbrainsignals.

evidence that neural networks in vivo operate close to a critical state (Touboul & Destexhe 2017, Wilting & Priesemann 2019). Whether these latter power laws are related to the power laws suggested from the observed PSDs of the brain signals is not clear.

The focus in the present section is on filtering and possible power laws in the PSD of extracellular potentials inside the brain. We will in this context include frequencies above the typical LFP frequency range of a few hundred hertz but will for simplicity still refer to the extracellular potential as the LFP. Note that the same mechanisms determining the filtering of the LFP also affect the frequency filtering of single-neuron variables, such as the membrane potential and membrane currents as well as systems-level measures like EEG and MEG. For a discussion of putative power laws for these other measurement modalities, see Pettersen et al. (2014).

8.4.2.1   Intrinsic Dendritic Filtering

Intrinsic dendritic filtering is due to the spatial spread of membrane currents as described in Section 8.4.1 and depends on the distance of the recording electrode from the neuron. Effects on the LFP from this type of filtering alone are shown in Figure 8.15 for a ball-and-stick model. Here, the synaptic currents are modeled as Dirac delta functions, and the incoming spikes have a stationary Poisson statistics. Both delta functions (Table F.1) and stationary Poisson processes have flat ("white") power spectra. This means that both $\hat{h}_{\mathrm{syn}}(f)$ and $\hat{v}(f)$ have a flat power spectrum so that $\hat{V}_{\mathrm{e}}(f) \propto \hat{h}_{\mathrm{dend}}(f)$ (Appendix F). Thus, the shape of the PSD ($|\hat{V}_{\mathrm{e}}(f)|^2$) is simply determined by $|\hat{h}_{\mathrm{dend}}(f)|^2$.

For the case where the synaptic current arrives at the soma, $|\hat{V}_{\mathrm{e}}(f)|^2 \propto 1/f$ immediately outside the soma for frequencies f above 100 Hz or so (Figure 8.15A), implying that $|\hat{h}_{\mathrm{dend}}(f)| \propto 1/f^{1/2}$ in this frequency range. Far away from the soma (10 000 μm), we instead have that $|\hat{V}_{\mathrm{e}}(f)|^2 \propto 1/f^2$, implying that $|\hat{h}_{\mathrm{dend}}(f)| \propto 1/f$ in this frequency range. These high-frequency power laws in the near-field and far-field limits can be analytically derived for the ball-and-stick model, if the stick is assumed to be infinitely long (Pettersen et al. 2014). Note also that the same difference of 1 in the power-law exponent for $|\hat{V}_{\mathrm{e}}(f)|^2$ between the near-field and far-field power laws was observed in Chapter 7 (see equation (7.6) and equation (7.10)).

The intrinsic dendritic filtering effect is specific to extracellular potentials as it reflects how the contributions from membrane currents sum up extracellularly according to volume-conductor (VC) theory (Chapter 4). Another more subtle dendritic contribution to power laws in brain signals is the spatial distribution of input currents driving the neurons (Pettersen et al. 2014).

If the synaptic input currents are distributed uniformly across the dendrite rather than positioned only on the soma, the exponents of the high-frequency power laws will be reduced by (approximately) 1/2 (Figure 8.15). For an infinite stick, it has indeed been shown analytically that the exponent of the high-frequency power law of the soma-current PSD is reduced from 1 to 1/2. Since the LFP recorded immediately outside the soma essentially is determined by this soma current, the power-law exponent for the LFP PSD is also reduced from 1 to 1/2. Likewise, for uniformly distributed synaptic currents, the power-law exponent for the dipole moment is reduced from 2 to 3/2 (Pettersen et al. 2014, table 2). These power-law behaviors are indeed seen in the numerical example in Figure 8.15B. The $1/f^{3/2}$ power law is observed in the far-field PSD, as this measure indeed reflects the dipole moment. However, the PSD immediately outside the soma (black line) is seen not to follow the soma-current prediction (orange line) for the highest frequencies. This reflects that membrane currents from the dendritic stick in practice also contribute to the LFP for these higher frequencies for the present example for LFPs 10 μm outside the soma center.

The inherent mechanism behind the fractional power-law exponents, namely the uniformly distributed input sources, may have applications for describing power laws beyond the field of neuroscience. For a discussion of this, as well as power laws predicted for the membrane potential and soma membrane currents, see Pettersen et al. (2014).

8.4.2.2   Synaptic Filtering

Real synaptic currents are not described by delta functions as assumed in the previous section but rather last at least some milliseconds after onset. This implies that a synaptic filter also will affect the LFP power spectra. For the situation where the incoming spike train is stationary Poissonian so that the firing-rate PSD is flat, the PSD power law for the LFP is given by

$$\mathrm{PSD}_{\mathrm{e}}(f) = |\hat{V}_{\mathrm{e}}(f)|^2 \propto |\hat{h}_{\mathrm{dend}}(f)\hat{h}_{\mathrm{syn}}(f)|^2 \,. \tag{8.37}$$

When the synaptic current is given by a delta function $h_{\mathrm{syn}}(t) \propto \delta(t)$, $\hat{h}_{\mathrm{syn}}(f)$ is a constant. When the synaptic current decays exponentially after onset ($h_{\mathrm{syn}}(t) \propto \exp(-t/\tau_{\mathrm{syn}})$, equation (3.10)), we instead have $\hat{h}_{\mathrm{syn}}(f) \propto 1/(f_0 + jf)$, where

$f_0 = 1/(2\pi\tau_{\text{syn}})$ (Appendix F) and j is the imaginary unit. Thus, in the high-frequency limit ($f \gg f_0$), we have $|\hat{h}_{\text{syn}}(f)|^2 \propto 1/f^2$. Note that other choices for temporal synaptic kernels listed in Section 3.1.4.1 will result in different filter properties.

With synapses as described here, we can thus get (within a certain frequency regime) a power-law exponent of 2 from the synaptic currents in addition to the exponent from dendritic filtering. This is in accordance with observations in the example in Figure 8.13F (grey curve), where a power-law decay close to $1/f^2$ is observed for frequencies around 100 Hz or so where the contribution from dendritic filtering is small. For larger frequencies where dendritic filtering also has an effect, the decay of the signal with frequency is even steeper.

With other choices of synaptic current functions, other power laws can be obtained. For example, with the synaptic current chosen to be a so-called α-function where $h_{\text{syn}}(t) \propto t \exp(-t/\tau_{\text{syn}})$, we find that $|\hat{h}_{\text{syn}}(f)|^2 \propto 1/f^4$ in the high-frequency limit (Appendix F).

8.4.2.3   Effect of Firing-Rate Spectrum

The LFP power spectra for the situations with a stationary Poissonian spike input in Figure 8.13 are relatively flat for frequencies less than a few tens of hertz. For this example model, neither the intrinsic dendritic filtering nor the synaptic filtering have a sizable effect on these lower frequencies. Experimentally observed LFPs typically have strongly varying PSDs also for these lower frequencies, and a key missing ingredient in these model PSDs is the effect of firing-rate modulations.

As an example of such effects, we consider a square-wave modulated firing rate, mimicking switching between cortical up-states and down-states (Figure 8.13D–E). The amplitude of this function in frequency space is given by $\hat{v}(f) \propto \sin(\pi f T)/f$, where T is the period of the square wave (see Table F.1). For low frequencies for which both synaptic and dendritic filtering effects are absent, we have

$$|\hat{V}_e(f)|^2 \propto |\hat{h}_{\text{dend}}(f)\hat{h}_{\text{syn}}(f)\hat{v}(f)|^2 \propto |\hat{v}(f)|^2 \propto \sin^2(\pi T f)/f^2 . \qquad (8.38)$$

The numerator $\sin^2(\pi T f)$ will cause the PSD to oscillate as a function of frequency, while the denominator f^2 attenuates these oscillations with increasing frequency. In Figure 8.13F, the depicted PSD is filtered to reduce the noise from the noisy spiking input, and as a consequence, the oscillatory effect from the factor $\sin^2(\pi T f)$ is essentially filtered out. Thus, for frequencies up to about 10 Hz, a $1/f^2$-dependence of the PSD is observed, in accordance with the prediction $|\hat{V}_e(f)|^2 \propto 1/f^2$. Incidentally, the approximate $1/f^2$ behavior seen for low frequencies was observed experimentally in Baranauskas et al. (2012), where the effect of switches between up-and-down states on the LFP was studied.

For the case with square-wave modulated input with delta-like synapse currents (Figure 8.13D), the PSD follows the PSD for intrinsic dendritic filtering alone for large frequencies (yellow and black curves in Figure 8.13F). This reflects that the square-wave modulated firing rate is accompanied by a sizable stationary background firing rate. For the example in Figure 8.13, the square-wave modulated part thus dominates the PSD for low frequencies while the stationary Poissonian part, which has a flat frequency spectrum, dominates the high frequencies. For the case with square-wave modulated input with exponential synapses (Figure 8.13E), the PSD instead follows the PSD for the case with stationary Poissonian input with exponential synapses for large frequencies (red and grey curves in Figure 8.13F).

8.4.2.4   Channel Filtering

Synaptic currents are not the only membrane-current inputs that can generate LFPs. The stochastic nature of currents crossing ion channels may also contribute to the LFP and thus to shaping LFP power spectra (Pettersen et al. 2014). The effect on the PSD power from an ion-channel current without any synaptic or firing-rate filtering is given by

$$PSD_e(f) \propto |\hat{h}_{dend}(f)\hat{h}_{chan}(f)|^2 , \qquad (8.39)$$

where $\hat{h}_{chan}(f)$ is the Fourier transform of the channel current. For some potassium channels, it has been observed that current noise typically follows $|\hat{h}_{chan}(f)| \propto 1/f^{1/2}$ (Derksen & Verveen 1966, Fishman 1973, Siwy & Fuliński 2002), and in Pettersen et al. (2014), it was shown that power laws observed in high-frequency tails of PSD spectra of LFPs and other electric brain signals could be explained by the combined filtering action of channels and dendrites. In the latter study, a prerequisite for reproducing experimentally observed power-law exponents was the assumption of uniformly distributed ion channels over the dendrites.

8.4.3   Frequency Content of Population LFPs

The model used to analyze population LFPs in Section 8.3 represented a population of layer 5 pyramidal neurons with passive membrane dynamics and current-based synapses. Such models are linear in the sense that if the weights of the synapses are doubled, the resulting population LFP will be doubled as well. In Section 8.3, we used the model to study the total LFP signal with all its frequency components included. However, the linearity of the model implies that we can use it to study each frequency component separately (Łęski et al. 2013). The example shown in Figure 8.16 illustrates how the spatial reach of four selected frequency components in the LFP depends on correlations in the synaptic input.

Correlations in the single-neuron LFP contributions can be induced in several ways. In Figure 8.16, correlations were induced through the choice of synaptic inputs on the neurons in the population. Specifically, the presynaptic spike trains driving the n_{syn} synapses on each neuron were drawn from a finite pool of n_{pool} premade Poissonian spike trains (without return so that each spike train could only be drawn once for a particular cell). For the population, the finite size of the pool implies that two neurons shared on average a fraction n_{syn}/n_{pool} of their inputs, inducing what is known as shared-input correlation. The smaller the size of the pool, the larger the correlations will be, and in Figure 8.16, the correlations were quantified by the input correlation

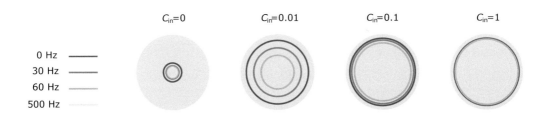

Figure 8.16 Frequency dependence of the spatial reach of LFP. Results are shown for a population of layer 5 pyramidal cells taken from Mainen & Sejnowski (1995) receiving basal input for different values of correlations of synaptic input c_{in}. Colored lines denote parts of the whole population (grey, radius = 1 mm) that contribute 95 percent of LFP amplitude at each given frequency in the middle of the population, at the soma level. Note that "0 Hz" corresponds to all frequencies $f \in [0, 30\,\text{Hz})$, "30 Hz" to $f \in [30\,\text{Hz}, 60\,\text{Hz})$, "60 Hz" to $f \in [60\,\text{Hz}, 90\,\text{Hz})$, and "500 Hz" to $f \in [470\,\text{Hz}, 500\,\text{Hz})$. The model setup was the same as in Figure 8.9, and LFPs were computed with the direct summation formula in equation (8.11). Figure adapted from Łęski et al. (2013).

coefficient $c_{in} = n_{syn}/n_{pool}$. This approach was the one used in Lindén et al. (2011) and Łęski et al. (2013).

In the example in Figure 8.16, the population of model neurons receives basal synaptic input for different values of c_{in}, from the uncorrelated situation with $c_{in} = 0$ to a strongly correlated situation with $c_{in} = 1$. A larger value of c_{in} implies a larger value of the correlation c_{Ve} of single-cell LFP contributions, but the quantitative connection must be computed numerically. For the population of basally activated pyramidal neurons considered in Figure 8.16, $c_{in} = 1$ corresponds to a correlation coefficient c_{Ve} of about 0.1 (Lindén et al. 2011, figure 8G).

The circles depicted in Figure 8.16 show the spatial reach for different frequency components of the population LFP. In the uncorrelated case, the spatial reach is relatively small for all frequencies considered, in accordance with the results in Section 8.3.1. However, the reach is largest for the component with the lowest frequency (denoted "0 Hz") and shrinks with increasing frequency.

As expected from the findings in Section 8.3.1, the spatial reach increases with increasing values of c_{in} for all considered frequency components. However, the increase is largest for the low-frequency components. For example, for $c_{in} = 0.01$, a plausible value for cortical networks (Lindén et al. 2011, figure 6), the spatial reach is increased from about 200 μm to about 800 μm, close to the population radius, for the 0 Hz component. For the 60 Hz component, the spatial reach is only about 400 μm for $c_{in} = 0.01$.

8.4.3.1 Frequency-Resolved Analytical Model

The frequency-dependent population LFPs can be well accounted for in an analytical population-LFP model similar to that derived in Section 8.3.1. For the frequency-resolved case, it can be shown that the power spectral density (PSD) is given by (Łęski et al. 2013, Hagen et al. 2016)

$$\mathrm{PSD_e}(f, R) = (1 - \hat{c}_{\mathrm{Ve}}(f))\mathrm{PSD_{e,u}}(f, R) + \hat{c}_{\mathrm{Ve}}(f)\mathrm{PSD_{e,c}}(f, R),\qquad (8.40)$$

where

$$\mathrm{PSD_{e,u}}(f, R) = 2\pi\rho \int_0^R r F(r; f)^2 \, dr \qquad (8.41)$$

and

$$\mathrm{PSD_{e,c}}(f, R) = \left(2\pi\rho \int_0^R r F(r; f) \, dr\right)^2 \qquad (8.42)$$

are the PSDs in the uncorrelated and fully correlated cases, respectively. Equation (8.40) is obtained by generalizing equation (8.24) to the frequency-resolved case, with equation (8.41) and equation (8.42) being analogous to equation (8.19) and equation (8.25), respectively. In the frequency-resolved equations, the single-neuron shape function $F(r; f)$ is frequency dependent and must be separately computed for each value of f. Likewise, the correlation coefficient in equation (8.40) has been replaced with a frequency-dependent *coherence* $\hat{c}_{\mathrm{Ve}}(f)$ (Appendix F.6). This quantity corresponds to the population-averaged coherence between single-cell LFP contributions and can, as the mean pairwise correlation c_{Ve} in Section 8.3.1, be estimated from numerical simulations of the population LFPs (see Łęski et al. (2013) for details).

The frequency dependence of these key factors is illustrated in Figure 8.17. Panel A shows, for three selected frequency components, the squared shape functions $F(r; f)^2$ at the soma level for the situation where a layer 5 pyramidal cell receives basal synaptic stimulation. As expected from the intrinsic dendritic filtering effects studied in Section 8.4.1, the high-frequency LFP components decay faster with distance than the low-frequency components. This leads to the power spectra seen in panel B.

The frequency dependence of the single-neuron shape function $F(r; f)$ can be quantified by making the cut-off distance r_* (panel C) between the short-distance ($F(r) \sim 1/\sqrt{r}$) and long-distance regimes ($F(r) \sim 1/r^2$) in equation (8.9) frequency dependent. As expected, the frequency-dependent cut-off distance, denoted $\hat{r}_*(f)$, decays monotonically with increasing frequency (panel D).

Like the cut-off distance, the coherence also decreases with frequency (Figure 8.17E). For the three nonzero coherence values considered in Figure 8.17, $\hat{c}_{\mathrm{Ve}}(f)$ drops by about a factor of hundred from 0 Hz to 500 Hz. This may be understood on biophysical grounds by considering the dendritic morphology of the cell: for high-frequency synaptic input, the return currents will be closer to the positions of the synaptic input because of filtering in the dendritic cable. For the present example with basal stimulation of layer 5 pyramidal neurons, the resulting dipoles will be aligned along the set of short basal dendritic segments. These point outwards from the soma in all directions, leading to large cancellations of LFPs and a small coherence. However, for low-frequency input, some of the synaptic input current will return through the apical dendrite, and the orientation of the effective dipoles will be more similar between cells, leading to a higher coherence (Łęski et al. 2013).

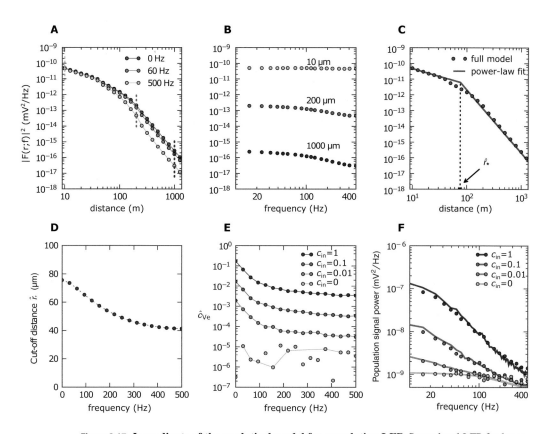

Figure 8.17 Ingredients of the analytical model for population LFP. Soma-level LFP for layer 5 pyramidal cells taken from Mainen & Sejnowski (1995) receiving basal synaptic input. **A**: Spatial decay in the lateral direction for the squared single-cell shape functions $|F(r; f)|^2$ for three different frequencies, $f = 0$, 60, and 500 Hz. See caption of Figure 8.16 for explanation of frequency bands. **B**: Single-cell LFP PSD $|F(r; f)|^2$ for three different lateral distances from the soma (dotted vertical lines in A). **C**: Log-log plot of the squared near-DC (~ 0 Hz), shape function $|F(r; 0)^2|$ (dots) approximated by a piecewise-linear function with cut-off distance r_* (dotted line; see Section 8.2.5). **D**: Frequency dependence of the cut-off distance $\hat{r}_*(f)$. **E**: Population-averaged LFP coherence $\hat{c}_{Ve}$ for different input correlation levels c_{in}. The dots not connected with lines for $c_{in} = 0$ have negative values for $\hat{c}_{Ve}$, and $|\hat{c}_{Ve}|$ is plotted instead. **F**: PSD for population LFP ($R = 1$ mm). Dots correspond to results from numerical simulation, and the lines correspond to predictions from the analytical model in equation (8.40) based on $\hat{r}_*$ and $\hat{c}_{Ve}$ given in D and E, respectively. Figure adapted from Łęski et al. (2013).

The frequency-dependent cut-off distance $\hat{r}_*(f)$ can be inserted into equation (8.41) and equation (8.42) to obtain frequency-dependent expressions for the uncorrelated and correlated LFP PSDs, respectively. These can next be used together with the numerically computed, frequency-dependent coherence $\hat{c}_{Ve}(f)$ to obtain predictions for the PSD for each frequency component for any level of coherence using equation (8.40). The resulting power spectral density (PSD) for our example situation is in excellent agreement with numerical simulation results (Figure 8.17F).

The results in Figure 8.17F show the overall strong effect of input correlations on the PSD. For the lowest frequencies, the LFP PSD for the highly correlated case ($c_{in} = 1$) is a factor of a hundred larger than for the uncorrelated case ($c_{in} = 0$). The results also show that the decay of the PSD with frequency is much more pronounced for the correlated cases. For the uncorrelated case where the only filtering comes from intrinsic filtering effects on the single-neuron shape function $F(r; f)$, there is little low-pass filtering for frequencies less than 100 Hz. In contrast, for the correlated case, the low-pass filtering effect kicks in at much lower frequencies.

For the case of uncorrelated single-neuron LFP contributions ($\hat{c}_{Ve}(f) = 0$), the general formula from equation (8.33) including dendritic, synaptic, and firing-rate filtering still applies so that the PSD is given by

$$\text{PSD}_e(f) = \text{PSD}_{eu}(f, R) \propto |\hat{h}_{dend}(f)\hat{h}_{syn}(f)\hat{v}(f)|^2 . \tag{8.43}$$

As we reasoned in Section 8.4.2, summing contributions from uncorrelated single-neuron LFPs with similar frequency content should not change the overall shape of the PSD. For the example in Figure 8.17, there is no synaptic or firing-rate filtering since the synaptic currents are delta functions,[6] and input spikes are generated by a stationary Poisson point process. Thus, only intrinsic dendritic filtering is present so that $\text{PSD}_e(f) \propto |\hat{h}_{dend}(f)|^2$. The PSD of the LFP at the population center for this $c_{in} = \hat{c}_{Ve}(f) = 0$ case is, for the example in Figure 8.17F, essentially flat for frequencies up to about 100 Hz. This is in qualitative accordance with the effects of intrinsic dendritic filtering seen at the single-neuron level in Figure 8.14C.

When the correlated contribution dominates in equation (8.40), we instead have

$$\text{PSD}_e(f) = |\hat{c}_{Ve}(f)|\text{PSD}_{ec}(f, R) \propto |\hat{c}_{Ve}(f)| |\hat{h}_{dend}(f)\hat{h}_{syn}(f)\hat{v}(f)|^2 , \tag{8.44}$$

with $\text{PSD}_{ec}(f, R)$ given in equation (8.42). When synaptic- and firing-rate filtering effects are absent, we thus have $\text{PSD}(f) \propto |\hat{c}_{Ve}(f)||\hat{h}_{dend}(f)|^2$. For the largest coherence ($c_{in} = 1$) considered in Figure 8.17, the PSD over the frequency interval between 10 and 100 Hz roughly follows a $1/f$ power law and is reduced by a factor of ten. This is consistent with a reduction of about a factor of ten of $|\hat{c}_{Ve}(f)|$ over the same frequency interval (Figure 8.17E). Thus, in the low-frequency part of the spectrum, correlations in single-neuron LFPs impose a substantial filtering that is absent for uncorrelated inputs.

The frequency dependence of $\hat{c}_{Ve}(f)$ also explains the frequency dependence of the spatial reach seen in Figure 8.16. As described in detail in Section 8.3.1, the spatial reach increases with increasing correlation c_{Ve}, and the same will hold for increasing coherences $\hat{c}_{Ve}(f)$. As the coherence will be largest for the lowest frequencies, so will the spatial reach.

The quantitative results in Figure 8.16 and Figure 8.17 are particular to the layer 5 neuron considered in Łęski et al. (2013). However, in the same study, it was found that the qualitative insights gained from these results generalize to other neuron types with non-uniform synaptic input patterns. While not discussed here, Łęski et al. (2013) also

---

[6] Implemented as alpha-functions with a very brief time constant of 0.1 ms.

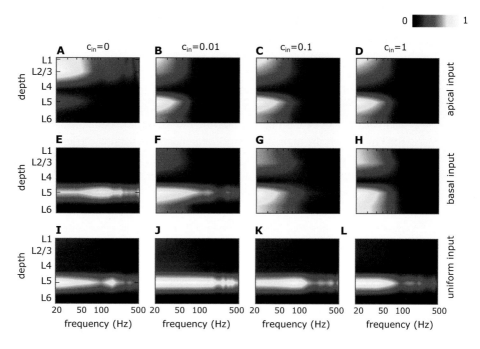

Figure 8.18 Depth dependence of LFP PSD. Results are shown for the LFP in the center of a population of layer 5 pyramidal cells taken from Mainen & Sejnowski (1995) for different input correlation levels ($c_{in} = 0, 0.01, 0.1, 1$) and different patterns of synaptic input (apical, basal, uniform). Population radius: $R = 1000$ μm. Values in each panel are normalized separately. Note that this normalization hides that the PSDs resulting from uncorrelated (left column) or uniform inputs (bottom row) may be several orders of magnitude smaller than those of correlated and non-uniform inputs (see Łęski et al. (2013) for details). Figure adapted from Łęski et al. (2013).

investigated the frequency dependence of the LFP decay outside an active population, similarly to how we in Section 8.3.2 studied the decay of the total (not frequency-resolved) LFP. Also, in that case, it was found that the numerical observations could be well accounted for by a frequency-resolved analytical model (equation (8.40)).

8.4.3.2   Depth Dependence

Whereas we focused on the population LFP calculated at the soma level of neuronal populations earlier, we here move on to investigate how the LFP varies with depth. This can also be studied using the analytical model in equation (8.40), which has been shown to reproduce salient features of the depth dependence well when compared to "ground-truth" numerical simulations (Łęski et al. 2013, figure 14). For the neural simulation used in Figure 8.16 and Figure 8.17, it predicts a depth dependence of the LFP as shown through the power spectral densities (PSDs) in Figure 8.18.

As for the soma-layer LFP, the level of correlations is also a crucial parameter at other depths (Figure 8.18). For the case of uncorrelated ($c_{in} = 0$) non-uniform synaptic inputs, prominent LFP PSDs are essentially evoked only near the inputs (superficial layers in panel A, deep layers in panel E).

For highly correlated non-uniform synaptic inputs (Figure 8.18D, H) the LFP (for frequencies less than about 60 Hz) has clear "dumbbell patterns," with two poles, one at each end of the dendritic structure of the neuron. This dumbbell structure is not present in the case of uniform inputs (panel I–L). Panels B–D, G, and H also reveal that correlations in the input boosts the low-frequency components ($\lesssim 100$ Hz) of the PSD more than the high-frequency components, as we saw earlier in Figure 8.17F. For uniform or uncorrelated inputs, on the other hand, there is no such "boosting bias," and less relative attenuation of the higher-frequency components.

The similarity in PSD magnitudes between superficial and deeper layers in Figure 8.17C–D, G–H can be explained by (i) the population being large (with R much greater than the cut-off distance r_*) and (ii) the inputs being correlated. As we argued in Section 8.3.1, the LFP in the center of large populations receiving non-uniform correlated input will be dominated by the summed contributions from distal neurons ($r \gtrsim r_*$). As indicated in Figure 8.7, the shape functions for different depths converge towards similar values in the large-r limit.

The almost-zero PSD halfway between the apical and basal dendrites reflects the dipolar spatial structure of the single-neuron LFP, which changes sign at this halfway point.

Note that a dipolar dumbbell structure as seen in Figure 8.18 does not directly imply that the current sources can be represented by a single (current-) dipole (Section 4.4) or a two-monopole (two-compartment) configuration (Section 3.2.4). After all, the current sources underlying the dumbbell structure in Figure 8.18 are spatially distributed and not localized in points.[7]

For uncorrelated ($c_{in} = 0$) or uniformly distributed input, the coherences between the individual LFP contributions are very small, and the neurons close by ($r \lesssim r_*$) will dominate. Then, the PSD will generally be much larger in the layer receiving the most synaptic inputs than in other layers, such as especially seen in Figure 8.18A, E.

8.5   LFP Contributions from Active Ion Channels

The membrane of a typical neuron contains a multitude of different active ion channels (or conductances). Arguably, the star among them is the Na^+ ion channel, which opens when the membrane potential is depolarized above a certain threshold and leads to the runaway depolarization known as the action potential (AP). The extracellular signature of an AP was referred to as a *spike* in Chapter 7. In the current section, we will use the more specific term *sodium spike* to distinguish it from so-called *calcium spikes*, which reflect another type of runaway depolarization of the neural membrane due to the activation of certain types of Ca^{2+} ion channels, and so-called *NMDA spikes*, which reflect runaway depolarization due to the activation of NMDA synapses. While it is

---

[7] For a study of the applicability of the dipole approximation to model LFPs, see Thio et al. (2022).

typically assumed that LFPs are dominated by subthreshold synaptic input currents and the associated return currents, possible LFP contributions from sodium, calcium, and NMDA spikes are briefly discussed in Section 8.5.2, Section 8.5.3, and Section 8.5.4, respectively.

Besides spike-generating Na^+ and Ca^{2+} and NMDA conductances, neurons can also possess a collection of "less dramatic" ion channels that are activated at voltage levels below the AP threshold. Although these *subthreshold* ion channels do not give rise to runaway responses, they may still affect how the neuron integrates synaptic inputs (Stuart et al. 2007, chapter 9). While LFPs from sodium and calcium spikes with some right can be regarded as specific, isolated events, the subthreshold ion channels can rather be seen as providers of modulations of the synaptic return currents. Modeling studies have demonstrated that the effects of such modulation on synaptically generated LFPs may be substantial (Sinha & Narayanan 2015, Ness et al. 2016, Ness et al. 2018, Sinha & Narayanan 2021). We therefore put an extra focus on presenting these modulatory subthreshold effects on LFPs and start this section with that. As before, we use the Hay model (described in Section 3.2.6) for our studies.

8.5.1   Subthreshold Active Ion Channels

Unlike the conductance of passive leak channels, the subthreshold conductances of active ion channels vary with the membrane potential. This implies that the responses of the dendrites to current input in principle will be nonlinear. When active conductances are present, the frequency components of the LFP will no longer be fully decoupled, as should be kept in mind when analyzing them.

8.5.1.1   Linearity

Fortunately (at least from a modeler's perspective), the subthreshold response of neurons may still be quite linear in practice. One example is shown in Figure 8.19, where the Hay model neuron is stimulated with step-current injections. The resulting somatic peak membrane potential is seen to grow linearly with the stimulus amplitude over a wide range. The only substantial deviation from linearity is for suprathreshold current input leading to the generation of an action potential. Such linearity in the presence of active conductances implies that either the active conductances remain unactivated or that their activation results in a roughly linear response over the relevant voltage regime.

The approximately linear responses suggest that linear "quasi-active" models of the nonlinear membrane currents can be used (see equation (6.36)). This has two benefits. Firstly, by making the system linear, one may study each frequency component of the LFP separately as in Section 8.4.2. Secondly, the number of model parameters describing the active currents can be reduced while preserving the key dynamical properties of the system (Remme & Rinzel 2011).

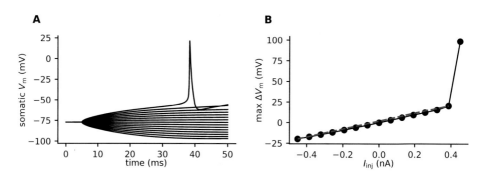

Figure 8.19 Subthreshold response to somatic current injection is close to linear. A: Somatic membrane potentials following step-like somatic current injections I_{inj} to the Hay neuron model. Results for 15 step-current amplitudes, from -45 nA to 0.45 nA, are shown. For the largest step-current value, an action potential is generated. **B:** Largest deviation from the resting membrane potential following the somatic step-current injections shown in panel A. The curve is almost perfectly linear for subthreshold current injections – that is, for injections that do not lead to the generation of a spike. The red dashed line is a straight line between the first and last point where the neuron is in the subthreshold regime. Code available via www.cambridge.org/electricbrainsignals.

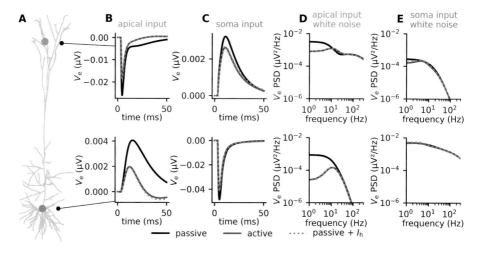

Figure 8.20 Illustration of the effect of I_h on single-cell LFPs. Extracellular potential V_e outside passive and active versions of the Hay model neuron receiving apical or somatic excitatory synaptic input. **A:** Input positions are marked by colored dots: apical (cyan), soma (green). Recording positions next to the input positions are marked with black dots. **B:** V_e next to the apical dendrite (top row) and soma (bottom row) for apical synaptic input. V_e responses are smaller for the active model (red curves) than the passive model (black curve). A model version that includes I_h but is otherwise passive (dashed curves) gives essentially the same V_e as the active model. **C:** Same as panel B but with somatic synaptic input. **D–E:** LFP power spectral densities (PSDs) for the same situations as in panels A and B, respectively, but with the synapses replaced with white-noise current injections. Code available via www.cambridge.org/electricbrainsignals.

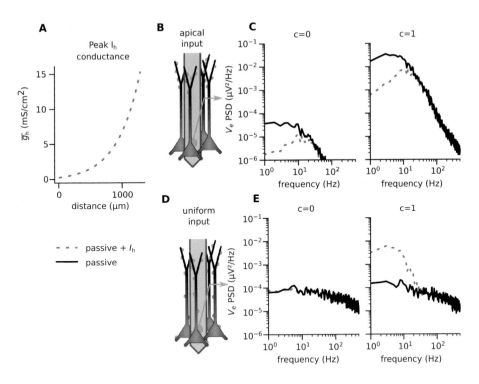

Figure 8.21 Effect of input correlations on population LFP with I_h. Two versions of the Hay model: one with all active conductances removed, except I_h, and one completely passive version. **A**: Maximal I_h conductance ($\bar{g}_h$) along the apical dendrite as function of distance from soma. **B–C**: LFP-PSD for apical synaptic input (panel B) with different levels of correlation (panel C, columns). **D–E**: Same as panels B–C but for uniformly distributed synaptic input. The blue dots in panels B and D illustrate the position of the LFPs depicted in panels C and E. Redrawn based on data from Ness et al. (2018).

8.5.1.2   Effects of the Subthreshold Current I_h

We can explore the effects that active subthreshold conductances have on the LFP by comparing LFPs produced by an active cell model to those produced by a passive version of the same model (where all active conductances have been removed). Such a comparison is done for the Hay model in Figure 8.20.

The difference between the LFPs produced by the active and passive model versions depends strongly on where the synaptic input is placed, and this difference is more pronounced for apical input (Figure 8.20B) than for somatic input (Figure 8.20C).

Ness et al. (2016) points to a key contribution to the LFP from the hyperpolarization-activated cation current I_h. The I_h conductance has been found both in cortical and hippocampal pyramidal neurons, predominantly in the distal dendrites (Magee 1998, Harnett et al. 2015, Kalmbach et al. 2018). This is also the case in the Hay model where the I_h conductance increases strongly with distance from the soma along the apical dendrite (Figure 8.21A). As demonstrated in Figure 8.20, the full Hay model and a model with I_h as the only active ion channel included give essentially identical LFPs. This suggests that the LFP contribution from I_h dominates over the contributions from other active currents in this subthreshold regime.

With an apical input, a biphasic LFP signal – that is, a signal that changes sign over its time course – is observed close to the soma for the active model (Figure 8.20B, lower panel). In frequency space, this corresponds to band-pass behavior as seen clearly when white-noise input is applied (Figure 8.20D, lower panel): for the case including I_h (passive + I_h), the power spectral density (PSD) of the LFP recorded outside the soma is dramatically decreased at low frequencies, with the result that the PSD has a (global) peak value for a frequency of about 10 Hz. However, the peak is absent when the h-type conductance is completely removed (passive).

For apical input, the I_h-induced peak is more prominent in the LFP close to the soma (Figure 8.20D, lower panel), reflecting that the I_h conductance modifies the membrane current in the soma region more than in the apical region close to the synaptic inputs. A qualitative picture explaining this effect is that the return current from the apical input has to pass through a long stretch of dendrite with high density of I_h channels before it leaves the neuron in the soma region. The I_h conductance has a much smaller effect on the LFP when the input is in the soma, as it is only weakly expressed in the basal and proximal apical dendrites in the model (Figure 8.20E). A further observation is that the PSD exhibits a notably steeper falloff with frequency when measured on the side opposite to the synaptic input. This is not predominantly an effect of I_h but rather of intrinsic dendritic filtering as described in Section 8.4.1.

The salient effects of I_h on the single-neuron LFP PSD are qualitatively reproduced on the population level, provided that neurons receive uncorrelated synaptic inputs (Figure 8.21C). This is as expected since contributions from uncorrelated single-neuron LFPs with similar frequency content should not change the overall shape of the PSD, only the magnitude (Section 8.4.3.1).

In some cases, correlations of the synaptic input can lead to a boosting of the LFP (Figure 8.21). For passive neurons, this boosting-by-correlation effect only takes place when the synaptic input is non-uniformly distributed on the cellular membranes (Figure 8.21C). In contrast, when the neurons have a non-uniform I_h conductance in the dendrite, the boosting takes place both for non-uniformly and uniformly distributed synaptic inputs (Figure 8.21C, E). However, the details of the PSD differ between the two input scenarios. For apical synaptic inputs, I_h acts to decrease the boosting of the LFP power at frequencies lower than $\approx 10\,\text{Hz}$ compared to the passive case (compare the solid and dashed curves in panel C). For uniform synaptic input, the presence of I_h instead boosts the LFP PSD at frequencies less than $\approx 20\,\text{Hz}$ (compare the solid and dashed curves in panel E). This boosting stems from breaching the uniformity of the electrical membrane properties of the neuron by the I_h conductance (Figure 8.4H).

The key take-home message from Figure 8.21 is that boosting of LFPs by correlations presupposes either a non-uniform spatial distribution of synaptic inputs or, at least for the Hay model, a non-uniform spatial distribution of the I_h conductance.

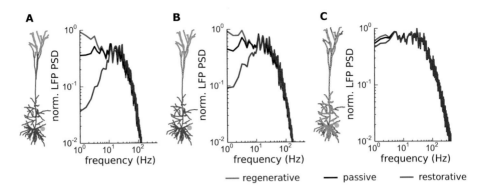

Figure 8.22 Band-pass properties of LFP with restorative active conductances. A: Passive Hay model, with customized quasi-active (linear) conductances added. The model expresses a single linearly increasing (with distance from the soma) quasi-active conductance that is either restorative (blue), passive-frozen (black), or regenerative (red). One thousand excitatory conductance-based synapses (green circles) are distributed across the distal apical tuft more than 900 μm from the soma. The synapses are activated by independent Poisson processes, and the PSD is computed from the LFP close to the soma (cyan dot). **B:** As in panel A but with synapses distributed above the main bifurcation, 600 μm from the soma. **C:** As in panel A but with synapses distributed uniformly across the neuron. Redrawn based on data from Ness et al. (2016).

8.5.1.3   Restorative versus Regenerative Conductances

The I_h conductance stands out as maybe the most important subthreshold conductance in terms of its putative effects on the LFP for two reasons: (i) its unusual electrical properties providing a hyperpolarization-activated depolarizing membrane current and (ii) its spatially varying expression in large pyramidal neurons in the cortex and hippocampus (Migliore & Shepherd 2002, Ness et al. 2016, Ness et al. 2018).

However, there are other active subthreshold conductances, and all can be divided into two classes: restorative and regenerative. The I_h conductance is an example of a restorative conductance, another being the conductance of the M-type slow potassium current (Hu et al. 2007). Restorative currents phenomenologically act as negative feedback and dampen low-frequency components of the membrane potential when responding to synaptic inputs (Remme & Rinzel 2011). As a consequence, the temporal duration of the response to a synaptic input is shortened (Figure 8.20). Regenerative currents instead act as positive feedback, amplifying the low-frequency components of the membrane-potential responses to synaptic inputs, thus making the membrane-potential response to a synaptic input broader. Examples of regenerative conductances are those of the persistent sodium current I_{NaP} and the low-voltage-activated calcium current I_{Ca_LVA} (Remme & Rinzel 2011).

The key differences between restorative and regenerative conductances in terms of the PSD of LFP signals are illustrated in Figure 8.22, where all the active conductances in the Hay model have been replaced with a single quasi-active conductance (equation (6.36)) whose expression increases linearly with distance from the soma.

In accordance with previous findings for uncorrelated synaptic inputs, a distinct peak in the PSD of the LFP close to the soma is observed only in the case of apical synaptic input (Figure 8.22A, B). For uniformly distributed input, this peak is almost absent (Figure 8.22C), as previously also observed for a neuronal population in Figure 8.21E. Analogously, for the regenerative model, the boosting of the low frequencies in the PSD is only reliably observed for the models receiving non-uniform input (panels A and B). The qualitative picture is very similar for the case of correlated apical inputs, though somewhat different and more complex for uniform inputs (see Ness et al. (2018, figure 4)).

To summarize, modeling studies suggest that the most striking effect of active ion channels on synaptically driven LFPs is a sizable peak in the PSD outside the soma produced by restorative conductances in the case of apical synaptic inputs. This suggests that restorative conductances like I_h can play a unique role in shaping the LFP in a way that depends on the positions and correlation of synaptic inputs (Ness et al. 2018). The peak in the LFP PSD essentially stems from the interaction between the low-pass filtering of the membrane and the active dampening of low frequencies by the restorative conductance, and this peak can also be seen in the membrane potential PSD (Hu et al. 2009, Remme & Rinzel 2011, Zhuchkova et al. 2013). As the peak in the LFP PSD reflects the membrane currents, the peak in the more low-pass filtered membrane potential PSD is expected to occur at a slightly lower frequency.

8.5.2   Sodium Spikes

Despite their brief duration, sodium spikes may contribute to extracellular potentials at frequencies as low as 100 Hz – that is, well within the LFP band (Pettersen et al. 2008, Ray & Maunsell 2011, Schomburg et al. 2012, Scheffer-Teixeira et al. 2013, Luo et al. 2018). We saw an example of this in Figure 7.3, showing the frequency spectrum for sodium spikes generated by the Hay model.

An action potential (AP) is not mediated by Na^+ alone. For one, the repolarization phase (downstroke) of the AP is mediated by K^+ ion channels that activate at voltage levels somewhere above the AP firing threshold. In addition, many neurons contain a collection of other *suprathreshold* ion channels, which may act to modify the shape of the AP or the membrane potential in the aftermath of the AP. Many of these suprathreshold ion channels have slower gating dynamics than the main AP-generating Na^+ and K^+ couple, and these channels may therefore increase the low-frequency components of the sodium spike waveform. Important players in this context are so-called afterhyperpolarizing K^+ ion channels, which can cause quite enduring spike afterhyperpolarizations (Buzsáki et al. 2012).

As described in Section 7.8.1, low-frequency components in sodium spikes can be boosted compared to the higher-frequency components if sodium spikes from different neurons are synchronized. Schomburg et al. (2012) explored this phenomenon in the context of fast oscillations and so-called ripple-wave generation in the hippocampus. A conclusion from their modeling study was that spikes contribute about half the signal power in the frequency range from 140–200 Hz, the other half being contributed by synaptic inputs. This large contribution from spikes might be a special case pertaining to a particular phenomenon in the hippocampus, however, and in any case, spikes only contribute to the high-frequency ("high-gamma") part of the LFP.

8.5.3   Calcium Spikes

There exist several types of voltage-gated Ca^{2+} channels whose activation can lead to neural calcium spikes. Many of these are preferentially located in neural dendrites, at least in some neuron types (Migliore & Shepherd 2002). In cortical pyramidal neurons, voltage-gated Ca^{2+} channels have been reported to be highly concentrated around the main bifurcation point of the apical dendrite (Larkum et al. 2009, Hay et al. 2011), as we have indicated in Figure 8.23A. The enduring membrane depolarization associated with dendritic calcium spikes (Figure 8.23B) can drive the somatic voltage level above the AP firing threshold and lead to bursts of action potentials. Reversely, a high level of somatic activity can also trigger dendritic calcium spikes (Larkum et al. 1999, Waters et al. 2005, Spruston 2008, Hay et al. 2011, Allken et al. 2014, Almog & Korngreen 2014). Since calcium spikes are thought to have an important physiological function (Spruston 2008, Larkum et al. 2009, Xu et al. 2012, Larkum 2013, Manita et al. 2015), their effect on, and their potential detectability from, the LFP is important.

As calcium spikes tend to be much broader than sodium spikes, they are prone to give more sizable contributions to the LFP. Indeed, Suzuki & Larkum (2017) reported that sensory-evoked dendritic calcium spikes gave LFP contributions of comparable magnitude as synaptic inputs. These findings have also been supported through computational modeling (Herrera et al. 2020, Næss et al. 2021, Herrera et al. 2022).

The Hay model neuron can exhibit dendritic calcium spikes, evoked either through coincident somatic and apical depolarization or in response to a rapid firing of somatic action potentials (Hay et al. 2011). As a demonstration of how calcium spikes can affect the LFP, we used the latter approach to evoke calcium spikes in the example in Figure 8.23. We gave the cell three strong excitatory synaptic inputs to the soma with an interval of five milliseconds, resulting in three somatic APs and a calcium spike in the apical dendrite (Figure 8.23B). The contribution to the extracellular potential was markedly different during the first somatic sodium spike (Figure 8.23D) than during the calcium spike (Figure 8.23E), with the sink in the latter case being located around the location of the calcium "hot" zone in the model (Figure 8.23A, red marker).

The same simulation but with Ca^{2+} channels removed gave a similar somatic membrane potential in response to the synaptic input, but it gave a weaker and briefer response in the membrane potential of the apical dendrite (Figure 8.23C).

We might expect that individual dendritic calcium spikes will give too-low LFP amplitudes to be above the noise level, and in the present example, the maximum amplitude was only on the order of $10\,\mu V$ (Figure 8.23E). We therefore also considered a population of pyramidal neurons, constructed on the basis of the single-neuron simulations with and without Ca^{2+} conductances.

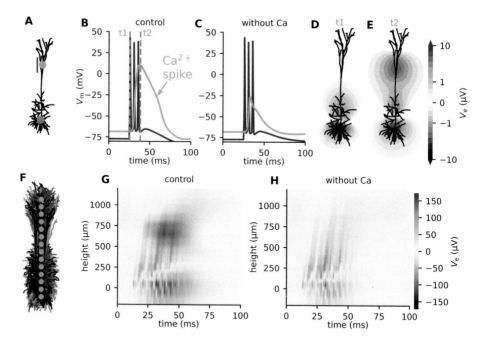

Figure 8.23   Contribution from dendritic calcium spikes to LFPs. A: The Hay model neuron, with the Ca^{2+} "hot" zone in the apical dendrite indicated with a red bar. **B**: Membrane potential at two locations (marked by dots in panel A) for three strong excitatory synaptic inputs to the soma, resulting in three somatic APs (blue) and a dendritic calcium spike (orange). **C**: The same as in panel B but with the Ca^{2+} conductances removed from the model. **D**: Extracellular potential during the first AP (at the time indicated by "t1" in panel B). **E**: Extracellular potential during the calcium spike (at the time indicated by "t2" in panel B). **F**: Population of 100 Hay model neurons, distributed uniformly in the horizontal plane and normally (with standard deviation 100 μm) in the depth direction within a cylinder with radius 100 μm. **G**: Laminar LFP computed by summing LFPs from 100 Hay models positioned as in panel F, each jittered (with a normal distribution) in time with standard deviation 10 ms. **H**: Same as in panel G but for the model without Ca^{2+} conductances (panel C). Code available via www.cambridge.org/electricbrainsignals.

Since sodium spikes are brief and have a biphasic extracellular waveform (Chapter 7) while calcium spikes are longer-lasting and have a monophasic extracellular waveform, we can expect more cancellation of sodium spikes than of calcium spikes when summing multiple (jittered) single-cell contributions. In this example, the additional sink from the calcium spike is clearly visible (compare panels G and H in Figure 8.23). These simulations seem to reproduce many of the features observed in the LFP measurements presented by Bereshpolova et al. (2007) and Suzuki & Larkum (2017). Note, however, that we do not here attempt to make any general claim about the relative importance of sodium versus calcium spikes to the LFP.

Since calcium spikes are broader than sodium spikes, calcium spikes will contribute to the LFP at a lower range of frequencies (Figure 1.2B). For the calcium spike in Figure 8.23, it can for example be verified that all its signal power is below 300 Hz (see code example available via www.cambridge.org/electricbrainsignals). In contrast, sodium spikes typically also have sizable power for frequencies larger than 500 Hz (Figure 7.3).

8.5.4   NMDA Spikes

NMDA synapses are glutamatergic synapses that are somewhat more complex than AMPA synapses. Unlike for AMPA synapses, binding to glutamate alone is typically not sufficient to activate NMDA synapses as they are often blocked by Mg^{2+} ions, which are only removed when the neuron is sufficiently depolarized. NMDA activation can thus act as a coincident detection between glutamatergic input and prior depolarization. Once activated, NMDA synapses can give rise to sustained plateau-like depolarizations, sometimes referred to as NMDA spikes. While calcium spikes are evoked in the thick apical dendrites of pyramidal neurons, NMDA spikes tend to occur in finer (basal, oblique, and tuft) dendritic structures (Schiller et al. 2000, Larkum et al. 2009).

In terms of amplitude and duration, NMDA spikes are comparable to calcium spikes and can give membrane potential amplitudes up to 40–50 mV and last for several hundred milliseconds (Antic et al. 2010). However, while calcium spikes are fairly global and involve larger areas of the thicker dendritic structures, NMDA spikes tend to be more local and can involve only a single branch in the dendritic tree (Antic et al. 2010). For that reason, NMDA spikes are probably less likely to contribute to LFPs than calcium spikes. However, simultaneous NMDA spikes in two nearby apical tuft dendrites have been shown to propagate to the trunk and trigger more global calcium spikes (Larkum et al. 2009). In such scenarios, effects from calcium spikes and NMDA spikes on LFPs may be hard to separate.

8.6   Slow Potentials

Fluctuations in the extracellular potential in the low-frequency range below about 0.5 Hz are often referred to as *slow potentials*.[8] Slow potentials have been known to be present since the early EEG recordings using galvanometers (Brazier 1963, Caspers et al. 1984). As measurements of slow potentials were often distorted by uncontrollable artifacts due to for example motion, changes in electrode polarizations, or changes in skin resistance, they originally received relatively little attention. However, slow potentials may also contain relevant information about physiological phenomena such as brain-state transitions and so-called readiness potentials, as well as pathological phenomena such as spreading depression, stroke, and epilepsy (Caspers et al. 1984, Kovac et al. 2018).

Slow potentials can only be recorded reliably using DC-coupled amplifiers (Caspers et al. 1984). Although recordings with DC-coupled amplifiers are not uncommon (Kraig & Nicholson 1978, Herreras & Somjen 1993, Richter & Lehmenkühler 1993, Somjen 2004, Pietrobon & Moskowitz 2014, Kovac et al. 2018, Herreras & Makarova 2020), most extracellular recordings are – for various technical reasons – done using AC-

---

[8] Other names used in the literature are ultraslow potentials, infra-slow potentials, steady potentials, standing potentials, sustained potentials, or DC potentials.

coupled amplifiers. These often apply a lower cut-off frequency set somewhere between 0.1 and 1 Hz (Einevoll et al. 2007, Einevoll, Lindén et al. 2013, Kovac et al. 2018) and thus discard some of the information available in the low-frequency part of the signal.

According to Speckmann et al. (1994), the origin of slow potentials is typically localized changes in neuronal or glial membrane potentials, causing sustained membrane-potential gradients along the cellular extensions. Such membrane potential gradients imply the presence of sustained current loops, with intracellular and extracellular currents running between the affected and unaffected membrane areas along sustained potential gradients.

There are several ways in which such a membrane-potential gradient can come about. One possibility is that it arises due to a change in input conditions, such as an increased average frequency of synapse activations and excitatory postsynaptic potentials (EPSPs) arriving in apical dendrites – for example, due to increased thalamic input (Caspers et al. 1984, Birbaumer et al. 1990, Speckmann et al. 1994). In this case, the membrane-potential gradient would share origin with the "conventional" synaptic LFP. Another possibility is that the membrane-potential gradient arises due to a change in the extracellular K^+ concentration. A K^+ gradient across, for example, the cortical depth will coincide with a gradient in the K^+ reversal potential, which in turn will lead to membrane-potential gradients and sustained current loops as those described in the previous paragraph (Speckmann et al. 1994, Somjen 2004, Herreras & Makarova 2020).

Extracellular K^+ gradients are associated with many pathological conditions and may arise due to intense neuronal activity – for example, during seizures or in the onset of spreading depression – or due to impaired homeostasis – for example, after a stroke. Due to their strong responsiveness to changes in K^+ concentrations, astrocytes are believed to be a major contributor to K^+-related slow potentials via a process called glial K^+ buffering, where K^+ is taken up by the glial syncytium in regions where the extracellular concentration is high and is released in regions where the concentration is low (Orkand et al. 1966, Gardner-Medwin 1983, Chen & Nicholson 2000, Somjen 2004, Halnes et al. 2013).

In principle, slow potentials may also originate from intrinsic neural properties. As a neuron's soma and dendrites typically do not share the same set of membrane mechanisms, it does not seem unlikely that membrane potential gradients and corresponding current loops could be maintained even in a neuron at rest. Such a loop is, for example, present in the Hay model (Hay et al. 2011) that we throughout this book have used as our default neuron model. In the resting state, the Hay model by design has a membrane potential gradient along the apical tree with a slope of approximately 10 mV/mm, with higher values in the distal dendrites compared to the soma (Kole et al. 2006, Hay et al. 2011). This essentially makes the resting Hay model a stationary (current-) dipole, with a static extracellular DC potential of about 0.5 µV over the neural extension (for a single cell). For large populations, this can add up to a substantial DC potential (see the code example available via www.cambridge.org/electricbrainsignals).

The slow potentials associated with the sources and sinks along neural membranes are not principally different from other extracellular potentials that can be computed with the MC+VC scheme (Section 2.6.2). However, as concentration effects are typically not accounted for in this scheme, the membrane depolarization that depends on them will not be present in MC+VC simulations. Understanding slow potentials associated with K^+ gradients, occurring for example under pathological conditions, will therefore require a simulation scheme accounting for intra- and extracellular ion-concentration dynamics. Such a scheme was for example applied in the simple and idealized electrodiffusive model by Sætra et al. (2021), which illustrates how slow extracellular potentials depend on three main components: (i) slow current loops due to local changes in membrane potentials in neurons, (ii) slow current loops due to local changes in membrane potentials in glial cells, and (iii) an extracellular diffusion potential, which depends solely on the extracellular concentration gradients, and not on the presence of any cellular activity per se, although such gradients typically are a product of cellular activity. While the two first components are the same kind of source- and sink-generated extracellular potentials that we compute with the MC+VC scheme, the diffusion potential belongs to a different category. It depends on currents due to extracellular diffusion of ions and cannot be understood using standard volume-conductor theory, where all extracellular currents are assumed to be ohmic. Diffusion potentials will be treated in further detail in Chapter 12.

The mechanisms discussed in the current section are probably not contributing much to the LFPs measured in most experiments, since the use of a lower cut-off frequency in most LFP recordings are likely to filter out their contributions. However, it cannot be ruled out that these slow-potential-generating mechanisms in some scenarios can contribute to the LFP above the lower cut-off frequency.

8.7   Network LFPs

Sections 8.2–8.6 primarily focused on principled issues for how LFP signals are generated by neurons. Here, we change gears and instead give a brief overview of the relatively short history of biophysical modeling of LFPs generated by neural networks.

In an early study, Pettersen et al. (2008) investigated the LFP generated by a single population of layer 5 neurons, mimicking the LFP generated in the rat barrel cortex following sensory stimulation. Here, only feedforward connections from the thalamus to the cortical neurons were included. More principled studies followed using feedforward models to study LFPs from populations of passive (Lindén et al. 2011, Einevoll, Lindén et al. 2013, Łęski et al. 2013) (Section 8.3.1, Section 8.3.2, and Section 8.4.3) and active neurons (Ness et al. 2018) (Section 8.5.1), as well as a more experimentally oriented study of the unitary LFP generated by a single thalamocortical neuron impinging on neuron populations in layer 4 of the sensory cortex (Hagen et al. 2017).

The first study of LFPs generated by a comprehensive network of recurrently connected neurons was conducted by Reimann et al. (2013), who simulated a model of a rodent neocortical column comprising more than 12,000 neurons. In the same vein, Głąbska et al. (2014) computed LFPs for the so-called Traub model for a similar column (Traub et al. 2005) in an in vivo situation, and later, Ness et al. (2015) computed the

LFP for the same network in an in vitro setting where LFPs are recorded by multielectrode arrays. Tomsett et al. (2015) likewise simulated LFPs mimicking multielectrode in vitro recordings from macaque cortical tissue. Later studies of LFPs for recurrently connected cortical networks include the two-population studies in Hagen et al. (2018) and Hagen et al. (2022), LFPs generated by the comprehensive network model for mouse visual cortex (Billeh et al. 2020, Rimehaug et al. 2023), and LFPs at the cortical surface (ECoG, Chapter 10) generated by the Blue Brain Project's model of a neocortical column (Baratham et al. 2022).

Hippocampal LFPs have been investigated in models, both to explore the role of the I_h current in determining spike-LFP relationships (Sinha & Narayanan 2015, Taxidis et al. 2015) and setting up the salient oscillations observed in hippocampal LFPs (Ferguson et al. 2015, Ferguson et al. 2017, Chatzikalymniou & Skinner 2018, Skinner et al. 2021) as well as to explore the generation of LFPs as such (Sinha & Narayanan 2021). Large-scale network models have further been used to explore LFPs and their dependence on behavioral state in the mouse primary motor cortex (Borges et al. 2022, Dura-Bernal et al. 2023). Models like this have also been used to explore LFPs in the macaque auditory cortex (Dura-Bernal et al. 2023).

These studies have all used the full MC+VC scheme. The approximate "hybrid LFP" scheme where LFPs are computed based on previously computed (or measured spikes) – see Chapter 6 – has been used to compute cortical LFPs from the so-called Potjans-Diesmann model (Potjans & Diesmann 2014, Hagen et al. 2016) as well as the Brunel model (Brunel 2000, Skaar et al. 2020). Based on this hybrid scheme, Mazzoni et al. (2015) identified simplified proxy formulas for computing LFPs in recurrently connected networks without having to resort to explicit LFP calculations based on MC models.

However, these are what we believe to be early days for network LFP modeling. We expect that, in the coming years, large-scale network models will increasingly aim to also compute LFPs and other population measures of activity, not only spikes and membrane potentials from individual neurons.

9 Electroencephalography (EEG)

Electroencephalography (EEG) is among the oldest methods for investigating the electric activity of the brain, having been in use for almost a century (Berger 1929). Still today, EEG is one of the most important techniques for studying brain function and dysfunction, with important neuroscientific and clinical applications (Nunez & Srinivasan 2006, Lopes da Silva 2013, Biasiucci et al. 2019, Ilmoniemi & Sarvas 2019). In EEG measurements, electric potentials are recorded non-invasively with electrodes placed across the exterior surface of the scalp. Such measurements thus have relatively little interference on subjects performing tasks (like, for example, taking a selfie, as in Figure 9.1).

While EEG measurements are disadvantaged by having a relatively low spatial resolution and a low signal-to-noise ratio, their advantages include a high temporal resolution, ease of use, cost-effectiveness, non-invasiveness, and the ability to investigate the interplay between brain regions (Tivadar & Murray 2019). EEG measurements are commonly used to investigate cognitive processes, characterize brain states, diagnose disease, and estimate functional connectivity.

The EEG signal is affected in known characteristic ways by various brain conditions, like sleep or attention (Klimesch et al. 1998, Palva & Palva 2011, Siegel et al. 2012), or by brain disorders like epilepsy and schizophrenia (Niedermeyer 2003, Light & Näätänen 2013, Freestone et al. 2015, Mäki-Marttunen, Kaufmann et al. 2019). Therefore, it seems clear that the EEG signal contains valuable information about the underlying neural function or dysfunction. However, as pointed out by Cohen (2017), we know "shockingly little" about the actual link between underlying neural activity and the EEG signal, even when it comes to well-studied EEG signatures such as alpha, beta, and gamma oscillations, or the stereotypical EEG shapes in response to sensory stimuli (event-related potentials). There is thus a strong motivation to develop mechanistic models that can better our understanding of this link (Uhlirova et al. 2016, Cohen 2017, Mäki-Marttunen, Kaufmann et al. 2019).

An important topic in EEG signal analysis is *source localization*. The aim is then to infer the location of the underlying neural current sources from measured EEG signals. This problem, known as the EEG inverse problem, is a major technical challenge. However, it is not a problem of insufficient understanding of the underlying biophysics. Rather, it is a problem of data analysis, as the inverse problem in electrostatics is inherently ill-posed, meaning that it has no unique solution (Helmholtz 1853). Other constraints are thus necessary to estimate source distributions using inverse modeling. As

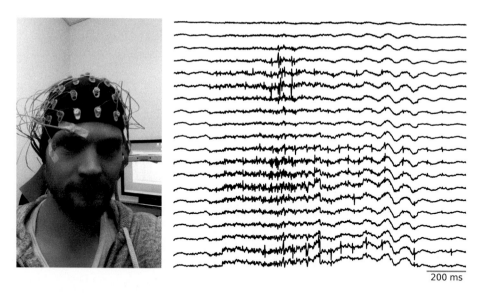

200 ms

Figure 9.1 EEG recording in progress. EEG electrodes are distributed over the surface of the scalp and held in place by an EEG cap. The EEG electrodes measure the electric potential stemming from brain-wide neural activity while the patient or research subject (in this case, one of the authors) is typically performing some mental task. To the right is an example of the raw EEG signal, which here also contain muscle artifacts that would typically be removed during the analysis.

the theme of this book is not data analysis, we refer readers interested in the EEG inverse problem to other literature sources (Nunez & Srinivasan 2006, Jatoi & Kamel 2017, Darbas & Lohrengel 2019, Ilmoniemi & Sarvas 2019, Sanei & Chambers 2021). In the current chapter, we will be dealing with forward modeling of EEG signals.

9.1   Forward Modeling of EEG Signals

Similar to the LFP (Chapter 8), the EEG signal is believed to predominantly originate from large numbers of synaptic inputs to populations of geometrically aligned pyramidal neurons (Nunez & Srinivasan 2006, Pesaran et al. 2018). EEG forward models can therefore be based on the same kind of two-step procedure as used for intracranial extracellular potentials (Section 2.6.2): one first simulates the neural activity in an independent step (step 1) and next combines the neural output with a volume-conductor model to compute the resulting EEG signal (step 2). However, since the EEG signal is measured outside the scalp, EEG forward models differ from models of LFPs and spikes in two main ways.

Firstly, when modeling intracranial extracellular potentials (Chapters 7–8), we relied on formulas like the point-source approximation,

$$V_{en}(\mathbf{r},t) = \frac{1}{4\pi\sigma_t} \frac{I_n(t)}{\|\mathbf{r} - \mathbf{r}_n\|} , \tag{9.1}$$

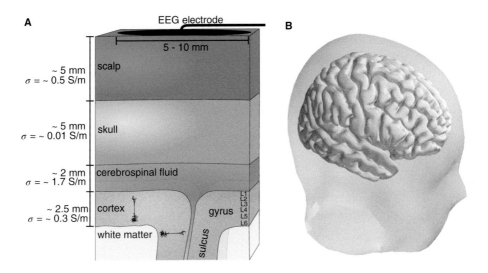

Figure 9.2 The spatial scale of EEG recordings. A: An illustration of the spatial scale involved in human EEG recordings, with typical thicknesses and electric conductivities of the different segments of the head. Note that there is a large variability in these parameters (see, for example, McCann et al. (2019)). **B**: An illustration of the human brain with its folded cortical surface and its distance from the scalp, based on the New York head model (Huang et al. 2016).

to compute the contribution from a membrane current $I_n(t)$ at position $\mathbf{r}_n$ to the extra-cellular potential at position $\mathbf{r}$. However, since the EEG signal is recorded far away from the neural activity it originates from, it reflects the activity of large populations of neurons, possibly spanning entire brain regions. At such large distances from the neural sources, details of the spatial distributions of membrane currents in individual neurons become less important. In EEG forward models, the neural membrane currents are therefore typically replaced by equivalent (current-) dipoles[1] (see Beltrachini (2019) for an exception). This is essentially a matter of convenience (further motivated in Section 9.5), as the membrane currents are still the origin of EEG signals. For a single dipole (moment) $\mathbf{P}_n(t)$ at position $\mathbf{r}_n$, the equation analogous to equation (9.1) is

$$V_{en}(\mathbf{r}, t) = \frac{1}{4\pi\sigma_t} \frac{\|\mathbf{P}_n(t)\| \cos\theta_n}{\|\mathbf{r} - \mathbf{r}_n\|^2} \, , \tag{9.2}$$

where θ_n is the angle between the vectors $\mathbf{P}_n$ and $\mathbf{r} - \mathbf{r}_n$ (Section 4.4).

Secondly, the analytical formulas listed here were derived under the assumption of a constant tissue conductivity σ_t. However, the EEG signal is strongly affected by the conductivity variations between different parts of the head, like the grey matter, skull, and scalp (Figure 9.2). EEG forward models thus call for a more complex volume-conductor model ($\sigma_t(\mathbf{r})$), often referred to as the *head model*, which in the most general case will

---

[1] When we speak of "dipoles" here, we refer to "current dipoles." Charge dipoles are not encountered in this book.

be an inhomogeneous and anisotropic conductivity tensor. Various head models used in forward modeling of EEG signals are introduced in Section 9.2.

For a non-homogeneous $\sigma_t(\mathbf{r})$, one can normally not derive an analytical expression for the relationship between the sources and extracellular signals. Instead, we can think of an EEG forward model $F_{\text{EEG}}(\mathbf{r}, \mathbf{r}_n, \mathbf{P}_n(t))$ more generally as any kind of function or numerical algorithm that provides a mapping

$$V_{en}(\mathbf{r}, t) = F_{\text{EEG}}(\mathbf{r}, \mathbf{r}_n, \mathbf{P}_n(t)) \tag{9.3}$$

from a time-dependent dipole $\mathbf{P}_n(t)$ at position $\mathbf{r}_n$ to the EEG signal at position $\mathbf{r}$. In many cases, such an algorithm can be precomputed for a given dipole position and set of recording positions to construct the so-called *lead-field matrix* $\mathbf{M}(\mathbf{r}, \mathbf{r}_n)$, which can be applied with a matrix multiplication as

$$V_{en}(\mathbf{r}, t) = \mathbf{M}(\mathbf{r}, \mathbf{r}_n)\mathbf{P}_n(t) . \tag{9.4}$$

See Section 6.3.6 for additional details.

In the following three sections, we simply postulate the presence of dipolar sources $\mathbf{P}_n$ and use them to examine how the EEG signal depends on the choice of head models (Sections 9.2 and 9.3) and on the degree of correlations between the dipoles (Section 9.4).

While the forward link between known dipoles and the resulting EEG signal is relatively well understood, the same is not true for the relationship between different types of neural activity and the equivalent dipoles themselves. This relationship is the topic of the last two sections of this chapter (Sections 9.5 and 9.6).

9.2   Head Models

Because the EEG signal is measured outside the head, it is affected by the different conductivities of the brain, cerebrospinal fluid (CSF), skull, and scalp (Figure 9.2A). Further, the human cortex is strongly folded, and the contribution to the EEG signal from a neural population will depend on whether the population is located in a *sulcus* ("groove") or a *gyrus* ("ridge") (Figure 9.2B). An integral part of an EEG forward model is therefore the head model – that is, the volume-conductor model describing the spatial variation of the conductivity $\sigma_t(\mathbf{r})$ of the head. Head models tend to be more complex than volume-conductor models that are used to simulate extracellular potentials inside the brain. Conveniently, the conductivity of the head seems approximately frequency independent (Pfurtscheller & Cooper 1975, Nunez & Srinivasan 2006, Ranta et al. 2017, Ilmoniemi & Sarvas 2019), as is also the case for the conductivity of brain tissue (Section 5.4).

9.2.1   Simplified Head Models

The first attempt to model the source of the EEG signal in terms of dipoles was done by Mary A. B. Brazier in 1949 (Brazier 1949, Brazier 1966). The electric potential

Region	Radius (cm)	σ (S/m)
Brain	8.9	0.276
CSF	9.0	1.65
Skull	9.5	0.01
Scalp	10.0	0.465

Table 9.1. Radii and electric conductivities used in the four-sphere model when applied to humans. The radius of each spherical shell in the four-sphere model, with σ denoting the respective electric conductivities. These parameters were taken from Huang et al. (2013).

produced by a dipole was calculated under conditions where the head was approximated either as an infinite homogeneous electric conductor or as a homogeneous spherical conductor. This allowed for a first investigation of how the shape and conductivity of the head model affected the EEG signal. The model was later expanded, first by including a spherical shell outside the main sphere, making it a two-sphere model (Geisler & Gerstein 1961), and later to make three-sphere (Rush & Driscoll 1969) and four-sphere models (de Munck & Peters 1993, Srinivasan et al. 1998, Næss et al. 2017).

Typical parameters for the sizes and electric conductivities of the different spheres are provided in Table 9.1. These spherical models all provide analytical formulas linking dipoles with EEG signals. As such, they remain valuable for studying and building intuition about generic features of EEG signals. For clinical purposes, however, they may be less useful because source localization based on spherical head models can lead to large localization errors when addressing the inverse problem (Akalin Acar & Makeig 2013).

The analytic solution to the four-sphere model is quite complex and cumbersome to work with (de Munck & Peters 1993, Srinivasan et al. 1998, Næss et al. 2017). However, open-source implementations are available (Næss et al. 2017, Hagen et al. 2018), and a simple minimal example with corresponding code can be found in Figure 9.3.

9.2.2 Detailed Head Models

In the past decades, advanced imaging techniques have become available for replacing the spherical head models with detailed head reconstructions (Dale et al. 1999), and subject-specific head models are now commonplace. Constructing subject-specific and anatomically accurate head models requires knowledge of both anatomy and the corresponding electrical properties. The first step in creating a detailed head model typically involves segmentation of the different tissue types of the head, based on data from magnetic resonance imaging (MRI) (Huang & Parra 2015), and the resulting head models often have an anatomical resolution of $1\,\mathrm{mm}^3$ or below (Huang et al. 2013, Huang et al. 2016). The second step is assigning a conductivity value to each tissue type, which can either be chosen from literature or experimentally estimated through a variety of methods (McCann et al. 2019).

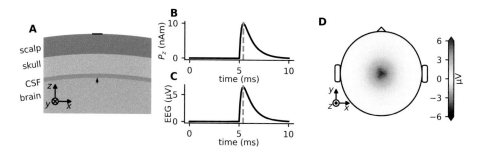

Figure 9.3 Minimal example of the use of the four-sphere model. A: The four-sphere model with its different concentric spheres. The black arrow marks the position and orientation of the dipole in the present example. **B**: The z-component of the dipole moment, used as the basis to calculate the EEG signal. **C**: The simulated EEG signal at the top of the head when the dipole moment in panel B is inserted into the four-sphere head model. **D**: Color plot of the EEG signal from the four-sphere head model at the time of the maximum dipole moment (dashed vertical line in panels B and C). The parameters for the four-sphere model are supplied in Table 9.1. Code available via www.cambridge.org/electricbrainsignals.

As imaging technology and computational methods have improved, increasingly detailed head models have been presented. Initially, the detailed head models would typically only include the brain, skull, and scalp (Vorwerk et al. 2014), but such models now frequently also incorporate CSF and anisotropic white matter (Wolters et al. 2006, Bangera et al. 2010) and might further subdivide the skull into its soft and hard constituents (skull *spongiosa* and *compacta*). For a review of the importance of including each of the different brain tissue types, see Vorwerk et al. (2014).

The accuracy of source localization from EEG recordings is critically dependent on the head model (Dale et al. 1999, Bangera et al. 2010, Akalin Acar & Makeig 2013, Vorwerk et al. 2014, Darbas & Lohrengel 2019, McCann et al. 2019). Source localization based on spherical head models has typically resulted in localization errors around 1–3 cm (Akalin Acar & Makeig 2013), while modern subject-specific head models typically give localization errors of less than 1 cm (Akalin Acar & Makeig 2013, Biasiucci et al. 2019).

Subject-specific head models can be constructed in different ways, with different levels of complexity, depending both on the problem at hand and on the method chosen for solving the EEG forward modeling problem. For EEG forward models based on the boundary element method (BEM), the boundaries between different head regions are represented as two-dimensional meshed surfaces, and the volumes between the surfaces are treated as distinct tissue types (Gramfort et al. 2010, Darbas & Lohrengel 2019). EEG forward models based on BEM simulations are substantially more accurate than the analytic spherical head models but have some limitations – for example, the surfaces must be reasonably smooth while the conductivity within all regions of each tissue type must be isotropic and homogeneous (Akalin Acar & Makeig 2013, Ziegler et al. 2014, Huang et al. 2016, Darbas & Lohrengel 2019).

EEG forward models based on the finite element method (FEM) can in principle incorporate arbitrary complex head models, since the entire volume is meshed instead of only the boundaries as in BEM (Ziegler et al. 2014, Darbas & Lohrengel 2019). FEM is computationally demanding, but with the steady increase in available computational power, it is emerging as the gold standard for EEG forward modeling based on detailed MRI reconstructions of the head.

Assuming that an appropriate forward model (equation (9.3)) has been constructed, applying it can still often be computationally demanding. This can however sometimes be aided by the so-called *reciprocity principle* (Rush & Driscoll 1969, Ziegler et al. 2014). The reciprocity principle essentially states that the electric potential at an electrode at position 1 resulting from a current source at position 2 will be the same as the electric potential at position 2 with the current source at position 1. When applied to the case of EEG signals, the reciprocity principle implies that, from a simulation of a current injection through a given EEG electrode (flowing to a ground electrode), we can – typically through FEM simulations – calculate the electric field at any position in the brain, which can be translated into a mapping that will give the EEG signal at the given EEG electrode from a dipole at any position in the brain. This can be done separately for each electrode to construct the lead-field matrix, **M** (see Section 6.3.6 for more technical details), from which we can find the EEG signal V_e at position **r** from an arbitrary set of dipoles $\mathbf{P}_n(t)$ at positions $\mathbf{r}_n$:

$$V_e(\mathbf{r}, t) = \sum_n \mathbf{M}(\mathbf{r}, \mathbf{r}_n)\mathbf{P}_n(t) . \qquad (9.5)$$

Using the reciprocity principle massively simplifies calculations since it means that solutions to the forward problem can be precomputed for a given head model through computationally demanding methods like FEM, after which the EEG signal at all electrodes from any combination of dipoles can be found by simple (lead-field) matrix multiplications. A lead-field matrix is available for the New York head model (Huang et al. 2016) and can be used as a stand-in EEG forward model in the absence of individual MRI data. A minimal example of EEG signals computed for the New York head model is shown in Figure 9.4. For a more complete treatment of the reciprocity principle applied to EEG signals, we refer to Rush & Driscoll (1969).

9.3 Effect of Head Models on EEG Signals

To illustrate how simulated EEG signals may depend on head model choice, we consider a cortical, radially oriented dipole (Figure 9.5A) and examine how the amplitude of the extracellular potential V_e falls off with distance from the source in the radial direction for two simple head models. With the head model being an infinite homogeneous medium, it follows from equation (9.2) that the amplitude of V_e falls off with distance as $1/r^2$ (grey line in Figure 9.5B). When using the inhomogeneous four-sphere model, which takes into account the different conductivities of the brain, CSF, skull, and scalp, as well as the finite volume of the head, the distance profile of the decay becomes more complex (black line in Figure 9.5B).

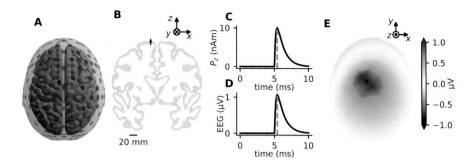

Figure 9.4 Minimal example of the New York head model. A: The New York head model (Huang et al. 2016), seen from the top, showing the cortical surface, the outline of the head, EEG electrodes (grey), and dipole position (black dot). **B**: A cross-section of the cortex from the New York head model, seen from the front. Each grey point represents a possible dipole position (the number of points is so high it looks continuous). The dipole position is marked by a black arrow. **C**: The z-component of the dipole moment, used as the basis to calculate the EEG signal. **D**: The EEG signal calculated from the dipole moment in panel C, inserted into the New York head model. Only the EEG signal at the electrode closest to the dipole position is shown. **E**: Color plot of the EEG from the New York head model, seen from the top, at the time of the maximum current dipole moment (dashed vertical line in panels C and D). Code available via www .cambridge.org/electricbrainsignals.

The spatial spread of the potential through the head from a cortical dipole is further illustrated by comparing a cross-section of the potential obtained for an infinite homogeneous medium (Figure 9.5C) with the four-sphere model (Figure 9.5D). In particular, notice that the EEG signal (the potential at the surface of the scalp) is more smeared out in the four-sphere model in that there is a larger separation between the contour lines (the spatial gradient of the potential is reduced). This effect can also be illustrated by considering how the EEG signal at the top of the head changes when the dipolar source is moved tangentially along the cortical surface (Figure 9.5E). For the infinite homogeneous head model, the EEG amplitude is initially higher than for the four-sphere head model, but the amplitude falls off more steeply as the population is moved further away tangentially to the cortical surface (Figure 9.5F). This spatial smearing (also called spatial low-pass filtering) effect of the EEG signal is well-known and stems from the low conductivity of the skull (Srinivasan et al. 1998, Nunez & Srinivasan 2006, Nunez et al. 2019).

9.3.1   Effect of Dipole Position on EEG Signals

As we saw in Figure 9.5B, the electric potential decays with distance from its source. Although subcortical sources in some cases can contribute (Krishnaswamy et al. 2017, Seeber et al. 2019, Piastra et al. 2021), the EEG signal is typically assumed to be dominated by cortical sources since they are closest to the EEG electrodes.

Intuitively, one might expect that neurons in the upper cortical layers will dominate the EEG because they are closer to the EEG electrodes than neurons in the lower cortical

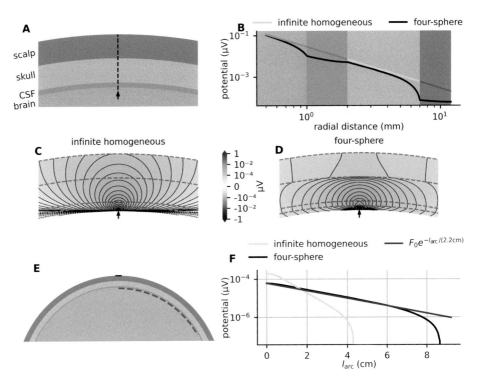

Figure 9.5 Decay of electric potential with distance from the source is strongly affected by head model. A: The four-sphere head model with the dipole position marked by the black arrow. **B**: Electric potential as a function of radial distance from the dipole (along the black dashed line in panel A). Results are shown for two different head models, an infinite homogeneous medium (grey line) and the four-sphere head model shown in panel A (black line). **C**: Contour plot of the electric potential from the dipole in panel A for an infinite homogeneous medium. **D**: Contour plot of the electric potential from the dipole in panel A for the four-sphere head model. **E**: The four-sphere head model where the dipole is moved along an arc tangentially to the cortical surface (red dashed line) while the EEG signal is calculated at the top of the head. The dipole is always radially directed, relative to the head model. **F**: The amplitude of the EEG signal at the top of the head as a function of dipole position along the arc (red dashed line in panel E) for an infinite homogeneous medium (grey line) or the four-sphere head model (black line). The decay with arc distance for the four-sphere head model appears approximately exponential, with an exponential-decay parameter of 2.2 cm (equation (9.15), blue) The parameters of the four-sphere model are listed in Table 9.1, while for the infinite homogeneous head model, we used the conductivity of the brain from Table 9.1, which is 0.276 S/m. Note that we here just assume the existence of a dipole, while in reality, the dipole approximation might not be justified close to the neural signal generators in panels A–D (see Section 9.5.1). Code available via www.cambridge .org/electricbrainsignals.

layers. However, relative to the dimensions of individual neurons or the thickness of the human cortex, the EEG electrodes are also far away from cortical neural sources (Figure 9.2). Together with the spatial smearing that we saw in Figure 9.5, the large distance implies that the EEG signal will be relatively insensitive to small shifts (on the order of millimeters) in the positions of the neural dipoles.

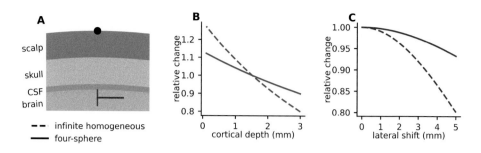

Figure 9.6 EEG signal is relatively insensitive to small movements of the dipole.
A: Illustration of the four-sphere head model, where the EEG signal from a single cortical dipole is calculated at the scalp surface (black dot). The dipole is moved either in the depth direction (red line, length 3 mm measured from the cortical surface) or in the lateral direction (blue line, length 5 mm measured from the lateral position of the EEG electrode). **B**: The relative change in the simulated EEG signal as a function of moving the dipole in the depth direction for either an infinite homogeneous head model (dashed line) or the four-sphere head model (full line). **C**: Same as in panel B but for lateral movement of the dipole. Code available via www .cambridge.org/electricbrainsignals.

Figure 9.6 illustrates how the EEG signal amplitude changes when the position of a single dipole is shifted along the depth direction of the cortex (Figure 9.6B) or up to a few millimeters in the lateral direction (Figure 9.6C). Such repositioning of the dipole changes the EEG signal amplitude quite substantially if the head is assumed to be an infinite homogeneous medium (dashed lines) but only by a few percent if the conductivity profile of the head is taken into account (full lines).

9.3.2 Effect of Dipole Orientation on EEG Signals

The EEG signal is highly affected by the orientation of the dipoles relative to the EEG electrodes. As alluded to also in earlier chapters (see, for example, Figure 8.4 and Figure 8.5), we can expect from radial symmetry that at the level of neural populations, the cortical dipoles will be aligned with the normal direction of the cortical surface (this is also illustrated later in Figure 9.11). However, in humans, the cortical surface has a highly folded geometry, with sulci and gyri (Figure 9.2). Dipoles within these struc-tures have different orientations relative to the EEG electrodes and will thus contribute differently to the EEG signal (Nunez et al. 2019).

The EEG signal contribution from a collection of dipoles is maximized if the dipoles are aligned in the same direction within a gyrus (Figure 9.7A) but will be practically non-existent if the same dipoles are randomly oriented up or down (Figure 9.7B). We can also expect a substantial EEG contribution from dipoles aligned in the same direction in a sulcus, but for this scenario, we expect a more dipolar EEG pattern above the dipole (Figure 9.7C). If the dipoles are aligned in the depth direction of the cortex but still distributed over a region that includes both a gyrus and a sulcus, we can expect something in between what we saw in panels A and C, as shown in Figure 9.7D. If the dipoles are divided into two opposing sulci, however, we can expect a very weak EEG contribution (Figure 9.7E).

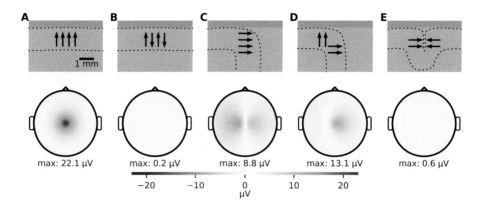

Figure 9.7 **Cortical geometry matters.** Different hypothetical folding patterns of the cortical surface are indicated with black dashed lines. The EEG signal is calculated from four (identical) dipoles with different orientations in the four-sphere head model. **A**: The dipoles are aligned in the same direction in a gyrus. **B**: The dipoles are pointing in opposite directions in a gyrus. **C**: The dipoles are aligned in the same direction in a sulcus. **D**: The dipoles (pointing towards the cortical surface) are divided between a gyrus and a sulcus. **E**: The dipoles (pointing towards the cortical surface) are divided between two opposing sulci. In all panels, the magnitudes of the dipole moments were 10 nAm, and the positions of the dipoles were at the centers of the depicted arrows. Code available via www.cambridge.org/electricbrainsignals.

Note that in Figure 9.7, panels A, C, D, and E could essentially represent the same ongoing neural activity – for example, basal excitatory input to a large population of pyramidal cells, causing cortical dipoles that are perpendicular to the (local) cortical surface. The entirely different EEG signals that we expect from these cases are introduced through the different assumed folding patterns of the cortex.

9.3.3   Comparison of Simple and Detailed Head Models

Given similarly positioned and oriented dipoles, we can expect EEG signals simulated using simple versus detailed head models to have the same qualitative properties but also some quantitative differences. This is illustrated in Figure 9.8, which compares the EEG signals obtained using the New York head model (panel A) to those obtained using the four-sphere head model (panel D) for two dipoles oriented radially or tangentially to the skull. The EEG signals were qualitatively similar in the two head models but of somewhat higher amplitude in four-sphere head model (see Figure 9.8, figure caption).

9.4   Effect of Dipole Correlations on EEG Signals

Like the LFP, the EEG signal is strongly affected by the level of correlation between the signal sources. To illustrate this, we consider a collection of dipoles evenly spread out over the cortical surface in a four-sphere head model. Following the approach used

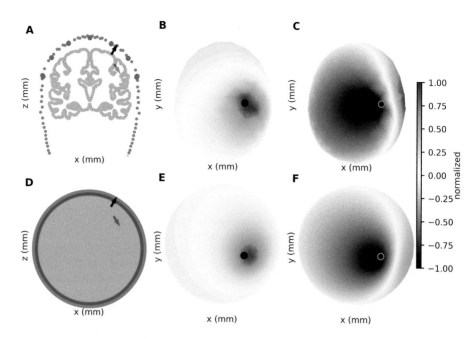

Figure 9.8 Comparison of EEG signals from simple and detailed head models. A: Two manually chosen dipole positions in the New York head model, either approximately radial (black) or approximately tangential (grey). The head model is seen from the back (x, z-plane). EEG electrode positions close to the chosen cross-section plane are marked in light blue, while the cortical cross-section is marked in light brown, and the outline of the head is marked by darker brown dots. The dipole moment is in all cases 10 nAm. **B**: Interpolated color plot of EEG signal from the radial dipole in panel A, seen from the top (x, y-plane). The plotted EEG signal is normalized, but the maximal value is 1.1 µV. **C**: Interpolated color plot of EEG signal from the tangential dipole in panel A, seen from the top (x, y-plane). The plotted EEG signal is normalized, but the maximal value is 0.7 µV. **D**: Similar dipole positions as in panel A but in the four-sphere head model. The head model is seen from the side (x, z-plane). **E**: Interpolated color plot of EEG signal from the radial dipole in panel D, seen from the top (x, y-plane). The plotted EEG signal is normalized, but the maximal value is 4.1 µV. **F**: Interpolated color plot of EEG signal from the tangential dipole in panel D, seen from the top (x, y-plane). The plotted EEG signal is normalized, but the maximal value is 1.7 µV. Note that although the dipole coordinates are similar in the two head models, the head shape and size, as well as the distance to the closest EEG electrode, will be somewhat different. Code available via www.cambridge.org/electricbrainsignals.

in Section 8.2.3, we assume the EEG contribution $V_{en}(t)$ from each dipole n to be a product of a position-dependent part $F_n(\mathbf{r}, \mathbf{r}_n, \overline{\mathbf{P}}_n)$ and a temporal function $\xi_n(t)$:

$$V_{en}(\mathbf{r}, t) = \xi_n(t) F_n(\mathbf{r}, \mathbf{r}_n, \overline{\mathbf{P}}_n) \,. \tag{9.6}$$

This formulation follows from equation (9.4) via the definitions

$$\mathbf{P}_n(t) = \overline{\mathbf{P}}_n \xi_n(t) \tag{9.7}$$

and

$$F_n(\mathbf{r}, \mathbf{r}_n, \overline{\mathbf{P}}_n) = \mathbf{M}(\mathbf{r}, \mathbf{r}_n, \overline{\mathbf{P}}_n) \,, \tag{9.8}$$

where $\overline{\mathbf{P}}_n$ is a constant vector in the same direction as $\mathbf{P}_n(t)$ and with a magnitude defined in the next paragraph.

The spatial component $F_n(\mathbf{r}, \mathbf{r}_n, \overline{\mathbf{P}}_n)$ determines the amplitude of the contribution to the EEG signal at $\mathbf{r}$ from the dipole positioned at $\mathbf{r}_n$. The temporal component $\xi_n(t)$ will in the following be assumed to have a (temporal) mean value of zero ($\langle \xi_n(t) \rangle_t = 0$) and variance of unity ($\langle \xi_n(t)^2 \rangle_t = 1$). The zero mean value implies a mean value of zero also for $V_{en}(t)$ so that the DC contribution to the EEG signal will be zero. The variance of unity implies that the constant vector $\overline{\mathbf{P}}_n$ must have magnitude $\|\overline{\mathbf{P}}_n\| = \sqrt{\langle \mathbf{P}_n(t)^2 \rangle_t}$.

Note that the assumption of spatiotemporally separable signals in equation (9.6) is valid provided that the head is frequency independent, which seems to be a good approximation (Section 9.2).

The EEG signal from the collection of dipoles is given by the sum of individual contributions

$$V_e(t) = \sum_n V_{en}(t) = \sum_n \xi_n(t) F_n , \tag{9.9}$$

where we have introduced the short-hand notations $V_{en}(t) \equiv V_{en}(\mathbf{r}, t)$ and $F_n \equiv F_n(\mathbf{r}, \mathbf{r}_n, \mathbf{P}_n)$ for readability.

As a measure of the magnitude of the EEG signal, we will in the following use its temporal variance denoted $V_{e\sigma}^2$. For a collection of dipoles as in equation (9.9), the variance is in Appendix G found to be

$$V_{e\sigma}^2 = (1 - c_P) \sum_n F_n^2 + c_P \left(\sum_n F_n \right)^2 . \tag{9.10}$$

Here,

$$c_P = \langle \xi_n(t) \xi_m(t) \rangle_t \tag{9.11}$$

is the Pearson correlation coefficient (equation (E.8)), which is assumed to be the same for all pairs of dipoles n and m where $n \neq m$.

9.4.1   Uncorrelated Dipoles

For the case where the dipoles are uncorrelated ($c_P = 0$), equation (9.10) simplifies to

$$V_{e\sigma}^2 = \sum_n F_n^2 . \tag{9.12}$$

Figure 9.9A–G shows how the compound EEG signal in this uncorrelated case grows with the number of dipoles, or more specifically with the area of the cortex containing dipole sources. Again, F_n determines the amplitude of the EEG contribution from a dipole n that in principle can have an arbitrary position and orientation. Figure 9.9A shows an example of how the EEG signal varies across the scalp for a small area of dipole sources (1 cm^2) at a randomly chosen time. The signal is computed with equation (9.9), and the (radially symmetric) patch of dipole sources is centered in the middle. Panel B shows the observed variances from this simulation, illustrating that the strongest EEG signals are found directly above the area containing dipole sources.

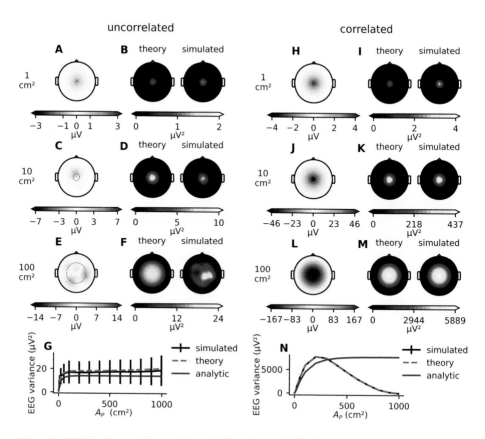

Figure 9.9 Effect of correlation and population size on EEG signals. The EEG signal was calculated from an uncorrelated or correlated population of dipoles in the four-sphere head model (parameters in Table 9.1). Each dipole moment had a maximum amplitude of 1 nAm, and the dipoles were evenly spread with a fixed density within a given patch of the (spherical) cortical surface with population area A_P. When the population area corresponded to the entire cortical surface, the number of dipoles was 5,000. **A:** The EEG signal at the time of the maximum amplitude for a completely uncorrelated population with radius $A_P = 1\,\text{cm}^2$. Dashed circles show the area of the population. **B:** The variance of the EEG signal in panel A, either calculated from theory (equation (9.12)) or directly from the numerical simulation in panel A. **C–D:** Same as panels A–B, but for $A_P = 10\,\text{cm}^2$. **E–F:** Same as panels A–B, but for $A_P = 100\,\text{cm}^2$. **G:** The EEG variance at the electrode on the top of the head from uncorrelated populations (panels A–F) as a function of population area, A_P, either calculated from the numerical simulation over 50 trials (black, error bars showing standard deviation), predicted from equation (9.12) (dashed grey), or the approximate analytic equation (9.16) (blue). **H–N:** Same as panels A–G, but for correlated populations. Code available via www.cambridge.org/electricbrainsignals.

However, moderately large EEG variances are also seen outside the patch of dipole sources. This spatial spread in the EEG signal resembles that seen for a single dipole in Figure 9.5F.

Figure 9.9B also shows that the variances predicted from equation (9.12) are in qualitative agreement with the simulation results. The agreement is not perfect though, since the "simulated" variance is computed from a simulation lasting a finite time. A perfect agreement would only be expected in the limit of infinitely long simulations.

Figure 9.9C–F show corresponding results when the area of dipole sources is larger ($10\,\text{cm}^2$ and $100\,\text{cm}^2$, respectively). The EEG signal is again by far the largest in the area containing dipole sources, and the formula in equation (9.12) gives a good qualitative account of the observations.

Figure 9.9G summarizes results for how the variance of the EEG signal above the center of the patch of current sources grows with the area of the patch. For patch sizes larger than about $100\,\text{cm}^2$, the EEG signal no longer increases with patch size. This is analogous to the observation in Section 8.3.1 that the population LFP saturates with increasing population sizes when the LFP sources are uncorrelated. A cortical area of $100\,\text{cm}^2$ corresponds to a maximal distance from the center of the dipole-source patch to the edge of the sources of about 6 cm. Hence, in analogy to the observation in Section 8.3.1 that the spatial reach of the LFP is a few hundred micrometers in the case of uncorrelated single-neuron LFP sources, the spatial reach of the EEG is in this example a few centimeters for uncorrelated dipole sources.

Figure 9.9G also shows that the theoretical expression for the variance in equation (9.10) generally gives predictions in close agreement with the simulation.

9.4.2 Correlated Dipoles

For the case where the dipoles are fully correlated ($c_P = 1$), we have from equation (9.10) that

$$V_{e\sigma}^2 = \left(\sum_n F_n\right)^2. \tag{9.13}$$

Figure 9.9H–N shows how the compound EEG signal grows with the area of fully correlated current sources, again using the four-sphere head model. Panels H, J, and L show examples of the EEG signals at a random time instance for small ($1\,\text{cm}^2$), medium ($10\,\text{cm}^2$), and large ($100\,\text{cm}^2$) patches of current sources, respectively, computed by use of equation (9.9). Panels I, K, and M show that the theoretically predicted variances using equation (9.13) are in excellent quantitative agreement with the simulation results. The most striking observations from these panels compared to the corresponding results for the uncorrelated case are (i) the much larger amplitudes of the EEG signal and (ii) the more regular spatial EEG patterns (in particular for the examples with $10\,\text{cm}^2$ and $100\,\text{cm}^2$ population areas). The EEG signal is again seen to be by far largest in the region with dipole sources.

Panel N in Figure 9.9 summarizes results for how the variance of the EEG signals above the center of the dipole sources grows with the area of dipole sources. A first observation is the much larger (more than a factor of 300) maximum EEG signal in this correlated case compared to the otherwise identical uncorrelated case in panels A–G. This observation mirrors the observation of much larger compound LFPs for correlated compared to uncorrelated single-neuron sources in Section 8.3.

Another difference from the uncorrelated case is that for dipole-source sizes larger than about $200\,\mathrm{cm}^2$, the EEG signal no longer increases; in fact, it decreases. This is due to cancellation effects: the distant dipoles will have EEG contributions with the opposite sign from the contributions from the dipoles close to the electrode. In fact, when correlated dipoles cover the whole brain sphere ($\sim 1000\,\mathrm{cm}^2$ in Figure 9.9N), the observed EEG signal is approximately zero.

Panel N in Figure 9.9 also shows that the theoretical expression for the variance in equation (9.13) generally gives predictions in excellent agreement with the simulation.

9.4.3   Analytical Theory

An approximate analytical theory gives further insights into the results so far in this section. The signal that we will be studying is the variance in the EEG signal directly above the center of a circular patch of dipole sources.

9.4.3.1   Uncorrelated Dipoles

To obtain analytical estimates for how the EEG signal increases with the area containing dipole sources, we approximate the expression for the EEG variance for uncorrelated dipoles in equation (9.12) with the integral

$$V_{e\sigma}^2 = \sum_n F_n^2 \approx 2\pi r_{\mathrm{brain}}^2 \rho_\mathrm{P} \int_0^{\theta_{\max}} F(\theta)^2 \sin\theta \, d\theta \,. \tag{9.14}$$

Here, θ is the polar angle between the z-axis and a radial axis from the center of the head to a position within the patch of dipoles, which spans from 0 (in the center of the patch) to $\theta_{\max}$ in the outer edge of the patch, while r_{brain} is the radius of the brain and ρ_P the density of dipoles per patch area.

Using equation (9.14), we assume that all individual dipoles have equal amplitudes so that their contribution to the EEG depends only on their positioning. $F(\theta)$ can be viewed as a dipole shape function, analogous to the single-neuron LFP shape function in Section 8.2.5. It can be computed for the four-sphere head model, and the result is shown in Figure 9.5F. With $r_{\mathrm{brain}}=8.9$ cm as in Table 9.1, the observed value of the signal decay from a dipole is seen to be well described by the function

$$F = F_0 e^{-l_{\mathrm{arc}}/a} = F_0 e^{-\theta r_{\mathrm{brain}}/a} \tag{9.15}$$

for arc distances along the cortical surface $l_{\mathrm{arc}} \lesssim 6$ cm when the parameter a is set to 2.2 cm (Figure 9.5F). An arc distance of 6 cm corresponds in this case to an angle $\theta = l_{\mathrm{arc}}/r_{\mathrm{brain}} = 6/8.9 \approx 0.67$ in radians – that is, 39 degrees.

Insertion of the shape function (equation (9.15)) into equation (9.14) now gives

$$V_{e\sigma}^2 = \sum_n F_n^2 \approx 2\pi r_{brain}^2 \rho_P F_0^2 \int_0^{\theta_{max}} e^{-2\theta r_{brain}/a} \sin\theta \, d\theta$$

$$= \rho_P F_0^2 \frac{\pi a^2}{2} \frac{1 - e^{-2\theta_{max} r_{brain}/a}(\cos\theta_{max} + 2(r_{brain}/a)\sin\theta_{max})}{1 + a^2/(4r_{brain}^2)}.$$

$$(9.16)$$

The area A_P of the brain with dipole sources is related to θ_{max} via

$$A_P = 2\pi r_{brain}^2 (1 - \cos\theta_{max}), \tag{9.17}$$

or conversely via

$$\theta_{max} = \arccos\left(1 - A_P/(2\pi r_{brain}^2)\right). \tag{9.18}$$

The use of equation (9.18) in equation (9.16) gives the curve labeled "analytic" in Figure 9.9G, which agrees qualitatively with the simulated results. For small areas, further analysis of equation (9.16) reveals that the EEG variance grows proportionally to the dipole-source area A_P in this regime, as seen in Figure 9.9G. For large dipole-source areas (large θ_{max}), the factor $\exp(-2\theta_{max} r_{brain}/a)$ becomes very small, and equation (9.16) can be approximated as

$$V_{e\sigma}^2 \approx \rho_P F_0^2 \frac{\pi a^2/2}{1 + a^2/(4r_{brain}^2)}. \tag{9.19}$$

As seen in Figure 9.9G, equation (9.19) is in good, but not perfect, agreement for the largest values of A_P as the equation underestimates the simulated results.

The key insight from this analytical exercise is that the EEG variance for the case of uncorrelated dipole sources covering a large cortical area is essentially proportional to the dipole density ρ_P, the square of the single-dipole contribution F_0, and the square of the exponential-decay parameter a from the single-dipole shape function in equation (9.15) (when $a \ll 2r_{brain}$ as in the numerical example in Figure 9.9). Specifically, the variance of the EEG signal is proportional to $\rho_P a^2$ – that is, proportional to the number of dipoles within a distance a from the electrode.

9.4.3.2 Correlated Dipoles

For correlated dipoles, the expression for the EEG variance in equation (9.13) is, in analogy to equation (9.20), approximated by the integral

$$\left(\sum_n F_n\right)^2 \approx \left(2\pi r_{\text{brain}}^2 \rho_P \int_0^{\theta_{\text{max}}} F(\theta) \sin \theta \, d\theta\right)^2$$

$$= 4\pi^2 r_{\text{brain}}^4 \rho_P^2 F_0^2 \left(\int_0^{\theta_{\text{max}}} e^{-\theta r_{\text{brain}}/a} \sin \theta \, d\theta\right)^2$$

$$= 4\pi^2 a^4 \rho_P^2 F_0^2 \frac{\left(1 - e^{-\theta_{\text{max}} r_{\text{brain}}/a}(\cos \theta_{\text{max}} + (r_{\text{brain}}/a) \sin \theta_{\text{max}})\right)^2}{\left(1 + a^2/r_{\text{brain}}^2\right)^2}.$$

$$(9.20)$$

With the use of equation (9.18) relating θ_{max} to the patch area A_P, equation (9.20) gives the curve labeled "analytic" in Figure 9.9N. The analytical formula is seen to qualitatively agree with the simulated results up to patch areas of about $200 \, \text{cm}^2$ but then plateaus out instead of becoming smaller with larger patches. This is as expected as the single-dipole shape function in equation (9.15) only applies close to the dipole and does not capture the change of sign of single-dipole EEG contributions underlying the cancellation effect observed in the simulations.

Nevertheless, the analytical formula gives a reasonable account for the maximum value of the EEG variance seen for patch areas up to around $200 \, \text{cm}^2$ (corresponding to an arc distance on the brain of about 8 cm). For large dipole-source areas (large θ_{max}), the factor $\exp(-\theta_{\text{max}} r_{\text{brain}}/a)$ becomes very small, and equation (9.20) can be approximated as

$$V_{\text{e}\sigma}^2 \approx \rho_P^2 F_0^2 \frac{4\pi^2 a^4}{\left(1 + a^2/r_{\text{brain}}^2\right)^2}.$$

$$(9.21)$$

The key insight from this analytical exercise is that the EEG variance for the case of correlated dipole sources covering a large cortical area is essentially proportional to the square of the dipole density ρ_P, the square of the single-dipole contribution F_0, and the exponential-decay parameter a from the single-dipole shape function in equation (9.15) to the fourth power. The variance of the EEG signal is correspondingly proportional to $\rho_P^2 a^4$ – that is, proportional to the square of the number of dipoles within a distance a from the electrode (assuming $a \ll r_{\text{brain}}$ so that the denominator in equation (9.21) is essentially 1).

We close this section by pointing out that these insights were derived using a spherical head model and that the results do not necessarily directly transfer to detailed human head models with folded cortical surfaces.

9.5   Biophysically Detailed Modeling of Neural Activity for EEG Signals

The EEG forward models described earlier in this chapter take dipoles as input. In the rest of this chapter, the focus is on how we can derive these dipoles from simulated neural activity. We may derive the dipoles from biophysically detailed neurons, and this approach is introduced and used in this section. EEG signals are, however, the result of the combined neural activity from entire brain regions, and at present, simulating neural activity at this scale using biophysically detailed models is often impractical or infeasible. In Section 9.6, we, therefore, review simplified schemes for simulating large-scale neural network activity in the context of modeling EEG signals.

9.5.1   From Membrane Currents to Dipoles

The origin of extracellular potentials, including EEG signals, is spatially distributed neural membrane currents (sources and sinks). However, when modeling or analyzing EEG signals, it is more common to instead represent the sources in the form of dipoles. These are easily computed and compact representations of the neural sources, which still tend to give accurate predictions of EEG signals in humans due to the large distance between the neurons and the EEG electrodes.

We learned in Section 4.4 and Appendix B that V_e from an arbitrary set of current sources I_n can be expressed as a current-multipole expansion. We also learned that the monopole contribution is zero since no neuron is a net current source. The dipole moment is given by $\mathbf{P} = \sum_n I_n \mathbf{r}_n$ (equation (4.51)) and can be obtained from any set of simulated membrane currents by following the biophysically detailed approach to modeling neural activity outlined in Chapter 3. Calculating V_e from only the dipole moment is referred to as the dipole approximation. Since biophysically detailed cell models can have hundreds of compartments (and thus hundreds of sources I_n), representing them in terms of the dipole moment – a single time-dependent three-dimensional vector – is substantially more compact.

A brief discussion of the validity of the dipole approximation under different circumstances is in order. To validate the dipole approximation for calculating EEG signals, we can compare it to a "ground truth" EEG signal, which does not rely on making the dipole approximation. Since EEG forward models take dipoles as input, we calculate the "ground truth" EEG signal by using the MC+VC scheme implemented through the so-called *multi-dipole approximation* presented in Section 4.4.2. The validity of the dipole approximation can then be assessed by comparing its EEG signal predictions for a given neural simulation with those obtained with the MC+VC scheme.

Since the validity of the dipole approximation is dependent on the distance from the dipole to the EEG electrode, we evaluate it for two very different head sizes, representing either human or mouse (Figure 9.10A). We use eight different morphologies, corresponding to two example cell models from each of layer 2/3 and layer 5, both in mice and humans (Figure 9.10B). How much V_e predicted from the dipole approximation deviates from the ground truth (i.e. the MC+VC prediction) is very dependent on the synaptic input position, as illustrated for a few example cases in Figure 9.10C.

The EEG signal calculated for the mouse-sized head model resulted in EEG amplitudes that were about two orders of magnitude larger than for the human-sized head model with human cell models (Figure 9.10, panel D versus E). Note that this is caused by the different head models and not by the differences in the cell models: when comparing the EEG signal from mouse and human cells when inserted into the same head model (either mouse or human), the amplitude was essentially the same.

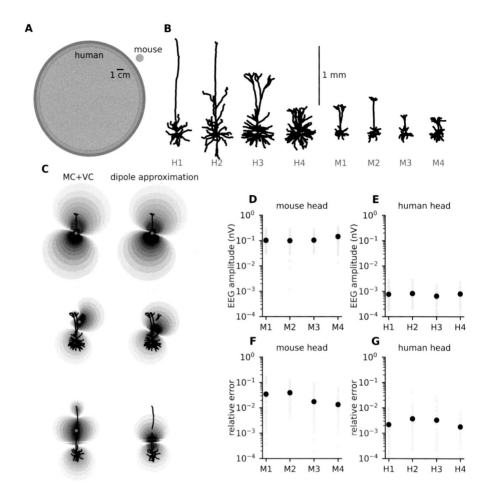

Figure 9.10 Evaluation of the dipole approximation for calculating EEG signals for human and mouse head models. A: Human- and mouse-sized versions of the four-sphere head model. Parameters were as in Table 9.1 but with the radii reduced by a factor of 15 for mice. **B**: Neuron morphologies taken from the Allen Brain Atlas (https://celltypes.brain-map.org/): two human layer 5 pyramidal cells (H1, H2), two human layer 2/3 pyramidal cells (H3, H4), two mouse layer 5 pyramidal cells (M1, M2), and two mouse layer 2/3 pyramidal cells (M3, M4). All cell models were passive and had the same passive parameters. **C**: Examples of V_e calculated from three morphologies: M2 (top), H3 (middle), and H1 (bottom). The neuron models received a synaptic input (blue dot). The resulting V_e as predicted using the MC+VC scheme (left) and the dipole approximation (right) is shown at the time of the maximum dipole moment. **D**: EEG signal amplitude in mouse head model. Signal predicted with MC+VC scheme directly above cell models M1–M4 receiving single randomly positioned synaptic inputs. Grey dots represent 100 trials, and black dots represent the mean. **E**: Same as panel D but for human cell and head models. **F**: Relative error (see code for definition) introduced when using the dipole approximation instead of the MC+VC scheme when predicting mouse EEG signals. **G**: Same as panel F but for the human head model. Code available via www.cambridge.org/electricbrainsignals.

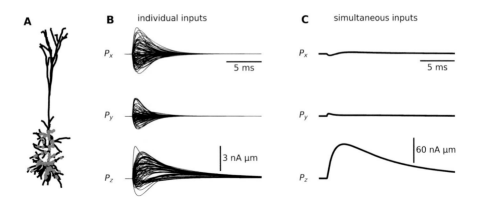

Figure 9.11 Only the component of the dipole moment that is aligned with the depth direction of the cortex remains for many synaptic inputs. A: Basal synaptic input positions are marked on the Hay model neuron. **B**: For individual synaptic inputs, the x-, y-, and z-component of the dipole moment **P** will often be similar in magnitude. **C**: If all synaptic inputs are instead activated simultaneously, radial symmetry tends to ensure that the x- and y-components cancel, while the same is not true for the z-component (defined to be the vertical direction in panel A). Code available via www.cambridge.org/electricbrainsignals.

Across all used morphologies and all tested synaptic positions, the relative error introduced by using the dipole approximation with the mouse head model was typically below 10 percent but was occasionally higher (Figure 9.10F). The relative error introduced by using the dipole approximation with the human head model was typically below 1 percent and did not in any cases tested here exceed 5 percent (Figure 9.10G). In conclusion, it therefore seems that the dipole approximation is well-justified for the purpose of calculating human EEG signals, but more caution is called for when calculating rodent EEG signals.

Reducing a large number of current sources I_n to a single time-dependent vector **P** is a convenient and compact representation of the underlying neural activity. Further, as briefly discussed in Section 9.3.1, we can often disregard the components of **P** that are not oriented along the depth direction of the cortex, leaving only a single time-dependent component P_z, as illustrated in Figure 9.11. The same principle has previously been demonstrated for a large neural population (Næss et al. 2021, figure 6), and we can therefore expect the EEG signal to be generated predominantly by dipoles that are oriented parallel to the apical dendrites of pyramidal neurons – that is, in the normal direction of the cortex.

Due to linearity (see Section 2.6.1), we can compute the total EEG signal by summing the contributions from each individual dipole. Furthermore, if such dipoles are grouped close together relative to the distance to the EEG electrodes ($\approx$ 1 cm in humans, Figure 9.2), we can in principle calculate the vector sum of the dipoles first and then calculate the EEG signal from the resulting single population dipole. This provides an interpretation of what it is one estimates when addressing the inverse source-localization problem: when a cortical dipole is localized, it is effectively a population dipole, built up from a large population of cells.

As mainly the components in the depth direction matter, the EEG contribution from a specific neural process can be characterized in terms of how it affects P_z (see, for example, Næss et al. (2021)). Calculation of P_z from morphologically detailed neuron models has been used to study the EEG and MEG contribution of spiking single cells (Murakami & Okada 2006) and to study how synaptic input positions affect the dipole (Lindén et al. 2010, Ahlfors & Wreh II 2015, Næss et al. 2021).

9.5.2   Differences and Similarities between LFP and EEG Signals

As for the LFP, we expect the main source of the EEG signal to be synaptic inputs to populations of geometrically aligned pyramidal neurons. It is therefore natural to ask how much these different brain signals have in common. In rodents, Bruyns-Haylett et al. (2017) observed an EEG signal that looked like a downscaled version of the LFP in the uppermost part of the cortex (Bruyns-Haylett et al. 2017). This is what we would expect if the dominant EEG sources are close to the LFP electrode and if the local cortical surface is radially oriented, like in rodents or in human gyri. This is illustrated in Figure 9.12, which shows the same simulated neural activity as in Figure 8.1 for the LFP but with the additional calculation of EEG signals. However, such a clear relationship between the LFP and the EEG will generally not be present.

The EEG signal is often observed to be more dominated by low frequencies than LFP signals. There may be several underlying reasons for this. Firstly, it may partly be due to the effect of intrinsic dendritic filtering becoming larger at larger distances from the sources. Secondly, low-frequency components of neural activity may be correlated on larger spatial scales than high-frequency components so that extracellular potentials evoked by larger neural populations are more prone to be dominated by the low-frequency components (Srinivasan et al. 1998, Dubey & Ray 2019, Nunez et al. 2019). One example of this effect is visible in Figure 9.12, where high-frequency spikes are clearly visible in the extracellular potential around the soma region of the population (panel C) while completely absent in the corresponding population dipole moment (panel D). Finally, further low-pass filtering of the EEG signals may occur due to the spatial smearing effects from the head (see section 9.3 and Nunez et al. (2019)).

The fact that the source of the EEG signal can be represented in terms of single (time-dependent) vector components P_z instead of the corresponding multitude of membrane-current sources (see Section 9.5.1) implies that the LFP signal in general will contain more information about the local current sources and sinks than the EEG signal. While the laminar LFP in some cases may be highly dependent on the exact target zone of the synaptic input, the EEG signal will tend to be much less sensitive to such details, as illustrated in Figure 9.13.

9.5.3   Cell-Type Specific EEG Contributions

As previously discussed in the context of LFPs (Chapter 8), not all cell types have the same capacity to produce strong extracellular potentials directly. Interneurons and layer

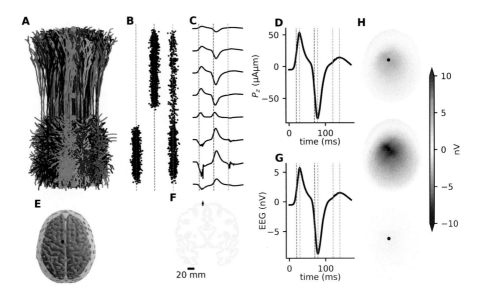

Figure 9.12 LFP and EEG signals from volleys of synaptic input to a neural population.
A: Example population with 1,000 Hay model neurons receiving excitatory synaptic inputs.
B: Depiction of spatiotemporal activation of synaptic inputs with three distinct epochs: basal inputs, apical inputs, and uniform inputs. Within each epoch, the input times are normally distributed with the means marked by the dashed colored lines. **C**: Extracellular potential computed by the MC+VC scheme at positions of electrode contact shown in panel A. Panels A–C are the same as in Figure 8.1. **D**: Population dipole moment, calculated by summing all single-cell contributions. The colored dashed lines correspond to the epochs in panel B, and the following grey dashed lines correspond to the times when the effect of each epoch on the dipole moment is largest (which is slightly delayed compared to the synaptic input). **E**: The New York head model (Huang et al. 2016) seen from the top, with EEG electrodes (grey) and position of population dipole (black dot). **F**: A cross-section of the cortex from the New York head model, seen from the front. The position of the population's dipole is marked by a black arrow. **G**: EEG signal calculated at the EEG electrode closest to the neural population. **H**: The EEG signal at the times of the maximum amplitude (grey lines in panel D), following basal synaptic input (top), apical synaptic input (middle), and uniformly distributed synaptic input (bottom). Code available via www.cambridge.org/electricbrainsignals.

4 stellate cells are often relatively spherically symmetric around the soma. Thus, they do not to the same extent as pyramidal neurons harbor distinct dendritic regions that can be asymmetrically targeted by synaptic input. Therefore, synaptic input to these cell types tends to cause randomly oriented dipole moments that will mostly cancel at the population level (Næss et al. 2021, figures 4 and 6). Pyramidal neurons, on the other hand, are both the most numerous type of neurons in the cortex, have distinct dendritic regions (apical and basal dendrites) that can serve as target zones, and are geometrically aligned perpendicular to the cortical surface. Pyramidal cells are therefore thought to be the main generators of the EEG (and LFP) signal.

 Pyramidal cells are found in most cortical layers, and there are substantial layer-specific differences in cell density, morphology, and physiology. It is therefore hard to make general statements about which class of pyramidal cells contribute the most to

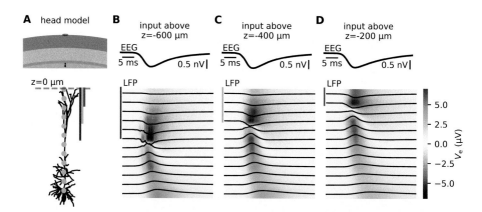

Figure 9.13 **EEG signals are less sensitive than LFP signals to the exact distributions of synaptic inputs.** **A**: A neuron model receiving different volleys of apical synaptic inputs, where the laminar LFP is calculated at different depths in the vicinity of the neuron, and the EEG signal is calculated at the top of the four-sphere head model. **B**: The EEG signal (top) and laminar LFP (bottom) resulting from a volley of apical excitatory synaptic inputs restricted to the top 600 μm of the cell model (depth marked by the purple line in panel A). The signals are calculated as the sum over 10 trials, to get smoother signals. **C**: Same as in panel B, but the synaptic inputs are restricted to the top 400 μm of the cell model (depth marked by the light green line in panel A). **D**: Same as in panels B and C, but the synaptic inputs are restricted to the top 200 μm of the cell model (depth marked by the red line in panel A). Code available via www.cambridge.org/electricbrainsignals.

the EEG signal. As mentioned earlier, the increased distance to the EEG electrodes for neurons in deeper layers seems of relatively little importance, at least in humans (see Figure 9.6).

9.5.4   Applications of MC+VC Scheme to EEG Signals

The forward modeling of EEG signals from given dipole moments (the VC part) have a long history (Nunez & Srinivasan 2006). However, examples of computation of such dipole moments from neural activity (the MC part) to account for specific experiments are few and far between. While forward modeling of neuron-based dipole moments in the context of magnetic signals started around the year 2000 (see Chapter 11), the application to EEG signals is more recent. Mäki-Marttunen, Krull et al. (2019) and Rosanally et al. (2023) used the scheme to explore putative effects of genes and gene expressions associated with schizophrenia on EEG. Further, Herrera et al. (2022) used the scheme to interpret EEG signals from a macaque visual cortex.

9.6   Simulating Large-Scale Neural Activity and Resulting EEG Signals

In the previous section, we showed how we could compute dipole sources for EEG forward models from biophysically detailed cell models. This works well if we study

EEG contributions from single cells or moderately small neural populations. However, constructing networks of biophysically detailed cell models that are large enough to model realistic EEG signals would require massive computational resources since the EEG signal is expected to reflect the neural activity of entire brain regions, containing millions of neurons. Biophysically detailed modeling of neural activity at this scale is therefore typically not an option, and simplified approaches to modeling neural activity are more commonly used. We here only give a brief overview of such approaches and refer readers who are interested in more in-depth treatments of such approaches to other sources (Deco et al. 2008, Sanz-Leon et al. 2015, Glomb et al. 2022).

9.6.1   Kernel-Based Approaches

As described in Section 6.4.3, an approximation to the EEG signal can be calculated from firing rates through kernel-based approaches. The advantage of this approach is that the firing rates can stem from simulations using simplified schemes for simulating the neural activity (or in principle even be estimated from experimental data).

In kernel-based approaches, the main challenge is to find suitable kernels H_{YX} for every connection pathway between presynaptic population X and postsynaptic population Y. In the context of predicting EEG signals, each kernel describes in essence the average contribution to the dipole moment component P_z from the postsynaptic population, given a spike in the presynaptic population (Section 6.4.3). Examples of such kernels are shown in Figure 6.6B–E, and an example prediction based on kernels and firing rates from the network simulation in Section 6.4.1 is shown in Figure 9.14.

Given suitable kernels, the total dipole moment – summed over all pre- and post-synaptic populations – can, analogous to equation (6.38), be found from

$$\check{P}_z(t) = \sum_X \sum_Y (\nu_X * H_{YX})(t), \tag{9.22}$$

where ν_X denotes the firing rate of the presynaptic population X. The estimated dipole $\check{P}_z(t)$ can then be used in combination with an EEG forward model as described previously in this chapter.

Even though finding a suitable kernel is generally challenging, obtaining a kernel for a network of biophysically detailed neuron models is in principle straightforward: every kernel H_{YX} can be found from the neural network parameters in combination with a single simulation of a representative cell model from the postsynaptic population X. In this scenario, it has been demonstrated that the kernel method can accurately reproduce the original population dipole moment from the MC+VC scheme (see Figure 6.6F) and consequently also the EEG signal.

Determining kernels for computing EEG signals from networks simulated with more simplified schemes is more challenging. The kernel-based approach will, in that case, rely on obtaining representative biophysical cell models and estimating the relevant network parameters, including the spatial distributions of synapses on the postsynaptic neurons, connection probabilities, and synaptic delays. Estimates of these biological parameters can be hard to come by, but the predicted EEG signal itself will be entirely

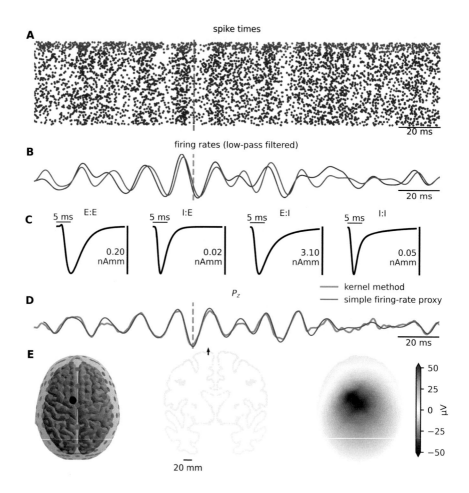

Figure 9.14 Illustration of the kernel method used to calculate EEG signals. A: Spike times from the simulation presented in Figure 6.6. Red dots correspond to spikes from the inhibitory population, and blue dots show spikes from the excitatory populations. **B**: Normalized population firing rates, calculated by pooling and low-pass filtering the data in panel A. **C**: The kernels for each synaptic pathway in the model, as also shown in the bottom row of panels B–E in Figure 6.6. **D**: The grey line shows dipole moment P_z calculated by the kernel method (equation (9.22)), as also shown in the bottom row of Figure 6.6F. The red line is a simple proxy, made by inverting and shifting by 5 ms the low-pass filtered firing rate of the inhibitory population that is shown in panel B (blue line). **E**: The dipole moment found by the kernel method (grey line in panel D) can be combined with a head model (Huang et al. 2016) to calculate the resulting EEG signal, similar to Figure 9.4. Code available via www.cambridge.org/electricbrainsignals.

determined by them. For example, the importance of the spatial distribution of synapses on the EEG signal is illustrated in Figure 9.12.

9.6.2   EEG Proxies

Other methods for simulating EEG signals that are still grounded in biophysically detailed modeling of neurons are the so-called proxy methods, validated against the hybrid scheme described in Section 6.4.2 (Mazzoni et al. 2015, Hagen et al. 2017, Martínez-Cañada et al. 2021).

When using proxy methods, neural network activity is first simulated in a point-neuron network (Section 3.5.2) before the spiking activity is replayed onto biophysically detailed neuron models for calculating EEG signals or dipole moments as described in Section 6.4.4. This gives us a "ground truth" signal, which is used to validate various proxies.

A proxy is, generally speaking, a set of mathematical operations that allows one to approximate the "ground truth" signal by relying exclusively on data that be directly extracted from the point-neuron network simulation. Proxies are typically found through extensive parameter searches, and the aim is to find proxies that mimic the "ground truth" signal as closely as possible.

Once established, proxy methods allow for EEG signals to be calculated directly from point-neuron network simulations (Martínez-Cañada et al. 2021). It has been demonstrated that a weighted combination of excitatory and inhibitory synaptic currents, which are available directly from the point-neuron network simulations, can give good predictions for the LFP (see figure 6.8 and Mazzoni et al. (2015)) or EEG signal (Martínez-Cañada et al. 2021).

9.6.3   Minimally Sufficient Biophysical Models

Another approach for modeling the neural activity at the level of individual neurons while still aiming for a high level of tractability is through using so-called minimally sufficient biophysical models (Murakami et al. 2002, Murakami et al. 2003, Jones et al. 2007, Jones et al. 2009, Herrera et al. 2020, Neymotin et al. 2020, Herrera et al. 2022, Kohl et al. 2022). As earlier noted, a minimal spatial structure of the neuron models is required for producing extracellular potentials (Figure 3.1), and the underlying assumption when using minimally sufficient biophysical models is that a few compartments are sufficient to reproduce the main features of electric brain signals like the EEG. To limit the parameter space, it is common to include only a few cell types with relatively simple connection rules in the simulations (Neymotin et al. 2020, Kohl et al. 2022).

Methods have been developed for automatically reducing the complexity of biophysically detailed neuron models (Amsalem et al. 2020, Wybo et al. 2021), which can potentially be used for predicting EEG signals. Note, however, that even though these reduced models have been demonstrated to be reasonably accurate in preserving the membrane potential dynamics of the original complex models (which is typically what they have been optimized for), it is not immediately clear if they will also faithfully reproduce the dipole moments calculated with the original models.

Simulating neural activity with minimally sufficient biophysical models massively reduces the parameter space and computational demands compared to using networks with biophysically detailed models of several cell types and experimentally recon-structed connection patterns. This means that the models can be simulated, explored, and constrained efficiently. A good example of this approach in use is the open-source Human Neocortical Neurosolver (HNN) software (Neymotin et al. 2020). This software allows researchers to quickly link measured and simulated EEG/MEG signals through an interactive graphical user interface, where the neural activity is simulated based on a pre-defined canonical neocortical column template network. The simulations will typically only use relatively small populations of neurons, where the predicted dipole moments are scaled by a scaling factor to match recorded EEG amplitudes and the corresponding estimated dipole moments. This allows for direct numerical comparison of measured and simulated EEG signals, which can be helpful in investigating the neural origin of measured evoked potentials or neural oscillations.

9.6.4   Neural Mass and Neural Field Approaches to Modeling EEG Signals

In the approaches to modeling EEG signals described so far in this section, we have relied on using simplified models of individual neurons. However, since the EEG signal reflects neural activity at the level of entire brain regions, it is common to instead model the underlying neural activity at a higher level of abstraction (Nunez 1974, Jansen & Rit 1995, Jirsa et al. 2002, David & Friston 2003, Deco et al. 2008, Kiebel et al. 2008, Bojak et al. 2010, Sanz-Leon et al. 2013, Sanz-Leon et al. 2015). The aim is then to model the evolution of coarse-grained variables such as the mean membrane potential or the firing rate of neural populations (Glomb et al. 2022). Such high-level approaches to modeling neural activity were briefly presented in Section 3.5.3. As they drastically reduce the parameter space and the computational burden of the simulation, they can be used to study the interplay between entire brain regions, and run whole-brain simulations.

Since these high-level approaches do not directly model the individual neurons, they are not in a principled way able to predict dipole moments from the simulated neural activity. They therefore typically rely on the assumption that the dipole moments will be proportional to the simulated firing rates or mean membrane potential (Jirsa et al. 2002, Kiebel et al. 2008, Bojak et al. 2010, Ritter et al. 2013, Sanz-Leon et al. 2015). Like the dipole moments predicted previously in this chapter, the dipole moments predicted from these high-level simulations of neural activity can be used as sources in EEG forward models.

The assumption that dipole moments are approximately proportional to firing rates or the mean membrane potential is in many ways in disregard to the complex bio-physical origin of the dipoles. However, as shown by Martínez-Cañada et al. (2021), it can still in many cases be a reasonable approximation for the purpose at hand. An example can be seen in Figure 9.14D, where the EEG signal calculated by the kernel method (validated against "ground truth" simulation data) is quite well approximated by a simple proxy corresponding to the inverted (upside-down), low-pass filtered, and slightly shifted (+5 ms) firing rate of the inhibitory population. This example illustrates that it can be meaningful to assume that dipole moments are approximately proportional to simulated firing rates. However, we do not claim that the EEG signal in general will follow the inhibitory firing rate as closely as in this example.

Despite their lack of biophysical detail, the unmatched ability of high-level approaches when it comes to modeling entire brain regions or even the entire brain might still make their EEG-signal predictions more realistic than those that can presently be achieved using biophysically detailed cell models. An interesting opportunity for further development of these high-level approaches is to compare the dipole moments used in these approaches with those predicted using kernel-based approaches (described in Section 9.6.1). This could potentially be used to either improve or better validate the estimated EEG signals.

The Virtual Brain (TVB) is an example of software that is aimed towards whole-brain network simulations with corresponding EEG predictions (Ritter et al. 2013, Sanz-Leon et al. 2013, Sanz-Leon et al. 2015), which combines simulated neural activity with detailed and potentially personalized head models. Further, tractography-based methods can be applied to identify the patient-specific functional connectivity between brain regions (Sanz-Leon et al. 2013). The resulting models can then be used to aid in the localization of brain pathologies and to plan brain surgery to avoid interfering with vital neural pathways.

10 Electrocorticography (ECoG)

Electrocorticography (ECoG) refers to the measurement of electric potentials on the cortical surface and is the oldest reported electrophysiological method (Caton 1875). In animal studies, ECoG signals are commonly recorded using grids of micro-electrodes ($\sim$10 µm in diameter) embedded in insulating plastic sheets (Khodagholy et al. 2015, Donahue et al. 2018, Hill et al. 2018, Baratham et al. 2022, Thunemann et al. 2022), but larger macroelectrodes ($\sim$2–3 mm in diameter) are also used (Dubey & Ray 2019). ECoG recordings used to identify the locus of seizures in drug-resistant epileptic patients before brain surgery typically apply macroelectrodes.

The starting point for modeling ECoG signals is the CSD equation (equation (2.31)), which was also the starting point for modeling spikes, LFP, and EEG signals. However, since ECoG electrodes are placed on the cortical surface, one cannot a priori assume that the distance between the recording positions and neural sources is so large that the dipole approximation[1] applies, as one does when modeling the EEG (Chapter 9). Further, due to the pronounced discontinuity of the conductivity at the cortical surface, one cannot assume the conductivity to be constant over the relevant region, as one often does when modeling spikes (Chapter 7) or the LFP (Chapter 8). As an additional complication, the electrodes used in ECoG recordings – whether they are large metal electrodes or micro-electrodes embedded in plastic sheets – typically affect the measured signal significantly, and their effects must generally be included in the modeling.

To explore the accuracy of various methods for modeling the ECoG signal, we first assume idealized point electrodes and thus neglect any effects that the recording devices may have on the recordings. We can then use MC+VC simulations together with the four-sphere head model introduced in the EEG chapter as the "ground truth." By comparison against this ground truth, we evaluate the accuracy of ECoG predictions based on simpler methods, including the method of images (Section 10.1) and the dipole approximation (Section 10.2). Finally, in Section 10.3, we outline how effects from the ECoG recording device can be included in the modeling scheme.

---

[1] When we speak of "dipoles" here, we refer to "current dipoles." Charge dipoles are not encountered in this book.

10.1   Method of Images (MoI)

ECoG electrodes are positioned at an interface between the grey matter and the much-more conductive cerebrospinal fluid (CSF). The CSF is further covered by a poorly conductive skull, which in turn is covered by the scalp with a conductivity similar to grey matter. This conductivity profile, which is incorporated in the four-sphere model, strongly affects the potential at the cortical surface. As we saw in Figure 9.5B, the potential decays sharply in the cortex close to the highly conductive CSF, an effect not captured by the homogeneous-conductivity approximation (grey line in Figure 9.5B). The same sharp decay in potential when approaching the cortical surface was seen in Figure 5.9D, where the cortex was assumed to be covered by an "infinitely" good conductor.

Since the ECoG (and EEG) signals are affected by the spatial variations in conductivity, computing them is more cumbersome than computing spikes and LFPs (in homogeneous tissue). However, the method of images (MoI) (Section 5.7.1) offers a mathematical trick that allows electric signals in inhomogeneous tissues to be computed with similar ease as in homogeneous tissue, provided that we can assume our medium to be a sandwich structure of planar homogeneous slabs.

In Figure 10.1, we explore the use of MoI in the modeling of ECoG signals. A two-layer sandwich structure is assumed, where a slab with conductivity σ_{cortex} represents the cortical grey matter containing the neuronal sources while another slab with conductivity σ_{top} approximates the net effect of the CSF, skull, and scalp covering the cortex. For this structure, the ECoG signal can be computed by means of equation (5.32), which is nearly as simple as computing the electric signal in a homogeneous medium (Hagen et al. 2018).

The two left columns of panels in Figure 10.1 compare ECoG signals from a human cortical cell receiving an excitatory synaptic input when predicted with MoI versus the four-sphere model. Three different synaptic positions are considered. When the synaptic input is close to the top (panel A) or halfway up the apical dendrite (panel B), the four-sphere and MoI predictions are almost in perfect agreement when σ_{top} is set equal to σ_{CSF} (black and grey curves overlap). In contrast, if one does not account for conductivity variations but use the infinite homogeneous volume-conductor approximation (assuming that $\sigma_{top} = \sigma_{cortex}$), the predicted signal amplitude is almost a factor of four too large (red curve). This reflects that the highly conductive CSF layer above the cortex effectively short-circuits the potential at the cortical surface (Pettersen et al. 2006).

For a synapse positioned at the soma, MoI predicts a smaller signal amplitude (Figure 10.1C) than the four-sphere model. This implies that when the synaptic current is far away from the cortical surface, the short-circuiting effect of the thin CSF cover, which in turn is covered by the skull and scalp, is overestimated by MoI, which assumes an infinitely thick layer of CSF. The discrepancy can to some extent be remedied by assuming a smaller conductivity value ($\sigma_{top} < \sigma_{CSF}$) in MoI. This approach will however require a systematic study of how such a reduced conductivity value for σ_{top} should be chosen for the system at hand. Another approach for improving the agreement with the four-sphere model could be to expand the MoI model to include more than two electrically distinct layers (Ness et al. 2015, Rogers et al. 2020).

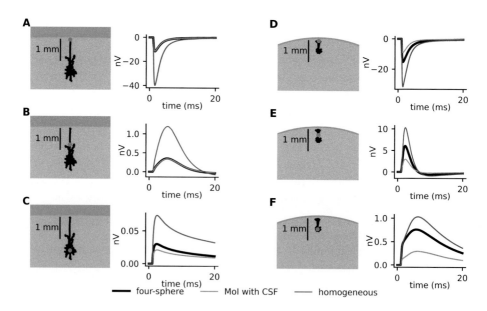

Figure 10.1 ECoG signals in humans and mice. **A**: ECoG signal from human layer 5 pyramidal cell model (morphology H2 in Figure 9.10) receiving apical input (blue dot) computed with the four-sphere model (black curve), method of images (grey curve), and the homogeneous volume-conductor model (red curve). **B**: Same as in panel A but with the synaptic input positioned halfway up the apical dendrite. **C**: Same as in panels A and B but with somatic input. **D–F**: Same as panels A–C but for a mouse layer 5 pyramidal cell (morphology M1 in Figure 9.10) and VC models representing the mouse brain. Parameters for the head models were as in Table 9.1 but with the radii reduced by a factor of 15 for mice. Code available via www .cambridge.org/electricbrainsignals.

The agreement of MoI with the four-sphere model predictions is much poorer for mouse ECoG signals (two right columns of panels in Figure 10.1). One reason for this is the much thinner CSF layer in mice, which reduces its electric short-circuiting effects. Another reason is the much-stronger curvature of the mouse head due to its much smaller size. Since MoI assumes a flat cortical surface, it is thus to be expected that the MoI predictions are better for humans than for mice. For the mouse, the MoI predictions underestimate the (ground-truth) ECoG amplitudes obtained with the four-sphere model, while the predictions obtained assuming an infinite homogeneous conductor overestimate the ECoG amplitudes (Figure 10.1D, E, F).

10.2   Dipole Approximation

As described in Chapter 9, using the dipole approximation instead of the MC+VC scheme greatly simplifies the computation of EEG signals. To explore whether the dipole approximation is warranted when computing ECoG signals, we have in Figure 10.2 compared its ECoG predictions with "ground-truth" predictions obtained

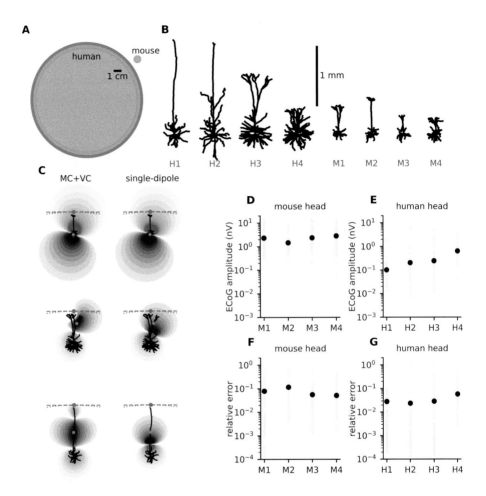

Figure 10.2 Evaluation of the dipole approximation for calculating ECoG signals for human and mouse head models. A: Human-sized and mouse-sized versions of the four-sphere head model – for parameters, see Table 9.1. **B**: The morphologies are the same as in Figure 9.10 and correspond to two human layer 5 pyramidal cells (H1, H2), two human layer 2/3 pyramidal cells (H3, H4), two mouse layer 5 pyramidal cells (M1, M2), and two mouse layer 2/3 pyramidal cells (M3, M4). **C**: V_e computed using the MC+VC scheme (left) and the dipole approximation (right) at the time of the maximum dipole moment of cell models M2 (top), H3 (middle), and H1 (bottom) receiving synaptic input (blue dot). The dashed and dotted curves illustrate the position of the cortical surface in the human and mouse models, respectively. The position of the ECoG electrode is indicated by the grey dot. **D**: Mouse ECoG signal amplitude calculated using the MC+VC scheme, for 100 randomly chosen synaptic locations (with area-weighted probability) in each mouse neuron. The ECoG amplitude from individual synaptic input is shown as small grey dots, and the mean is shown as a larger black dot. **E**: Same as panel D but for the human model. **F**: Relative error in ECoG signal (averaged over signal length and synaptic inputs) when using the dipole approximation instead of the MC+VC scheme for the mouse model. **G**: Same as panel F but for the human model. Code available via www.cambridge.org/electricbrainsignals.

with the MC+VC scheme implemented through the multi-dipole approximation as defined in Section 4.4.2.

The comparison is analogous to how we, in Figure 9.10, compared EEG signal predictions from the two schemes, and this comparison is done for both human and mouse heads, using four-sphere models together with a selection of biophysically detailed (MC) neuron models.

A first observation is that the ECoG signal amplitudes in the mouse and human are similar in magnitude, although the mouse signals are slightly larger (panels D, E). This reflects that the ECoG electrode is roughly as close to the neural sources in humans as in mice.

This is different from the corresponding EEG results in Figure 9.10, where the human EEG signal is much smaller than the mouse EEG signal. The amplitude of the mouse ECoG signal is only about a factor of 10 larger than the mouse EEG signal (see also panel D in Figure 9.10 and Figure 10.2). In contrast, the amplitude of the human ECoG signal is about 100–1,000 times larger than the human EEG signal (panel E). This reflects the much thicker skull in humans compared to mice, implying a much larger distance from the EEG electrode to the neural sources in humans.

The relative error in using the dipole approximation to compute ECoG signals is typically between 5 and 10 percent for the present example (Figure 10.2F, G). For mice, the error introduced by using the dipole approximation is only slightly larger for ECoG signals than for EEG signals, while for humans, the error in the ECoG signal is much larger than in the EEG signal, where it is typically less than 1 percent (Figure 9.10G). This reflects that the dipole approximation is valid in the far-field limit and thus gives much smaller prediction errors for the human EEG due to the large distance between the EEG electrodes and the neural sources.

In summary, caution is required if the dipole approximation is used for calculating ECoG signals both in humans and in mice (see also Næss et al. (2021)).

10.3   Electrode Effects

For ECoG signals recorded with small individual ECoG microcontacts, the computational schemes outlined in Section 10.1 and Section 10.2 may be used with the disk-electrode approximation (Section 4.5.2) to take the electrode size into account.

For grids of ECoG microcontacts embedded in plastic sheets (Khodagholy et al. 2015, Donahue et al. 2018), the effects of the electrically insulating plastic covering the cortex may, at least for human ECoG, be approximated by means of the method of images (MoI) with σ_{top} set to zero. A more comprehensive three-layer MoI model may also be used, letting the bottom layer represent the cortex, the middle layer the CSF (and other tissue sandwiched between the cortical surface and the recording grid), and the top layer the insulating plastic sheet with $\sigma_{top} = 0$ (Rogers et al. 2020). To account for the larger curvatures in rodent brains, one possibility is to use the four-sphere head model with $\sigma_{CSF} \approx 0$ to mimic the insulating effect of the plastic. Another alternative might be to keep a CSF layer in the model and instead set σ_{skull} to be approximately zero, in analogy to the MoI approach of Rogers et al. (2020). To account for the effects of the finite lateral extents of the plastic sheet, however, one must resort to more extensive numerical schemes like the finite element method (FEM) (Hill et al. 2018).

ECoG macrocontacts measuring millimeters across are so large that they, like the plastic sheets carrying micro-electrodes, sizably affect the electric fields around the electrode. A proper study of this requires FEM modeling, and Vermaas et al. (2020*b*) used FEM to explore how recorded ECoG signals depend on the electrode's surface impedance and position relative to the neural source. In their study, replacing an idealized point electrode with a macroelectrode could change the recorded potential by up to a factor of three. A qualitative conclusion was that the effect of the recording electrode on the surrounding electric field cannot be disregarded when the distance between the electrode and the neural source is equal to or smaller than the size of the electrode. This is in line with the qualitative conclusion found in Ness et al. (2015) regarding the modeling of spikes recorded in micro-electrode arrays (MEAs) (see Section 7.7.2).

11 Magnetoencephalography (MEG)

In earlier chapters, we have focused on how electric membrane currents in neurons give rise to electric extracellular potentials. However, it follows from Ampère's circuital law (equation (2.37)) that electric currents will also give rise to magnetic fields. The weak magnetic field produced by the cellular and extracellular currents in the brain can be measured non-invasively on the scalp. This type of measurement was first performed by Cohen (1968) and is known as magnetoencephalography (MEG), the magnetic cousin of EEG. The measurement quality was later dramatically improved by the use of sensitive SQUID (superconducting quantum interference device) detectors, first used in this context by Cohen (1972). SQUID detectors are also the most commonly used measurement device for MEG signals today.

Both EEG and MEG signals originate from cellular current sources (or equivalent dipoles) in the brain, and both can be computed using the two-step (MC+VC) framework described in Section 2.6.2 – that is, by (1) computing the neural activity in an independent simulation based on multicompartment (MC) models of neurons and (2) applying a forward volume-conductor (VC) model to compute the resulting electric and magnetic fields.

Although electric fields $\mathbf{E}$ and magnetic fields $\mathbf{B}$ originate from the same kind of neural processes, the challenges involved when deriving equations for the two differ. While originating from membrane currents, the extracellular field $\mathbf{E}$ ultimately reflects the local ohmic volume-current density in the tissue, which at any point in space is proportional to and parallel with $\mathbf{E} = -\nabla V_e$. Hence, computing V_e is essentially a matter of computing tissue volume currents.

In contrast, the magnetic field reflects both intracellular currents and tissue volume currents, as it follows from Maxwell's equation (2.37) that any current inside the brain will generate a magnetic field that spreads everywhere. However, since the intracellular and extracellular currents are interrelated through the principle of current conservation, it is for idealized cases – such as when the medium is assumed to be an infinite homogeneous- or spherically symmetric conductor – possible to express both V_e and $\mathbf{B}$ analytically as functions of cellular current sources alone. We have already seen in Chapters 4 and 9 that this is the case for V_e, and later we will see that it is also the case for $\mathbf{B}$.

Another difference between electric and magnetic fields is that the electric field evoked by current sources in a medium depends on the medium's conductivity, whereas the magnetic field analogously depends on the medium's magnetic permeability μ.

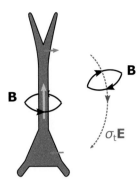

Figure 11.1 Electric versus magnetic fields. Extracellular electric fields (**E**) are proportional to extracellular volume currents. Magnetic fields (**B**) are evoked by both extracellular (red) and intracellular (yellow) currents.

The conductivity in the head varies abruptly when passing from brain tissue to the cerebrospinal fluid and then to the skull and scalp, so conductivity effects must be accounted for in the computation of the EEG signals. In contrast, the permeability of the entire head is more or less the same as for empty space, and permeability effects are therefore not an issue in the computation of the MEG signal (Hämäläinen et al. 1993).

A consequence of these discussed differences is that MEG and EEG signals are largely complementary: whereas the EEG signal predominantly reflects the radial component of an underlying dipolar source, the MEG signal predominantly (and in ideal cases, exclusively) reflects its tangential component relative to the scalp's surface. As we saw in Section 9.3, electric potentials measured on the scalp are strongly influenced by local conductivity inhomogeneities in the head. For this reason, the MEG signal often turns out to be superior to the EEG signal when it comes to spatial resolution (Hämäläinen et al. 1993), in particular if the region of interest is located in a sulcus.

As the main focus in this book is on electric fields, we will keep this chapter about MEG rather short. We put a focus on presenting simple cases, where the brain is treated either as an infinite homogeneous conductor or a spherically symmetric conductor, and comment only briefly on how MEG signals can be modeled in the case of a more general head model. Readers that desire a more comprehensive introduction to the physics and mathematics of MEG are advised to consult the book by Ilmoniemi & Sarvas (2019).

11.1 From Currents in the Brain to Magnetic Fields outside the Head

The starting point for computing magnetic fields from the brain is the Ampère-Laplace law, showing how a spatially distributed total current density (i_{tot}) induces a magnetic field **B** at location **r**:

$$\mathbf{B}(\mathbf{r}) = \frac{\mu_0}{4\pi} \iiint \frac{\mathbf{i}_{tot}(\mathbf{r}') \times (\mathbf{r} - \mathbf{r}')}{\|\mathbf{r} - \mathbf{r}'\|^3} \, dv' . \tag{11.1}$$

This equation is derived from Ampère's circuital law (equation (2.37)) and rests on the assumption that $\mathbf{B}$, and therefore $\mathbf{i}_{tot}$, must go to zero at infinity. The integral is taken over the entire volume where $\mathbf{i}_{tot}$ is nonzero. The permeability of the head is assumed to be the same as the permeability of free space μ_0.

In the MEG literature, it is common to decompose the total current density

$$\mathbf{i}_{tot} = \mathbf{i}_i - \sigma_t \nabla V_e \tag{11.2}$$

into a so-called *impressed current density* $\mathbf{i}_i$ and an ohmic macroscopic volume-current density $\mathbf{i}_t = -\sigma_t \nabla V_e$ (Chapter 4).

It is implicit in equation (11.2) that the current densities are defined on a coarse-grained spatial scale (as defined in Section 2.6), where it is meaningful to consider brain tissue as a macroscopic conductor with an average tissue conductivity σ_t. At this scale, a reference volume will span both intra- and extracellular space. We may think of $\mathbf{i}_i$ as accounting for all the cellular currents while $-\sigma_t \nabla V_e$ accounts for all the extracellular currents. A more formal definition of $\mathbf{i}_i$ is given in Section 11.2.1. We note that $\mathbf{i}_{tot}$, $\mathbf{i}_i$, V_e, and (generally) σ_t are all functions of the position $\mathbf{r}'$.

It is useful to write equation (11.1) in the form

$$\mathbf{B}(\mathbf{r}) = \mathbf{B}_i(\mathbf{r}) + \mathbf{B}_t(\mathbf{r}) \,, \tag{11.3}$$

where we have separated the contributions from the impressed currents,

$$\mathbf{B}_i(\mathbf{r}) = \frac{\mu_0}{4\pi} \iiint \mathbf{i}_i(\mathbf{r}') \times \frac{\mathbf{r} - \mathbf{r}'}{\|\mathbf{r} - \mathbf{r}'\|^3} \, dv' \,, \tag{11.4}$$

and the tissue volume currents,

$$\mathbf{B}_t(\mathbf{r}) = -\frac{\mu_0}{4\pi} \iiint \frac{\nabla' \sigma_t(\mathbf{r}') \times \nabla' V_e(\mathbf{r}')}{\|\mathbf{r} - \mathbf{r}'\|} \, dv' \,, \tag{11.5}$$

where ∇' is the spatial derivative with respect to $\mathbf{r}'$. The Ampère-Laplace law expressed as in equation (11.3), with $\mathbf{B}_i$ and $\mathbf{B}_t$ as in equation (11.4) and equation (11.5), is the foundation for computing magnetic fields resulting from neural activity.

Equations (11.3)–(11.5) follow from the insertion of equation (11.2) into equation (11.1). While the equation for $\mathbf{B}_i$ follows directly from insertion, getting the equation for $\mathbf{B}_t$ in the form of equation (11.5) requires some more work. A derivation of this equation is given in Appendix H.1.

As an alternative formulation, one may write the expression for $\mathbf{B}_t$ as

$$\mathbf{B}_t(\mathbf{r}) = \frac{\mu_0}{4\pi} \iiint \left(V_e(\mathbf{r}') \nabla' \sigma_t(\mathbf{r}') \right) \times \frac{\mathbf{r} - \mathbf{r}'}{\|\mathbf{r} - \mathbf{r}'\|^3} \, dv' \,. \tag{11.6}$$

This form was suggested by Hämäläinen et al. (1993) and is useful when computing magnetic fields numerically for detailed head models (see Section 11.3.3). A derivation of this alternative equation for $\mathbf{B}_t$ is found in Appendix H.2.

11.2   Sources of the MEG Signal

In the expression for the total current (equation (11.2)), the cellular sources of the MEG signal are expressed in terms of an impressed current density $\mathbf{i}_i$. As we know from previous chapters, the cellular sources are also the origin of the tissue volume currents represented by the term $-\sigma_t \nabla V_e$.

In the current section, we give a formal definition of $\mathbf{i}_i$ and briefly discuss how $\mathbf{i}_i$ is related to the so-called *primary current density*, which is sometimes used in its place in the MEG literature (Ilmoniemi & Sarvas 2019). We also show how the cellular sources can be expressed in terms of a dipole moment.[1]

11.2.1   Impressed Current Density

The *impressed current density* $\mathbf{i}_i$ in equation (11.2) is closely related to the current-source density C_t that we encountered in previous chapters (equation (2.31)).

We can see how $\mathbf{i}_i$ and C_t are related by requiring that the total current in equation (11.2) is conserved, so that $\nabla \cdot \mathbf{i}_{tot} = 0$. This gives us

$$\nabla \cdot \mathbf{i}_i = \nabla \cdot \sigma_t \nabla V_e , \tag{11.7}$$

which is consistent with the current-source density equation (equation (2.31)) provided that

$$C_t = -\nabla \cdot \mathbf{i}_i . \tag{11.8}$$

Hence, the divergence of $\mathbf{i}_i$ determines the membrane current-source density in the system.

For a formal definition of $\mathbf{i}_i$, we can revisit the derivation of equation (2.31) in Section 4.3.2. We then realize that $\mathbf{i}_i$ is equivalent with what we there defined as $\langle \mathbf{i}_\mu \rangle_c$. To recapitulate, $\langle \mathbf{i}_\mu \rangle_c$ is the microscopically defined current density averaged over the cellular domain of a suitable coarse-graining region of tissue. Although the cellular domain is defined so that it includes both the intracellular space and membranes of cells, the intracellular part is much greater than the volume of the (thin) membranes. It therefore seems safe to think of $\mathbf{i}_i$ as being predominantly intracellular. In an MC simulation, we can think of $\mathbf{i}_i$ as representing the axial currents inside neuronal compartments.

11.2.2   Impressed versus Primary Currents

When doing inverse modeling – that is, when we try to infer underlying neural sources from experimental recordings of $\mathbf{B}$ – it is often advised that we replace the impressed current density in equation (11.2) with the so-called *primary* current density $\mathbf{i}_p$, which is somewhat more cautiously defined (Tripp 1983, Hari & Ilmoniemi 1986, Ilmoniemi & Sarvas 2019).

---

[1] When we speak of "dipoles" here, we refer to "current dipoles." Charge dipoles are not encountered in this book.

In practice, replacing $\mathbf{i}_i$ with $\mathbf{i}_p$ amounts to admitting that possible inaccuracies are introduced when describing tissue with a constant, macroscopic (average) conductivity σ_t. As argued by Hari & Ilmoniemi (1986), a locally impressed current $\mathbf{i}_i$ will likely be accompanied by local changes in the conductivity near and inside the active cell membranes so that fluctuations in a measured V_e may reflect not only changes in $\mathbf{i}_i$, but also such microscopic conductivity changes not accounted for by σ_t. More technically detailed versions of the same argument were provided earlier by Geselowitz (1967) and Tripp (1983).

Hence, in inverse modeling, we are restricted to estimate the primary source $\mathbf{i}_p$, which includes both $\mathbf{i}_i$ and possible effects of how currents are being guided locally in the unknown microscopic conductivity landscape. According to Hari & Ilmoniemi (1986), this restriction is not severe since the distribution of microscopic conductivity effects coincides with the distribution of $\mathbf{i}_i$ and thus does not jeopardize source localization. In the literature, the two terms "impressed current" and "primary current" are sometimes used interchangeably. However, whereas $\mathbf{i}_i$ represents an actual current on the cellular domain, $\mathbf{i}_p$ is defined more loosely as *all current that is not explained by the macroscopic volume current defined as* $-\sigma_t \nabla V_e$ (Ilmoniemi & Sarvas 2019).

We emphasize that the distinction between $\mathbf{i}_i$ and $\mathbf{i}_p$ is relevant only when addressing the inverse problem. In forward modeling, there are no microscopic conductivity changes or other kinds of guiding of the tissue currents unless we explicitly put them into the model, which we normally don't. In forward modeling, there is thus no distinction between the primary and impressed current density, as the latter is the only driving battery for macroscopic volume currents. We will therefore stick with $\mathbf{i}_i$.

11.2.3   Dipole Sources

When computing MEG signals, it is custom to express the neural sources in terms of a (current-) dipole moment $\mathbf{P}$ (see Sections 4.4 and 9.5.1).

The relationship between the impressed current density and a dipole moment at location $\mathbf{r}_P$ can be expressed as (Hämäläinen et al. 1993)

$$\mathbf{i}_i(\mathbf{r}') = \mathbf{P}(\mathbf{r}')\delta^3(\mathbf{r}' - \mathbf{r}_P) . \tag{11.9}$$

As a simple example illustrating that equation (11.9) makes sense, consider the case where $\mathbf{i}_i$ is an axial current density in the z-direction in a neural segment of length d. Integrating equation (11.9) over the xy-plane then gives us the total axial current

$$\mathbf{I}_i(x_P, y_P, z') = \mathbf{P}(x_P, x_P, z)\delta(z' - z_P) \tag{11.10}$$

in the segment. Next, integration over the segment length gives us the standard definition of a (current-) dipole:

$$\mathbf{I}_i(x_P, y_P, z_P)d = \mathbf{P}(x_P, x_P, z_P). \tag{11.11}$$

11.3   Head Models

To compute the magnetic field resulting from a set of neural current sources, we need to specify the head model. By head model, we simply mean a specification of the spatial profile of the conductivity σ_t.

11.3.1   Infinite Homogeneous Head Model

It follows from equation (11.5) that volume currents will contribute to the magnetic field only if the conductivity σ_t varies in space so that $\nabla'\sigma_t \neq 0$. If we assume the head to be an infinite homogeneous volume conductor, the magnetic field is given by $\mathbf{B_i}$ alone and thus by equation (11.4).

If we express the source term in equation (11.4) in terms of a dipole moment (equation (11.9)), the equation simplifies to

$$\mathbf{B}(\mathbf{r}) = \frac{\mu_0}{4\pi} \iiint \frac{\mathbf{P}(\mathbf{r}') \times (\mathbf{r} - \mathbf{r}')\delta^3(\mathbf{r}' - \mathbf{r_P})}{\|\mathbf{r} - \mathbf{r}'\|^3} dv' = \frac{\mu_0}{4\pi} \frac{\mathbf{P}(\mathbf{r_P}) \times (\mathbf{r} - \mathbf{r_P})}{\|\mathbf{r} - \mathbf{r_P}\|^3}. \quad (11.12)$$

This is Ampère-Laplace's law for an infinite homogeneous volume conductor.

Equation (11.12) is the simplest way of approximating the magnetic signal resulting from a neural simulation. It can be used to estimate $\mathbf{B}$ at any position $\mathbf{r}$, provided that it is far enough away from the cellular sources for the dipole approximation to hold (see Section 11.6).

As equation (11.12) assumes the head to be an infinite homogeneous conductor, the scalp (or any other structure) is not in any way represented in this model. Using equation (11.12) to approximate MEG signals is therefore just a matter of inserting positional arguments $\mathbf{r}_{\mathrm{MEG}}$ that one lets represent the scalp.

11.3.2   Spherically Symmetric Head Model

In reality, the head is neither infinite nor homogeneous. An approximation that is a bit closer to reality is to assume that it is spherically symmetric. In a spherically symmetric head model, the conductivity can be set individually for each concentric layer so that one can account for conductivity differences – for example, between the brain, cerebrospinal fluid, skull, and scalp. This means that there will be spatial conductivity variations ($\nabla'\sigma_t \neq 0$) and, consequently, that volume currents will contribute to the magnetic field.

Although volume currents indeed contribute to the magnetic field, it has been shown that in the case of spherical symmetry, their contributions are independent of the actual conductivity profile $\sigma_t(r)$. This, in turn, means that the magnetic field can be expressed as a function of the impressed sources alone. This was first shown by Grynszpan & Geselowitz (1973) for the case of a (current-) dipole in a spherically symmetric head model, and simpler derivations were later presented by Ilmoniemi et al. (1985) and Sarvas (1987).

Without including the derivation, we here present the formulas derived by Sarvas (1987), which state that the magnetic field outside a spherically symmetric conductor is given by

$$\mathbf{B}(\mathbf{r}) = \frac{\mu_0}{4\pi} \frac{F_S \mathbf{P} \times \mathbf{r}_P - (\mathbf{P} \times \mathbf{r}_P) \cdot \mathbf{r} \, \nabla F_S}{F_S^2}, \tag{11.13}$$

where

$$F_S = a(ra + r^2 - \mathbf{r} \cdot \mathbf{r}_P)$$

and

$$\nabla F_S = (r^{-1}a^2 + a^{-1}\mathbf{a} \cdot \mathbf{r} + 2a + 2r)\mathbf{r} - (a + 2r + a^{-1}\mathbf{a} \cdot \mathbf{r})\mathbf{r}_P,$$

with $\mathbf{a} = \mathbf{r} - \mathbf{r}_P$, $a = \|\mathbf{a}\|$, and $r = \|\mathbf{r}\|$. We emphasize that equation (11.13) is only valid outside the head, where $\sigma_t = 0$.

It may appear paradoxical that equation (11.13) is independent of both σ_t and V_e despite the fact that the magnetic field it predicts indeed depends on volume currents. As explained by Ilmoniemi & Sarvas (2019), the reason why the volume-current term can be eliminated is that the different field components are not independent, but interrelated by Maxwell's equations. In the particular case of spherical symmetry, the tangential field components can be derived directly from the radial field components, which in turn can be computed as direct functions of the impressed current sources. Readers that would like to delve into the depth of this matter are advised to consult the derivation by Sarvas (1987).

11.3.3   Detailed Head Models

For more detailed head models – for example, constructed from MR-images of real heads – there are no available analytic solutions relating dipole sources to the MEG on the scalp. Numerical methods must then be used to compute the magnetic field.

A good starting point for such an endeavor is to write equation (11.1) in the form

$$\mathbf{B}(\mathbf{r}) = \frac{\mu_0}{4\pi} \iiint \left(\mathbf{i}_i(\mathbf{r}') + V_e(\mathbf{r}') \nabla' \sigma_t(\mathbf{r}') \right) \times \frac{\mathbf{r} - \mathbf{r}'}{\|\mathbf{r} - \mathbf{r}'\|^3} \, dv', \tag{11.14}$$

as suggested by Hämäläinen et al. (1993). Here, the first and last terms on the right follow from equations (11.4) and (11.6), respectively.

Once V_e is known, it should be relatively straightforward to solve this integral (numerically) for $\mathbf{B}$. The core of the problem is thus again to determine V_e by solving the CSD equation (equation (4.1)) numerically for the case of an inhomogeneous conductivity, as discussed in Section 5.7.

Numerical strategies for forward MEG modeling are normally based on assuming that the head is a medium with piecewise constant conductivity (Geselowitz 1970, Hämäläinen et al. 1993, Van Uitert et al. 2003). The theory for computing $\mathbf{B}$ and V_e for a piecewise constant medium is, for instance, presented in chapter 3.4 in the book by Ilmoniemi & Sarvas (2019).

11.4   MEG for Infinite Homogeneous and Spherically Symmetric Head Models

We have seen that if we approximate the head as being (i) an infinite homogeneous (Section 11.3.1) or (ii) spherically symmetric (Section 11.3.2) conductor, the MEG signal can be computed analytically as a direct function of the cellular sources or their (current-) dipole moments **P**. Hence, in simulations based on the two-step multicompartment plus volume conduction (MC+VC) scheme, the MEG signal can – like the extracellular potential – be computed from an analytical formula once the underlying neural sources are known.

To analyze the MEG signal, it is useful to decompose it into its components in the θ, ϕ, and r directions, defined in the spherical coordinate system (Figure 11.2). Some key insights can be obtained by examining how single dipoles, either radially or tangentially oriented relative to the scalp surface, contribute (or do not contribute) to the MEG signal in the two simple head models (i) and (ii). In Sections 11.4.1 and 11.4.2, we have computed the MEG signal from such dipoles on a spherical surface with radius $r_{MEG} = 10.5$ cm, with the dipoles placed 8.8 cm above the center of the sphere. Although we picture r_{MEG} as representing the radius of the external surface of the scalp, we note that the scalp radius does not in itself enter as a parameter into the equations for the magnetic field. Hence, mathematically r_{MEG} can represent any spherical surface where we have chosen to evaluate the magnetic field. The only restrictions are that, for both head models, the surface r_{MEG} must enclose the dipole and be far enough away from it for the dipole approximation to hold (see Section 4.4). For the spherically symmetric model, we have the additional restriction that r_{MEG} must be interpreted as being outside the head, since equation (11.13) is only valid in the region where $\sigma_t = 0$.

An important difference between the equations for the spherically symmetric (equation (11.13)) and infinite homogeneous (equation (11.4)) head model is that tissue volume currents (implicitly) contribute to the MEG signal in the former but not the latter.

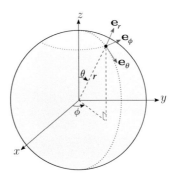

Figure 11.2 Illustration of spherical coordinate system. In the spherical coordinate system, locations are specified in r, θ, and ϕ, and directions with the unit vectors $\mathbf{e}_r$, $\mathbf{e}_\theta$, and $\mathbf{e}_\phi$.

11.4.1   Radial Dipoles

In the case of a radially directed dipole, the volume currents cancel so that the MEG signal in a spherically symmetric head model is always zero (Figure 11.3, bottom row). This follows from equation (11.13), since both terms in the equation contain the cross product $\mathbf{P} \times \mathbf{r_P}$, which is zero when the two vectors are parallel.

A qualitatively different result is obtained with an infinite homogeneous head model, where a radial dipole leads to an MEG signal with a zero radial component but a nonzero tangential component (Figure 11.3, top row). This result follows from the cross product $\mathbf{P} \times (\mathbf{r} - \mathbf{r_P})$ in equation (11.4), where we only need to consider the first term $\mathbf{P} \times \mathbf{r}$, since we already know (from earlier) that the last term $\mathbf{P} \times \mathbf{r_P}$ is zero for radial dipoles. If we compute the vector product between this first term and the three unit vectors $\mathbf{e}_r$, $\mathbf{e}_\theta$, and $\mathbf{e}_\phi$ using spherical coordinates, we confirm that the radial component B_r must be zero because $\mathbf{P} \times \mathbf{r} \perp \mathbf{e}_r$. The two tangential components can in general both be nonzero, and a zero polar component B_θ in our example is a consequence of our dipole $\mathbf{P}$ being directed along the z-axis. Then, the product $\mathbf{P} \times \mathbf{r}$, which is perpendicular to the plane

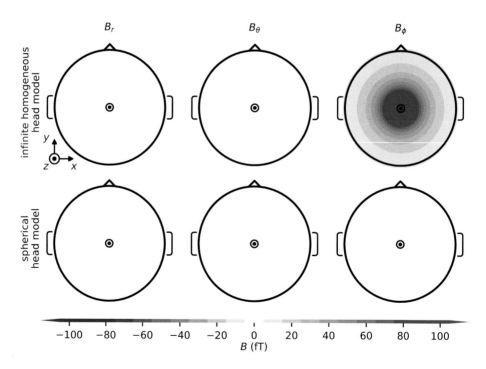

Figure 11.3 MEG signal from radial dipole. Computed for an infinite homogeneous head model equation (11.4) (top row) and a spherically symmetric head model equation (11.13) (bottom row). In both cases, a radial dipole with strength 1 nA m was positioned on the z-axis 8.8 cm above the origin. The MEG signal was estimated at 10,000 locations on a spherical shell at a distance 10.5 cm away from the origin. The three columns show the radial component B_r, the polar B_θ, and the azimuthal component B_ϕ, respectively. Code available via www.cambridge .org/electricbrainsignals.

spanned by $\mathbf{r}$ and $\mathbf{P}$, is also perpendicular to the $\mathbf{r}$-z-plane, and since $\mathbf{e}_\theta$ by definition is always in the $\mathbf{r}$-z-plane, we must have $B_\theta = \mathbf{B} \cdot \mathbf{e}_\theta = 0$.

Since an infinite conductor is also spherically symmetric, it is natural to ask why the zero-field prediction based on equation (11.13) does not apply also to the infinitely homogeneous conductor. The explanation is that equation (11.13) is only valid outside the conductor, and for an infinite homogeneous head model, we are never outside the conductor. Considering the spherical model as the more realistic among the two, the MEG produced by a radial dipole in the infinite homogeneous head model can be regarded as an artifact resulting from not having insulating boundary conditions on the scalp.

11.4.2   Tangential Dipoles

For a tangential dipole, all spatial components of the MEG are nonzero in both head models (Figure 11.4). As proven mathematically in Ilmoniemi & Sarvas (2019, section 3.6), the radial component of the MEG signal does not depend on volume currents. The radial component of the magnetic field is therefore identical in the two models (equation (11.13) and gives the same result as equation (11.4)).

However, the tangential components of the MEG signal differ between the models, although they do show some similarities. The differences are again explained by the volume currents, which are accounted for only in the spherical model.

11.4.3   Summary of Findings

The main insights for the MEG in an (i) infinite homogeneous and a (ii) spherically symmetric head models are:

- The radial component of the MEG signal does not depend on volume currents.
- The radial component of the MEG signal is (head-) model independent for all dipole sources and can be computed with the simple equation $B_r = \mathbf{B}_i \cdot \mathbf{e}_r$, with $\mathbf{B}_i$ as in equation (11.4).
- The radial component of the MEG signal is zero for a radial dipole.
- The tangential component of the MEG signal depends on volume currents in (ii) but not in (i).
- The tangential component of the MEG signal from a radial dipole is zero in (ii) but not in (i).

For general, more-detailed head models, one must turn to numerical schemes for computing the MEG, as described in Section 11.3.3.

11.5   Applications of the MC+VC scheme to MEG Signals

As for the EEG signal, the forward modeling of MEG signals from given dipole moments (the VC part) dates back to the development of the MEG technique in the

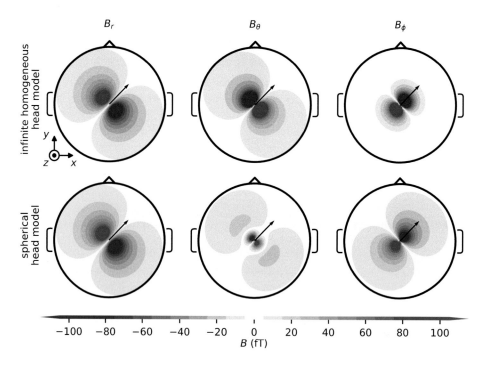

Figure 11.4 MEG signal from a tangential dipole. Computed for an infinite homogeneous head model equation (11.4) (top row) and a spherically symmetric head model equation (11.13) (bottom row). A tangential dipole with strength 1 nA m positioned on the z-axis 8.8 cm above the center of the sphere gives rise to a magnetic field outside the head. The MEG signal is estimated in 10,000 locations on a spherical shell a distance 10.5 cm away from the origin. The three columns show the radial B_r, the polar B_θ, and the azimuthal B_ϕ components, respectively. Code available via www.cambridge.org/electricbrainsignals.

1970s (Ilmoniemi & Sarvas 2019). However, the computation of such dipole moments from neural activity (the MC part) came later. In a series of pioneering papers, Murakami and co-workers computed dipole moments in the context of estimating magnetic signals from slice recordings, considering both stylized (Murakami et al. 2002, 2003) and biophysically detailed neuron models (Murakami & Okada 2006). Later, Jones and co-workers have in a series of papers (Jones et al. 2007, 2009, Sherman et al. 2016, Kohl et al. 2022) modeled cortical MEG signals based on circuits of stylized neurons, focusing their work on beta-oscillations.

11.6   Magnetic Fields inside the Brain

While equation (11.13) for a spherically symmetric head model is only valid outside the head where $\sigma_t = 0$, equation (11.12) for an infinite homogeneous head model and equation (11.14) for a general head model apply also on the inside.

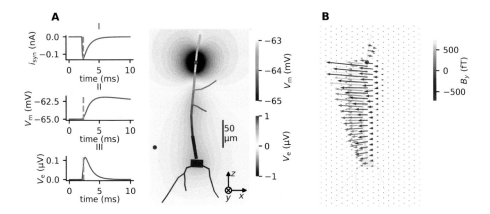

Figure 11.5 Magnetic field predictions from axial currents. A: Membrane voltages and extracellular potentials of an MC neuron. The passive neuron model lies flat in the xz-plane and receives a synaptic input current (inset axes I) at the location of the red dot marker. The membrane potential in the synaptic input location is shown in inset axes II. The contour plot shows the extracellular potential V_e, computed in a plane parallel with the xz-plane offset by $-10\,\mu m$ along the y-axis, at the time of maximum synaptic current magnitude (dashed grey line in insets). $V_e(t)$ computed at the location marked by the blue dot is shown in inset axes III. **B**: Magnetic field computed in the same plane as V_e. The arrow lengths and directions display the field components along the $x-$ and $z-$axis, while their colors denote the $y-$component. Code available via www.cambridge.org/electricbrainsignals.

The simplest way of approximating the magnetic field inside the brain is to (i) assume an infinite homogeneous head model, (ii) compute the dipole moment **P** from a neural simulation, and (iii) plug it into equation (11.12). However, this procedure will only allow us to compute the field at locations that are sufficiently far away from the cellular sources for the dipole approximation to hold.

If we wish to compute the magnetic field in closer proximity to the neurons, we can – instead of using the dipole approximation – compute $\mathbf{B_i(r)}$ from equation (11.4), letting $\mathbf{i_i}$ represent the axial intracellular currents in simulated neurons (Blagoev et al. 2007, Hagen et al. 2018). For a discrete set of constant axial currents I_{an}, running along neural compartments represented by vectors $\mathbf{d}_n$ with midpoint positions $\mathbf{r}_n$ in an MC neuron simulation, we can derive a modified version of equation (11.12), expressing **B** as a sum over the contributions from individual compartments:

$$\mathbf{B(r)} = \frac{\mu_0}{4\pi} \sum_n I_{an} \frac{\mathbf{d}_n(\mathbf{r}_n) \times (\mathbf{r} - \mathbf{r}_n)}{\|\mathbf{r} - \mathbf{r}_n\|^3}. \qquad (11.15)$$

Using equation (11.15), **B** can be computed in close proximity to the neural sources.[2]

---

[2] We must still be far enough away for the dipole approximation to hold for each dipole moment separately.

An example simulation using this strategy is shown in Figure 11.5, which shows the extracellular potential (panel A) and magnetic field (panel B) outside a neuron receiving a single synaptic input. Since an infinite homogeneous head model has been assumed, **B** depends exclusively on the intracellular axial currents (equation (11.15)) and not on the tissue volume currents associated with the extracellular potential. At this fine spatial scale, local features of the cell morphology influence the magnetic field, and **B** differs strongly from what would be predicted from an equivalent dipole.

12   Diffusion Potentials in Brain Tissue

Neurons generate electric signals by exchanging ions with their surroundings. For example, an action potential is essentially a brief depolarization of the membrane caused by an influx of Na^+ ions into the neuron, followed by a repolarization (or hyperpolarization) largely caused by an efflux of K^+. Knowing this, it may seem peculiar that many multicompartment (MC) models of neurons so lightheartedly assume that ion concentrations remain constant during neural activity.

As discussed in Section 3.4, the justification for neglecting short-term concentration changes is simply that the number of ions crossing the neuronal membrane during an action potential, or any brief period of neural activity, is too small to have a notable impact on the ion concentrations inside or outside a neuron. Short-term concentration changes are therefore believed to be negligible on the cellular level and above, at least when it comes to the most abundant ions, K^+, Na^+, and Cl^-.

Even if a few action potentials cause a negligibly tiny change in ion concentrations, we might imagine that many such tiny changes could add up to a substantial change over time in a population of vividly firing neurons. However, to prevent such accumulative concentration changes from occurring, neurons and glial cells possess a homeostatic machinery consisting of ion pumps and cotransporters that constantly work to reverse the ionic exchange occurring during neuronal signaling. An important player in the homeostatic machinery is the Na^+/K^+ pump, which uses energy to pump K^+ back into the neuron and Na^+ back out from the neuron, thus acting (in symphony with other homeostatic mechanisms) to reverse the ionic exchange that occurs during neural activity. Hence, ion concentrations tend to also remain close to constant on a longer time scale, at least during normal working conditions. In standard MC neuron models, the homeostatic machinery is therefore not explicitly modeled. Instead, ion concentrations are simply assumed to stay constant.

While the assumption of constant ion concentrations probably holds fairly well during normal cellular activity, there are scenarios when concentrations deviate quite dramatically from baseline levels, either because the homeostatic machinery is impaired – for example, due to lacking blood or oxygen supply – or because the neuronal activity is so intense that the homeostatic machinery fails to keep up with it. Concentrations may then diverge from baseline both in the intra- and extracellular space. Shifts in extracellular ion concentrations are a hallmark of many pathological conditions, such as spreading depression and epilepsy, but can also occur, although to a less-dramatic degree, under nonpathological periods of intense neuronal activity (Somjen 2004).

Local changes in ion concentrations imply that there will be changes in the transmembrane concentration gradients, which will affect neuronal reversal potentials (equation (3.50)). This can have dramatic effects on neuronal firing properties, as has been the topic of many computational studies (Cressman et al. 2009, Zandt et al. 2011, Øyehaug et al. 2012, Wei et al. 2014, Sætra et al. 2020). Local changes in ion concentrations will also cause changes in intracellular and extracellular concentration gradients. If intracellular concentration gradients become large, resulting longitudinal diffusion along cellular processes can be of comparable magnitude to electric drift currents, and standard MC models, accounting only for the latter, thus become inaccurate (Qian & Sejnowski 1989, Halnes et al. 2013, Sætra et al. 2020). Finally, and most importantly for the topic of this book, if extracellular concentration gradients become large, diffusive currents through the extracellular space can evoke extracellular *diffusion potentials* (Gardner-Medwin et al. 1981, Dietzel et al. 1989, Herreras & Somjen 1993). When present, these diffusion potentials will give contributions to the total extracellular potential (V_e) that are not accounted for by standard volume-conductor (VC) theory (Halnes et al. 2016, 2017, Solbrå et al. 2018, Sætra et al. 2021). The objective of this chapter is to briefly outline the theory for modeling diffusion potentials and to discuss if and when such diffusion potentials can be expected to contribute to measured local field potentials or if, as normally assumed, the local field potentials exclusively reflect the distribution of transmembrane cellular currents.

Before moving on, we emphasize that the concentration changes that we will be considering in the current chapter are those that develop fairly slowly on the spatial scale of tissue. We are not interested in the faster concentration changes that occur on very fine spatial scales, such as in dendritic spines or in synaptic clefts (Savtchenko et al. 2017, Jasielec 2021). Also, we are not considering the large and fast relative changes that may occur in intracellular $[Ca^{2+}]$ due to the very low intracellular baseline concentration (De Schutter 2009, chapter 4). While such more-local concentration changes are indeed important for many cellular processes, they are not likely to play any notable roles for extracellular diffusion potentials.

12.1   What Is a Diffusion Potential?

Before going into the mathematical description, let us establish an intuitive understanding of a diffusion potential by considering a simple model of two containers with different ionic solutions (Figure 12.1A). There is a high-concentration solution of NaCl in one container (x_1) and a low-concentration solution of NaCl in the other (x_2). Each container is modeled as a single compartment. and we assume that both containers initially are perfectly electroneutral so that $[Na^+] = [Cl^-]$ in each of them. The potential difference between them will then be zero. At $t = 0$, the two containers are set in contact so that they may exchange ions over a junction (dashed line in Figure 12.1). Apart from the junction, we assume that both containers have sealed (impermeable) walls.

When the two electroneutral solutions start interacting at $t = 0$, the initial system dynamics will be driven exclusively by diffusion. As intuition tells us, both Na$^+$ and

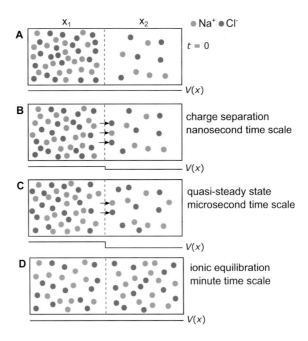

Figure 12.1 Diffusion potential in a simple two-compartment model of two ionic solutions interacting at a junction. A: Initial condition with a high-concentration NaCl in compartment x_1 and a low-concentration NaCl in compartment x_2. **B**: Since $D_{Cl} > D_{Na}$, the initial flux of Cl^- is higher than the flux of Na^+. **C**: The net charge transfer in panel B gives rise to a potential difference $V(x)$ between the compartments, which will prevent further charge separation. **D**: Eventually, the concentrations will equilibrate, and the voltage difference decay to zero.

Cl^- will diffuse rightwards towards the low-concentration container x_2. However, nature has it that not all ions diffuse equally fast. The diffusion constants for Na^+ and Cl^- in dilute solutions are $D_{Na} = 1.33 \times 10^{-9}$ m^2/s and $D_{Cl} = 2.03 \times 10^{-9}$ m^2/s, respectively (Table 5.1). The initial diffusive flux of Cl^- will therefore be larger than for Na^+, and diffusion will act as a charge-separation process (Figure 12.1B), resulting in an excess of positive charge in x_1 and an excess of negative charge in x_2.

The charge-separation process will "oppose itself" by causing a potential difference to develop between the two sides of the junction. This potential difference will act to drive anions towards x_1, thus reducing the flux of Cl^- towards x_2. Oppositely, it will drive cations towards x_2, thus increasing the flux of Na^+ towards x_2. The system will rapidly reach a so-called quasi-steady state, where the net fluxes of Na^+ and Cl^- have the same magnitude, and no further charge separation takes place (Figure 12.1C). Such a quasi-steady state will also occur in more-general systems when more than two ion species are present, and this state represents a *zero-current equilibrium* that occurs when the net electric current due to the diffusion of ions is exactly canceled by the net electric current due to the electric drift of ions. When we speak of the diffusion potential, it is the potential difference V_d associated with this equilibrium that we refer to. For dilute saline solutions (such as in the brain), it can be shown that it only takes about 10 ns to reach this quasi-steady state (Solbrå et al. 2018).

Note that the diffusion potential arises because the different ion species have different diffusion constants. If all ions had identical diffusion constants, there would be no charge separation and thus no diffusion potential.

When the system is in the quasi-steady state, V_d will still vary with time, hence the term "quasi." It eventually becomes zero when the system is equilibrated and the ion concentrations are identical on both sides of the junction (Figure 12.1D). However, this concentration-equilibration process occurs at a time scale that is several orders of magnitude slower than the time it takes to reach the quasi-equilibrium (see e.g. Solbrå et al. (2018, figure 2)). For practical purposes, it is therefore a good approximation to keep the two processes separate and assume that the system is *always* in a quasi-steady state, with V_d being an instantaneous function of the ionic concentrations on either side of the junction.[1]

Like in the example above, the diffusion potential is often discussed in the context of two different ionic solutions interacting at a junction. For that reason, the diffusion potential is sometimes called the *liquid-junction potential* (Aguilella et al. 1987, Sokalski & Lewenstam 2001). This term may be familiar for experimental neuroscientists, as the determination of liquid-junction potentials is often necessary when recording neural membrane potentials in patch clamp experiments (Barry & Lynch 1991, Neher 1992). The liquid-junction potential then causes some challenges when it comes to setting the reference point for membrane-potential recordings. The experimentalist normally wants to define the patch-pipette voltage to be zero just outside the target cell. However, as the pipette solution (normally made to mimic the cytosol solution) is different from the extracellular solution, a liquid-junction potential V_{LJ} will exist across its junction due to a process similar to that in Figure 12.1, only involving a larger number of different ion species. In the next step, when the experimentalist obtains the whole-cell recording configuration, the charge-separation layers at the pipette junction diffuse into the cytosol or pipette, as both contain similar ionic solutions, and the liquid-junction potential disappears (Neher 1992, Petersen 2017). For this reason, there will be an offset equal to V_{LJ} between the "true zero" and the "defined zero," and the same offset is inherited in all recordings of the membrane potential, V_m. This offset must be estimated and corrected for.

Other familiar examples of diffusion potentials are the ion-channel reversal potentials, or Nernst potentials (equation (3.50)), which are essentially single-species diffusion potentials across neural membranes. Diffusion potentials are also involved in determining the resting membrane potential and leakage reversal potential of the neuronal membrane. However, neural resting potentials and leakage potentials are not pure diffusion potentials, since they also are affected by homeostatic machinery involving the electrogenic Na^+/K^+ pump. This was discussed in further detail in Sections 3.4.4 and 3.4.5.

---

[1] The number of ions crossing the junction during the few nanoseconds it takes to reach the quasi-steady state will only constitute a fraction ($\approx 10^{-9}$) of the ions present within ~ 10 nm surrounding the junction (Aguilella et al. 1987), and this number practically has no impact on the ionic concentrations.

Compared to diffusion potential gradients over neural membranes or pipette junctions, diffusion potential gradients over intra- or extracellular bulk solutions tend to be less steep and are often neglected in neuroscience.

12.2   Theory for Computing Diffusion Potentials

The scenario depicted in Figure 12.1 is a simple example of an electrodiffusive process, and diffusion potentials are generally the result of such electrodiffusive processes. The fundamental equation governing electrodiffusive processes is the Nernst-Planck continuity equation,

$$\frac{\partial [k]}{\partial t} = \nabla \cdot \left[D_k \nabla [k] + \frac{D_k z_k [k]}{\psi} \nabla V \right] , \tag{12.1}$$

describing the interplay between a set of ion concentrations $[k]$ and the electric potential V. Equation (12.1) follows directly from combining the principle of particle conservation,

$$\frac{\partial [k]}{\partial t} = -\nabla \cdot \mathbf{j}_k , \tag{12.2}$$

with the Nernst-Planck flux density,

$$\mathbf{j}_k = -D_k \nabla [k] - \frac{D_k z_k [k]}{\psi} \nabla V , \tag{12.3}$$

which we encountered earlier in equation (2.21). We recall that $\psi = RT/F$ is defined by the gas constant (R), Faraday's constant (F), and the temperature (T), while D_k is the diffusion constant and z_k the valency of ion species k. The first term on the right (of equation (12.3)) accounts for the diffusive flux along the concentration gradient, and the second term accounts for the electric migration of ions along the voltage gradient. Equation (12.1) can in principle be applied everywhere (intracellularly, extracellularly, and inside the membrane), but the diffusion constants D_k will depend on the medium.

12.2.1   General Mathematical Frameworks

Equation (12.1) is a partial differential equation, continuous in both space and time. If the system in study contains K different ion species ($k \in \{Na^+, Ca^{2+}, Cl^-, \ldots\}$), we get one instance of equation (12.1) for each ion species, but the system is underdetermined since we are in need of an additional equation ($K + 1$) for the variable V that couples the dynamics of the individual ion species. The two main approaches to this problem are known as the *Poisson-Nernst-Planck* (PNP) and *electroneutral* frameworks. Among the two frameworks the PNP framework is the most physically detailed. Regardless of framework, the system of $K + 1$ equations must generally be discretized in both time and space and solved numerically using, for example, the finite element method (FEM).

12.2.1.1   Poisson-Nernst-Planck Framework

When using PNP (see e.g. Chen & Eisenberg (1993), Léonetti & Dubois-Violette (1998), or Mori (2006)), V is defined by Poisson's equation from electrostatics:

$$\nabla^2 V = -\rho/\epsilon \,, \tag{12.4}$$

with ϵ being the (local) permittivity of the medium. By expressing the local charge density ρ as a function of the ionic concentrations,

$$\rho = F \sum_k z_k [k] \,, \tag{12.5}$$

the PNP equations (equation (12.1) together with equation (12.4)) couple the dynamics of ion concentrations and electric potential.

12.2.1.2   Electroneutral Framework

An alternative to the PNP approach is to replace Poisson's equation with the approximation that the bulk solution is electroneutral, so that

$$\rho = F \sum_k z_k [k] = 0 \,. \tag{12.6}$$

The constraint determines what V must be at all points in space for there to be no charge accumulation anywhere in space. If a reference point ($V = 0$), is also set for the potential, this problem has a unique solution. If relevant, violations of electroneutrality associated with charge accumulation on neural membranes can be introduced as membrane boundary conditions (see e.g. Mori (2006) or Solbrå et al. (2018)). An example implementation of the electroneutral framework (applied to extracellular electrodiffusion surrounding active neurons) is outlined in Appendix I.

12.2.1.3   Framework Comparison

Combined with suitable boundary conditions, both the PNP and electroneutral frameworks can be solved for arbitrary complex geometries using some suitable numerical method (Mori 2006, Lopreore et al. 2008, Pods 2017, Solbrå et al. 2018, Song et al. 2018).

As we discussed when we considered the simple system in Figure 12.1, the electrodiffusive charge-separation process (illustrated in Figure 12.1B) typically lasts for only a few nanoseconds before the system reaches a quasi-steady state. In this quasi-steady state, the accumulated nonzero charge density is mainly concentrated on a very fine spatial scale at the junction between two solutions of different ionic composition. In neural tissue, nonzero charge densities are predominantly confined to nanometer thick layers around neuronal membranes (Grodzinsky 2011).

The physically detailed PNP framework explicitly accounts for the nanosecond-fast charge-relaxation processes (Figure 12.1B) and thus models the finely resolved accumulation of charge that sets up V (see equation (12.5)). To obtain stable and accurate PNP simulations, ionic concentrations must therefore be determined with hyperfine precision, requiring spatiotemporal resolution smaller than nanometers and nanoseconds.

This makes PNP simulations extremely computationally expensive and unsuited for estimating dynamics at the level of tissue. Applications in neuroscience have therefore been limited to studies of electrodiffusive processes taking place on a very tiny spatial scale, such as in dendritic spines or near and inside membranes (see Savtchenko et al. (2017) or Jasielec (2021) for reviews of applications in neuroscience).

Unlike the PNP framework, the electroneutral framework circumvents the charge-relaxation problem by assuming (and ensuring) that the system is always in the quasi-steady state (Figure 12.1C). The electroneutrality constraint (equation (12.6)) is then used to derive the potential V associated with this quasi-steady state. Although electroneutrality at a fundamental level is inconsistent with a nonzero V (see equation (12.4)), the electroneutral approach still gives accurate predictions of both $[k]$ and V under many circumstances (Feldberg 2000). In biophysical applications, it has been shown that the electroneutral approximation gives accurate results on spatiotemporal scales larger than micrometers and microseconds (Grodzinsky 2011, Pods 2017, Solbrå et al. 2018).

The advantage of the electroneutral approach is that it, unlike PNP, gives stable numerical solutions even for arbitrary coarse spatiotemporal resolutions, making the electroneutral framework much more computationally amenable than the PNP framework. However, the electroneutral framework is also too computationally heavy to allow for simulations of populations of neurons described with explicit geometries on today's computers. Ellingsrud et al. (2020) used a version of the electroneutral framework to simulate a small piece of tissue containing a bundle of nine axons described with idealized geometries.[2] To our knowledge, this is the largest system that at the time of writing has been simulated in 3D on a self-consistent scheme that couples intra- and extracellular electrodiffusion in tissue containing explicit cellular geometries (as in Figure 2.7D). However, whole-cell-level electrodiffusive models for single two-compartment neuron models have been developed for idealized scenarios considering a 1D extracellular space (Sætra et al. 2020, 2021).

Electrodiffusive models that account for further aspects of cellular geometries have also been constructed but using the MC+ED scheme (Figure 2.7B), where electrodiffusion was only considered for the extracellular dynamics while the neurodynamics was simulated with standard MC models (Halnes et al. 2016, Solbrå et al. 2018). Electrodiffusive simulations covering a larger spatial scale have so far only been performed using domain-type models (see Section 2.6.4), which do not account for any aspects of cellular geometries and are not suited to predict extracellular potentials.

12.2.2 Analytical Estimates of the Diffusion Potential

The general frameworks presented in Section 12.2.1 allow us to numerically compute how diffusion potentials vary in space and time. However, in many practical cases, we are content with finding the diffusion potential across a junction of two ionic solutions

---

[2] Axons were cuboid shaped with dimensions $0.1 \times 0.1 \times 390\ \mu m^3$, and separated with an $0.1\ \mu m$ intra-axonal distance.

(like in Figure 12.1) at a given instance in time, assuming that the system is in a quasi-steady state (i.e. as in the electroneutral framework). In such cases, we can avoid computationally heavy numerical solutions and instead make analytical estimates.

When two ionic solutions interact at a junction, the electrodiffusive flux is effectively one-dimensional and perpendicular to the junction, thus given by the 1D version of equation (12.3),

$$ j_k = -D_k \frac{d[k]}{dx} - \frac{D_k z_k}{\psi} [k] \frac{dV}{dx} . \tag{12.7} $$

Since we consider the quasi-steady state, the diffusion potential is the potential for which the net electric current across the junction is zero,

$$ i = F \sum_k z_k j_k = 0 . \tag{12.8} $$

By combining equation (12.7) and equation (12.8), the problem at hand can be written as

$$ \sum_k z_k D_k \frac{d[k]}{dx} = -\sum_k \frac{z_k^2 D_k}{\psi} [k] \frac{dV}{dx} . \tag{12.9} $$

Hence, we seek to find the potential for which the net electric drift current due to electric drift in one direction (right-hand side of equation) has the same magnitude as the net electric current due to diffusion (left-hand side of the equation) in the other direction.

If only a single ion species is included, as would be relevant for modeling an ion channel that is permeable only to a particular ion type, equation (12.9) reduces to equation (3.47), which was the starting point for deriving the Nernst potential in Section 3.4.1. When several ion species are involved, the problem is more challenging. Provided that the concentrations of all ions on both sides (call them x_1 and x_2) of the junction are known, it is still possible to integrate equation (12.9) over the junction to obtain analytical solutions for the diffusion potential. However, to do so, it is necessary to make certain assumptions regarding the physics within the junction itself. Three different approaches to this problem have been proposed by Planck (1890), Henderson (1907), and Goldman (1943) (see e.g. Rosenberg (1969) or Stanton (1983) for brief overviews), all based on the approximation of electroneutrality (Section 12.2.1.2). Most commonly used in neuroscience are the Henderson and Goldman equations, and here we will only consider those two.

In 1949, Hodgkin and Katz used the Goldman equation to describe the membrane potential of the resting squid giant axon (Hodgkin & Katz 1949). Since this work established its use in neuroscience, the Goldman equation is referred to by many as the Goldman-Hodgkin-Katz (GHK) voltage equation, as will we. Being simpler than the Henderson equation, and successful in fitting experimental measurements of membrane potentials, the GHK voltage equation has become the standard choice when the junction represents a (thin) biomembrane. For this reason, the GHK voltage equation has been coined the membrane equivalent of the Henderson equation (Stanton 1983). The Henderson equation is the common choice for computing other types of junction potentials, such as for liquid-junction potentials in patch-recording pipettes (Neher 1992) or even for diffusion potentials in the extracellular space – for example, between the different cortical layers (Dietzel et al. 1989). In the next sections, we will briefly introduce these equations and comment on how they differ.

12.2.2.1 Goldman-Hodgkin-Katz Voltage Equation

Goldman assumed, as did later Hodgkin and Katz, that the electric field is constant over the junction (membrane), or equivalently, that the voltage profile is linear. This led to the GHK voltage equation,

$$V_\mathrm{d} = \frac{RT}{F} \ln \left(\frac{\sum_{k+} P_{k+}[k^+]_{x_1} + \sum_{k-} P_{k-}[k^-]_{x_2}}{\sum_{k+} P_{k+}[k^+]_{x_2} + \sum_{k-} P_{k-}[k^-]_{x_1}} \right), \tag{12.10}$$

for the diffusion potential between two sides of the junction. The pluses and minuses indicate that the sums are taken over monovalent cations (+) and anions (−) separately.

When this equation is used, x_1 and x_2 often represent the extracellular and intracellular sides of a membrane, respectively, and p_k (with unit $\mathrm{cm^{-1}\,s^{-1}}$) is then the specific membrane permeability. As we discussed in Section 3.4.2, p_k is proportional to the diffusion constant D_k in the original Nernst-Planck equation.

We note that equation (12.10) is valid only in the case where all ions are monovalent. More complex relationships can be derived for cases when ions differ in valency (Pickard 1976, Stanton 1983), but we do not include such results here. The restriction to monovalent ions is normally not a serious limitation since the most abundant ions in biological systems are monovalent. We encountered the GHK voltage equation earlier for the special case when the involved ions on the two sides of the membrane were K^+, Na^+, and Cl^- (equation (3.52)).

12.2.2.2 Henderson Equation

Henderson solved equation (12.9) by making the simplifying assumptions that (i) the mobilities (or equivalently, the diffusion constants) of ions were constant over the junction and (ii) that the concentration profiles were linear across the junction. In the original derivation, the Henderson equation was expressed in terms of the mobilities of the ion species, but we here present it in the form

$$V_\mathrm{d} = \frac{RT}{F} \left(\frac{\sum_k z_k D_k([k]_{x_2} - [k]_{x_1})}{\sum_k z_k^2 D_k([k]_{x_2} - [k]_{x_1})} \right) \ln \left(\frac{\sum_k D_k z_k^2 [k]_{x_1}}{\sum_k D_k z_k^2 [k]_{x_2}} \right), \tag{12.11}$$

used for example in Alcaraz et al. (2009), where the mobilities have been replaced by diffusion constants, assuming that they are related via the Einstein relation (equation (2.23)).

12.2.2.3 Goldman-Hodgkin-Katz versus Henderson

Since the Henderson and GHK equations represent two alternative approaches to the same problem, it is natural to question why the GHK equation is used for membranes, while the Henderson equation is used for other kinds of junctions. As the two equations often give similar predictions (Rosenberg 1969), one could in principle estimate membrane potentials using the Henderson equation and other junction potentials using the GHK equation. However, later theoretical works, comparing predictions from the GHK and Henderson equations to numerical simulations using the PNP framework (Section 12.2.1.1), have concluded that the GHK and Henderson equations are, respectively, the thin and thick membrane (or junction) limits of the PNP equations (Perram & Stiles 2006). In this context, neural membranes and biomembranes do indeed qualify as thin, suggesting that the GHK equation is the better choice for membrane-related problems. In contrast, for problems not involving physical membranes but diffusion potentials arising at the boundary between two electrolytes, the Henderson equation is the common choice (Jasielec 2021), and it has been shown to give similar results as numerical solutions of the PNP equations for many cases of interest (Sokalski & Lewenstam 2001).

12.3   Can Diffusion Potentials Be Seen in Recorded LFPs?

Attempts to realistically simulate population level signals such as the LFP have so far been based on the MC+VC scheme (Figure 2.7A) or some of the (simulation-wise) less – detailed schemes presented in Section 6.4. None of these schemes account for possible effects of diffusion on extracellular potentials.

Whether diffusion potentials are likely to be "visible" in LFP recordings will depend on their magnitude and temporal development. The reason why the temporal aspect matters is that ion concentrations, and therefore diffusion potentials, tend to vary on a slow time scale compared to LFP fluctuations produced by electric neural activity. As most electrode systems used to record LFPs use a lower cut-off frequency of 0.1–1 Hz (Einevoll et al. 2007), the slow (DC-like) diffusion potentials might in principle be filtered out from such recordings, as we also discussed in Section 8.6.

To accurately account for concentration effects on extracellular potentials (in time and space), we would need to run numerical simulations using an electrodiffusive scheme (Figure 2.7C or D), based on either the PNP framework (Section 12.2.1.1) or the electroneutral framework (Section 12.2.1.2). Unfortunately, electrodiffusive numerical simulations are very computationally expensive, and attempts to model population-level LFPs mimicking realistic experimental scenarios using such schemes have so far not been made.

In the current section, we give a crude assessment of whether diffusion potentials are likely to show up in LFP measurements by following a simpler method (Brinchmann 2021). The idea of this method is to compare power-spectral densities of diffusion potentials (diffusion PSDs) with power-spectral densities of real, recorded LFPs (LFP PSDs). Using this method, in Section 12.3.3, we will compare the PSDs of a selection of previously published LFP recordings with PSDs of diffusion potentials estimated for three different scenarios representing (i) spreading depression (extreme concentration changes), (ii) neural hyperactivity (moderate concentration changes), and (iii) moderate neural activity (small concentration changes). If diffusion potentials indeed do not contribute to recorded LFPs, we would expect the diffusion PSDs to be smaller than the LFP PSDs for all frequencies included in the LFP recording.

The method is based on the following set of assumptions:

- The diffusion PSD is computed from the temporal development of the diffusion potential $V_d(t)$ between two fixed locations x_1 and x_2. The choice of positions is discussed further in Section 12.3.1.
- All ion concentrations $[k]$ at position x_1 remain at baseline concentrations $[k] = [k]_{bl}$ for all $t \geq 0$.
- Ion concentrations $[k] = [k]_{bl} + \Delta[k](t)$ at position x_2 deviate transiently from baseline. By assumption, each concentration has a peak deviance $\Delta[k]_{peak}$ at time $t = 0$ and decays exponentially towards the baseline with a time constant τ so that

$$\Delta[k](t) = \Delta[k]_{peak}e^{-t/\tau} , \text{ for } t \geq 0 . \tag{12.12}$$

- The temporal variation of the diffusion potential between x_1 and x_2 are given by the Henderson equation (equation (12.11)), which after the insertion of equation (12.12) can be written as

$$V_d(t) = \frac{RT}{F} \left(\frac{\sum_k z_k D_k^* \Delta[k](t)}{\sum_k z_k^2 D_k^* \Delta[k](t)} \right) \ln \left(\frac{\sum_k D_k^* z_k^2 [k]_{bl}}{\sum_k D_k^* z_k^2 ([k]_{bl} + \Delta[k](t))} \right) , \text{ for } t \geq 0 . \tag{12.13}$$

- The parameters $\Delta[k]_{peak}$ and τ are based on ion-concentration data from relevant experimental literature for the scenarios (i–iii), as explained in Section 12.3.1 and Section 12.3.2. The diffusion constants $D_k^* = D_k/\lambda_e$ are given by the diffusion constants D_k for dilute saline solutions (Table 5.1), divided by the tortuosity λ_e of the extracellular medium (equation (5.3)).[3]
- For a given scenario (i, ii, or iii), we will assume that all ion species have the same τ. Although the rate of change is not necessarily identical for all ion species, it can be expected to be of similar magnitude. After all, variations in the individual ion concentrations are not independent: due to the electroneutrality condition, they will covary in a way that always and everywhere adds up to a zero net charge density.
- The PSD of $V_d(t)$ computed with equation (12.13) will then give us the diffusion PSD, which we compare with LFP PSDs in Section 12.3.3.
- We ignore possible influences of osmotic pressures associated with ion-concentration gradients (Holter et al. 2017).

Before we move on, we note that $\Delta[k](t)$ as given by equation (12.12) has the same frequency content (for $f > 0$) as its mirror-image, the function

$$\Delta[k](t) = \Delta[k]_{d,peak}(1 - e^{-t/\tau}) \text{ for } t \geq 0 , \tag{12.14}$$

describing a temporal development from zero to $\Delta[k]_{d,peak}$. Hence, used with a τ reflecting the relevant process (rise or decay), equation (12.13) can be used to estimate PSDs of diffusion potentials both when they are rising (concentrations diverging from baseline)

---

[3] The tortuosity is the same for all ion species and can be eliminated from equation (12.13). D_k^* can thus be replaced with D_k.

or decaying (concentrations returning to baseline). To evaluate the maximal diffusion PSDs, we will in the following generally pick τ from the process that is fastest.

12.3.1   Magnitude of Diffusion Potentials

When we speak of a potential, we refer to a potential difference between two positions (x_1 and x_2). To compute the diffusion potential V_d between two positions, we plug ion concentrations recorded in these positions into the Henderson equation (equation (12.11)). This method has been previously used by, for example, Dietzel et al. (1989) to compute diffusion potentials between different cortical layers.

Using this method, the "junction" will consist of all the tissue between the recording positions (x_1 and x_2) and can in principle be quite thick. The assumption (implicit in the Henderson equation) that concentration profiles are linear over such a thick and complex junction is rather bold, and estimates of diffusion potentials obtained in this way will probably be inaccurate. However, assuming that the predicted values will still be in the right ballpark, we will adopt this strategy.

For simplicity, we will assume that the diffusion potential only depends on the four most abundant ion species in the extracellular space, K^+, Na^+, Cl^-, and HCO_3^-. Variations in these three ion concentrations are rarely recorded simultaneously within a given experiment. This means that we must make semi-qualified guesses about concentrations that are not included in the recordings. A constraint that helps us in this regard is the electroneutrality condition, which requires that

$$\Delta[K^+] + \Delta[Na^+] = \Delta[Cl^-] + \Delta[HCO_3^-] \, . \tag{12.15}$$

We assume that x_1 is a position where the ion concentrations remain at baseline, whereas x_2 is the position where the maximal deviances from baseline concentrations are recorded. This diffusion potential $V_{d,peak}$ at the time when deviances from baseline are at their highest will then probably be among the largest diffusion-potential differences within the system, but not necessarily the largest, since ion concentrations in principle can go both above and below baseline values (i.e. there could be a position x_3 where deviations from baseline went in the opposite direction compared to position x_2).

For the baseline concentrations, we assume a set of rather typical values $[K^+] = 3\,mM$, $[Na^+] = 147\,mM$, $[Cl^-] = 127\,mM$, and $[HCO_3^-] = 23\,mM$, which have been specified so that they sum up to a zero net charge (electroneutrality).

Many experiments performed in the cortex or hippocampus show concentrations that remain close to baseline in some layers, while they vary and deviate strongly from baseline in other layers. For example, ion-concentration variations during spreading depression have been observed to be restricted to specific layers both in the cortex (Richter & Lehmenkühler 1993, figure 3) and the hippocampus (Herreras & Somjen 1993, figure 1). Pronounced concentration profiles along the cortical depth have also been observed in experiments with less-dramatic concentration shifts (Cordingley & Somjen 1978, Dietzel et al. 1989). Hence, it seems reasonable to picture our two positions x_1 and x_2 as being associated with different cortical or hippocampal depths so that they are separated by distances on the order of a millimeter.

12.3.1.1 Spreading Depression

Among the most extreme concentration changes reported in the extracellular space of the brain are those recorded under the pathological condition called spreading depression. The pathology is associated with a strong increase in extracellular $[K^+]$ and decreases in extracellular $[Na^+]$ and $[Cl^-]$ (Kraig & Nicholson 1978, Herreras & Somjen 1993, Somjen 2004, Pietrobon & Moskowitz 2014).

Concentrations (for position x_1 and x_2) representing the scenario of spreading depression are listed in Table 12.1, along with the associated diffusion potential. These are taken from Kraig & Nicholson (1978), who found that $[K^+]$ increased by about 33 mM from baseline during spreading depression, while $[Na^+]$ and $[Cl^-]$ decreased by about 92 mM and 90 mM from baseline, respectively. They also reported an anion deficit, which we here assume to be HCO_3^-. Based on the recorded changes in the other ion species, it follows from equation (12.15) that we must have an increase in $[HCO_3^-]$ (in position x_2) by 31 mM. Plugging these concentrations into the Henderson equation gives a diffusion potential $V_{d,peak} \approx 6.8$ mV between the two positions.

We note that, during spreading depression, slowly varying electric potential differences can exceed 20 mV across the cortical (see e.g. Lauritzen & Hansen (1992)) or the hippocampal (see e.g. Herreras & Somjen (1993)) depth. As previously discussed in Section 8.6, these *slow potentials* are sometimes referred to as *DC potentials* and must typically be recorded with so-called *DC electrodes* (Kraig & Nicholson 1978, Herreras & Somjen 1993, Richter & Lehmenkühler 1993, Somjen 2004, Pietrobon & Moskowitz 2014, Herreras & Makarova 2020). We note that slow potentials arising during spreading depression are generally not explained by the diffusion potential alone as slow potentials also involve neuronal and glial current sources and sinks associated, among other things, with glial K^+ buffering (Herreras & Makarova 2020, Sætra et al. 2021). Cellular swelling, causing a pronounced decrease in the extracellular volume fraction and thus in the extracellular conductivity (Olsson et al. 2006, Ayata & Lauritzen 2015), is also likely to contribute to the large potentials observed.

12.3.1.2 Hyperactivity

During neural hyperactivity (such as seizures), the extracellular $[K^+]$ can increase by several millimolars, up to a ceiling level suggested to be around 8–12 mM. Concentrations exceeding this ceiling level are normally only observed during pathophysiological conditions such as anoxia or spreading depression (Larsen et al. 2016).

Concentration shifts during neural hyperactivity have been recorded in several experiments (Sypert & Ward 1974, Krív et al. 1975, Nicholson et al. 1978, Dietzel et al. 1982, Somjen et al. 1986, Dietzel et al. 1989, de Curtis et al. 2018). In most of these studies, the focus was on $[K^+]$, which is the most critical extracellular concentration due to its low baseline value. However, Dietzel et al. (1989, figure A2) estimated simultaneously the shifts in both $[K^+]$, $[Na^+]$, and $[Cl^-]$ and suggested that an activity-induced increase in $[K^+]$ was (electrically) compensated by changes in both $[Na^+]$, and $[Cl^-]$. The $[K^+]$, $[Na^+]$, and $[Cl^-]$ composition varied over time, but some representative values from this study, taken to represent the hyperactivity scenario, are listed in Table 12.1. When inserted into the Henderson equation, these values give a diffusion potential of

	Baseline	Spreading depression	Hyperactivity	Moderate
K^+ (mM)	3	36	10	4
Na^+ (mM)	147	55	143	146
Cl^- (mM)	127	37	130	127
HCO_3^- (mM)	23	54	23	23
τ (s)	$--$	15	2.5	0.85
Diffusion potential (mV)	0	6.8	0.13	0.035

Table 12.1. Local peak-deviances from baseline in extracellular ion concentrations during various conditions, and their time constants (i.e. how fast they change from baseline to peak deviance). The corresponding peak-diffusion potential is also included. $\Delta[k]_{peak}$ used in equation (12.12) is obtained by subtracting the baseline concentrations from the peak concentration.

0.13 mV. We note that, in the original study, Dietzel et al. (1989) estimated that even-larger diffusion potentials than this (up to maximally 0.4 mV) could occur under their experimental conditions.

12.3.1.3   Moderate Activity

Typical physiological neuronal activity has been estimated to cause relatively small transients in extracellular $[K^+]$, up to about 0.2–0.4 mM above baseline (Larsen et al. 2016). However, numbers higher than this have been reported. For example, in cats, the cortical $[K^+]$ concentration varied by about 2 mM between different cortical layers when moderate (not seizure-inducing) stimuli were applied in the thalamus (Cordingley & Somjen 1978). In the cat spinal cord, a rhythmic flexion of the knee joint was found to increase the extracellular $[K^+]$ by as much as 1.7 mM (Heinemann et al. 1990). Recordings from the cortex of anesthetized cats showed 0.5 mM fluctuations in extracellular $[K^+]$ even during the resting state (McCreery & Agnew 1983), and in vivo experiments have reported a 1 mM shift in $[K^+]$ when mice transitioned from quiescence to locomotion (Rasmussen et al. 2019).

Based on the (somewhat arbitrary) examples given above, we have in Table 12.1 assumed that moderate neural activity can give rise to a local 1 mM shift in $[K^+]$. Furthermore, we have assumed that this local concentration shift is caused by neurons exchanging K^+ for Na^+ during action potential firing, so that it is compensated with a 1 mM decrease in $[Na^+]$. These relatively small concentration shifts correspond to a diffusion potential $V_{d,peak} \approx 35\,\mu V$. This is not so small that it a priori can be neglected in recordings of extracellular potentials.

12.3.2   Temporal Development of Diffusion Potentials

Since V_d is a function of ion concentrations, its temporal development will depend on the time constant τ for the rate of change in ion concentrations. As mentioned previously, we assume the same τ for all ion concentrations and take it to be the time constants for the development of extracellular $[K^+]$. Values for τ representing the scenarios of

(i) spreading depression, (ii) neural hyperactivity, and (iii) moderate neural activity are given in Table 12.1 and these values are taken from relevant experimental articles (more information on this is given in Sections 12.3.2.1–12.3.2.3). For each scenario, we have chosen whichever is the fastest between the time constant of the rise (from baseline to peak deviance) and the time constant of the decay (from peak deviance towards baseline).

We note that the exponential form in equation (12.13) was chosen mainly because of its simplicity. Although [K$^+$] does indeed vary in a close to exponential fashion in many data sets (see e.g. Cordingley & Somjen (1978, figure 1), Somjen (2004, figure 8-3), Haj-Yasein et al. (2012, figure 4)), or Enger et al. (2015, figure 4), this is not the case in all data sets.

12.3.2.1 Spreading Depression

The temporal development of extracellular [K$^+$] during spreading depression has been recorded in many experiments. There is quite some variation between reported values, but in general, the rising phase is faster than the decay time (which often lasts more than a minute). We therefore use the value $\tau = 15$ s, which lies close to the time constant for the rise-phase seen in Pietrobon & Moskowitz (2014, figure 1b) and also seems representative for several other experiments (Kraig & Nicholson 1978, Hansen & Zeuthen 1981, Enger et al. 2015).

12.3.2.2 Hyperactivity

The literature on ion-concentration dynamics during neuronal hyperactivity or seizure activity is extensive, and a whole range of different rise and decay times for extracellular [K$^+$] can be found, depending on experimental conditions (Somjen 2004, Raimondo et al. 2015, Antonio et al. 2016). The data in Somjen (2004, figure 8-3) and Sypert & Ward (1974, figure 4B) indicate that time constants as fast as $\tau = 2.5$ s can occur under some conditions (both for the rise and decay phase).

12.3.2.3 Moderate Activity

Cordingley & Somjen (1978) used thalamic stimulus trains of various duration and frequency to evoke increases in cortical [K$^+$] and studied the temporal decay of [K$^+$] towards baseline following stimulus offset. In these studies, they found half-decay times as fast as ≈ 0.6 s (Cordingley & Somjen 1978, figure 3), which corresponds to a time constant $\tau \approx 0.85$ s.[4]

12.3.3   Power Spectra of Diffusion Potentials

The estimates in Table 12.1 suggest that the magnitude of diffusion potentials may be comparable to LFP amplitudes, while their temporal development is slow compared to the LFP frequencies. If they are too slow, they will not have any frequency content above

---

[4] τ is given by the half-decay time multiplied with ln 2.

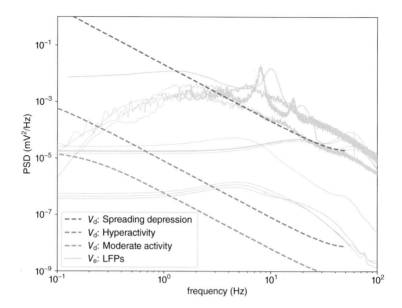

Figure 12.2 PSDs of diffusion potentials versus recorded LFPs. Diffusion PSDs (dashed curves) for conditions of spreading depression (blue), neural hyperactivity (orange), and moderate neural activity (green) versus PSDs of recorded LFPs (grey, solid curves). Diffusion potentials were computed with equation (12.13), with $\Delta[k](t)$ as in equation (12.12) and with parameters $\Delta[k]_{\mathrm{peak}}$ and τ as in Table 12.1. LFPs were taken from the literature, and the selection included LFPs from the hippocampus of rats performing open tasks (Mizuseki et al. 2009), the auditory cortex of rats responding to auditory input (Deweese & Zador 2011), the prefrontal cortex and hippocampus of rats performing memory tasks (Fujisawa et al. 2008), the basal forebrain of rats under resting conditions and during the exploration of an arena (Nair et al. 2018), the somatosensory cortex of anesthetized rats (Baranauskas et al. 2012), the anterior claustrum of freely moving rats (Jankowski et al. 2017), and the cerebral cortex of mice (unpublished data from subjects receiving no stimulus, obtained during the project by Thunemann et al. (2018)). All PSDs were computed using the periodogram function from SciPy's signal package (https://docs.scipy.org/doc/scipy/reference/generated/scipy.signal .periodogram.html), based on the Welch method. The same methods and LFP data were used in Brinchmann (2021), and more detailed information about the procedures can be found there.

the lower cut-off frequency, which for LFP recordings typically lie somewhere between 0.1 and 1 Hz (Einevoll et al. 2007). To explore whether it is conceivable that diffusion potentials are "visible" in the LFP, we have in Figure 12.2 compared their power spectra (dashed curves) with the power spectra of real LFP recordings (grey, solid curves).

With the insertion of equation (12.12) into equation (12.13), it is found that the diffusion potential $V_d(t)$ has approximately the same exponential temporal development as the ion concentrations $\Delta[k](t)$, and its PSD is approximately a $1/f^2$ power law[5] over most of the considered frequency interval. For the extreme case of spreading depression (blue curve), the diffusion potential has a higher power than most of the recorded LFPs at all frequencies. Hence, in subjects undergoing spreading depression, LFP recordings

---

[5] See Table F.1: the squared Fourier transform of an exponential function goes like $1/f^2$.

are likely to be dominated by concentration-dependent effects, and their PSDs would probably not resemble those seen for any of the real recorded LFPs in Figure 12.2.

Since most simulations and interpretations of recorded LFPs are made under the assumption that it is not at all affected by diffusive currents, it is reassuring that the scenario with moderate neural activity (green curve) has a very low diffusion PSD over most of the considered frequency range. It only exceeds the LFP PSDs that have the weakest power among those considered and then only for very low frequencies. In the case of neural hyperactivity (orange curve), the diffusion PSD is greater. It crosses the LFP PSDs with the weakest power at about 3 Hz as well as several of the LFP PSDs frequencies between 0.1 and 1 Hz.

The comparison in Figure 12.2 suggests that most of the frequency range in the LFP will be unaffected by diffusion potentials, except under extreme pathological conditions such as spreading depression. However, it also suggests that one cannot rule out the possibility that, in some experiments, diffusive effects can show up in the very-low frequency range of recorded LFPs. If they do, the lower frequencies in the LFP will not exclusively reflect transmembrane neural currents and cannot be simulated using standard MC+VC models. Similar conclusions were drawn in the computational study by Halnes et al. (2016), where diffusion potentials (under non-pathological conditions) were found to affect the LFP up to frequencies of at most a few hertz. By prediction, the diffusion potentials will then give rise to a $\sim 1/f^2$ power law in the low-frequency range of the PSD.

In the next section, we give a more theoretical discussion of the diffusion potential, its absence in standard MC+VC type models, and the possible consequences.

Before we go on, we emphasize that the diffusion PSDs and LFP PSDs that we have compared in this section do not represent the same experiments. Hence, the analysis does not give any direct clues as to whether diffusion was present in the actual LFP PSDs included in the comparison. The point was rather to explore whether diffusion PSDs in a general sense can be of magnitudes comparable to recorded LFPs. We also note that our analysis was by no means exhaustive. It included only a rather arbitrary selection of LFP PSDs and a few representative estimates of diffusion PSDs.

12.4   Extracellular Volume Conduction versus Electrodiffusion

When modeled using the standard MC+VC scheme, tissue currents are assumed to be purely ohmic so that diffusive extracellular currents are neglected. The extracellular potential (V_e) can then be computed as a direct function of the neural (and glial, if one wishes) membrane currents as determined from the current-source density (CSD) equation (equation (2.31)). The inaccuracy introduced when neglecting diffusive effects on V_e can be assessed by comparing the CSD equation obtained with the MC+VC scheme with an expanded CSD equation obtained with the MC+ED scheme depicted in Figure 2.7B. In the MC+ED scheme, the extracellular volume-conduction (VC) part of the MC+VC scheme has been replaced with a framework for extracellular electrodiffusion (ED).

The starting point for such a comparison is the principle of current conservation, as expressed through the continuity equation

$$\nabla \cdot \mathbf{i}_e = C_{all} , \qquad (12.16)$$

where C_{all} is the current-source density (cellular output) and $\mathbf{i}_e$ the extracellular current density. Equation (12.16) is analogous to equation (2.30), which we normally use in computations based on the MC+VC scheme. However, as we discussed in Section 4.3.3, standard applications of MC+VC simulations do not require us to make any assumptions regarding the pathways taken by tissue currents, which means that $\mathbf{i}_t$, featuring in equation (2.30), generally can account for both extracellular and intracellular pathways through the tissue. In contrast, the MC+ED scheme computes ionic fluxes (and thus currents) in the tissue from extracellular ion concentrations and thus exclusively accounts for the extracellular pathways. For the sake of comparing the schemes, we here therefore assume that the tissue-current density is purely extracellular ($\mathbf{i}_t = \mathbf{i}_e$) in both the MC+ED and MC+VC schemes. For this reason, we have also denoted the current-source density by C_{all}, as we implicitly assume that it accounts for "all" membrane currents present in the tissue.

If extracellular currents are purely ohmic, then $\mathbf{i}_e = -\sigma_e \nabla V_e$, with σ_e being the extracellular conductivity defined in Section 5.2. With this, equation (12.16) gives us the CSD equation,

$$\nabla \cdot (\sigma_e \nabla V_e) = -C_{all} , \qquad (12.17)$$

which is analogous to equation (2.31) but with $\mathbf{i}_t$ and σ_t replaced with $\mathbf{i}_e$ and σ_e.

The electrodiffusive counterpart to equation (12.17) is obtained by expressing the extracellular current density in terms of electrodiffusive ion fluxes. To get the extracellular current density from ionic fluxes, we multiply equation (12.3) by $F z_k$ and sum over all the ion species k to get

$$\mathbf{i}_e = F\alpha \sum_k z_k \mathbf{j}_k = -F\alpha \left(\sum_k z_k D_k^* \nabla [k]_e - \sum_k \frac{D_k^* z_k^2}{\psi} [k]_e \nabla V_e \right) , \qquad (12.18)$$

where the first term on the right-hand side is the diffusive current density, and the second term is the ohmic drift current density.

Before we go on, we should comment on the introduction of D_k^* and α in equation (12.18). Both these parameters come in because we consider electrodiffusion on the coarse-grained tissue scale described in Section 2.6. As we did earlier in equation (5.3), we have therefore introduced the effective diffusion constant $D_k^* = D_k / \lambda_e^2$ on the coarse-grained scale, defined by the diffusion constant in saline (D_k) and the tortuosity λ_e (≈ 1.6) of the extracellular medium, which accounts for the hindrances to free extracellular diffusion imposed by cellular structures in tissue (Syková & Nicholson 2008). The introduction of the extracellular volume fraction (α) is merely a matter of convention: in VC theory, the source term C is normally defined as neuronal output current per *tissue* volume, and extracellular current densities is defined as current per tissue cross-section area. In contrast, the extracellular ion concentration ($[k]_e$) is conventionally defined as the number of extracellular ions of species k (in mol) per *extracellular* volume, which is a fraction α (~ 0.2) of the tissue volume, as this gives us the "real" extracellular concentration (Syková & Nicholson 2008). Likewise, flux densities ($\mathbf{j}_k$) are conventionally defined relative to the extracellular cross-section area and must therefore be multiplied with α when converted to current densities per tissue cross-section area.

Since the last term in equation (12.18) is the ohmic current density, we must have

$$\sigma_e = \frac{\alpha F}{\psi} \sum_k D_k^* z_k^2 [k]_e \tag{12.19}$$

as we also defined it previously (see equation (5.1) and equation (5.4)). With this, we may express $\mathbf{i}_e$ more compactly as

$$\mathbf{i}_e = -\alpha \sum_k F z_k D_k^* \nabla [k]_e - \sigma_e \nabla V_e \,. \tag{12.20}$$

Finally, we can plug equation (12.20) into equation (12.16) to obtain a CSD equation that accounts for extracellular diffusion,

$$\nabla (\sigma_e \nabla V_e) = -C_{\text{all}} - F\alpha \nabla \cdot \left(\sum_k z_k D_k^* \nabla [k]_e \right) \,. \tag{12.21}$$

Comparing this with equation (2.31), we see that the difference is the last term on the right-hand side of equation (12.21), which accounts for the diffusive effects that are neglected in the standard theory. Hence, this last term can be seen as a correction term of the V_e predicted from standard VC theory.

To analyze the diffusive contribution separately, it is convenient to define the potential as (Solbrå et al. 2018)

$$V_e = V_{\text{VC}} + V_d \,, \tag{12.22}$$

where V_e is the "true" extracellular potential, while V_{VC} is the extracellular potential predicted from VC theory (standard CSD equation with sources C),

$$\nabla \cdot (\sigma_e \nabla V_{\text{VC}}) = -C_{\text{all}} \,, \tag{12.23}$$

and V_d is the additional diffusion potential evoked by concentration gradients in the extracellular space,

$$\nabla \cdot (\sigma_e \nabla V_d) = -F\alpha \nabla \cdot \left(\sum_k z_k D_k^* \nabla [k]_e \right) \,. \tag{12.24}$$

Again, we see that the diffusion potential V_d is a function of ion-concentration gradients alone. Although these gradients will generally be the result of the cellular activity over the last seconds or minutes, the diffusion potential in itself does not depend explicitly on the presence of cellular sources. The Goldman-Hodgkin-Katz voltage equation and Henderson equation that we considered in Section 12.2.2 are special-case solutions of equation (12.24).

As we showed in Chapter 4, the standard CSD equation (equation (2.31)) allows us to derive an analytical expression for the electric potential at all points in space once C_{all} is known. In comparison, equation (12.21) is not in itself so "user friendly," since it requires knowledge of all ionic concentrations $[k]_e$ at all points in space. To simulate extracellular concentration dynamics resulting from neural activity, one must solve the Nernst-Planck continuity equation (equation (12.1)) using finite element or finite difference methods, with separate source terms for each ion species as well as for capacitive membrane currents and with V_e computed from equation (12.21). A framework for doing this is summarized in Appendix I.

Simulations of diffusion potentials surrounding active multicompartment neurons have so far been limited to small systems, including only a single neuron in the case of full 3D simulations of the extracellular dynamics (Solbrå et al. 2018) or a small population of 10 neurons in simplified 1D simulations where $[k]_e$ and V_e were assumed to vary only in the direction perpendicular to cortical layers (Halnes et al. 2016, 2017). Due to the small population size and other model approximations, these models did not produce realistic LFPs. However, an examination of how much the diffusion potential contributed to the LFP in these models gave conclusions similar to those that came out of the analysis in Section 12.3.3. For frequencies above a few hertz, the LFP was always unaffected by the diffusion potential. For frequencies up to a few hertz, the LFP could be affected by diffusion potentials but only in simulations that produced large extracellular concentration gradients approaching (but not exceeding) pathological values. Hence, both the analysis in Section 12.3.3 and the cited computational studies suggest that, under nonpathological conditions, the LFP frequency spectrum will be unaffected by the (possible presence) of extracellular concentration gradients, except possibly for the lower range of frequency components ($\lesssim 1$ Hz).

Without solving for the concentration dynamics, the insight from equation (12.21) is simply that diffusion gives a "correction" of the potential predicted from standard VC theory. In CSD analysis, where the aim is to infer neural current sources from recorded extracellular potentials, the mathematical framework is typically based on equation (12.23). If, contrary to what the theory assumes, V_e contains contributions from extracellular diffusion, these will be interpreted by the CSD theory as additional (spurious) cellular current sources (Halnes et al. 2017).

13 Final Comments and Outlook

The first recordings of extracellular potentials date back a century or so (Caton 1875, Adrian 1928, Berger 1929). The biophysical foundation for understanding what these signals can tell us about the underlying neural activity was laid some decades later (Hodgkin & Huxley 1952*a*, Pitts 1952, Rall 1959, Rall 1962, Holt & Koch 1999) and has resulted in what we in this book refer to as the MC+VC scheme (Section 2.6.2). This scheme combines multicompartment (MC) models of neurons with volume-conductor theory (VC) to compute extracellular signals, and it is today's standard scheme for the forward modeling of extracellular signals.

Although proper modeling schemes have been in place for quite some time, interpretations of extracellularly recorded signals have often not been based on biophysical modeling, but rather on intuitions or "rules of thumb" that have been spread throughout the neuroscience community. Some commonly encountered rule-of-thumb interpretations can, on the basis of biophysical principles, be dismissed as fallacies. We give some examples of this in Section 13.1.

The MC+VC scheme that we have advocated throughout this book is not perfect, and in Section 13.2, we briefly summarize what we believe are its key limitations. Given the computational demand associated with using alternative, more-advanced schemes (Section 2.6.3), we still expect the MC+VC scheme to keep its position as the standard forward modeling scheme in years to come. We therefore conclude the book by discussing the future outlook of MC+VC-based forward modeling in Section 13.3.

13.1 Common Misconceptions about Extracellular Potentials

Below, we list three important learnings to clear up commonly encountered misconceptions regarding local field potentials (LFPs) and current-source densities (CSDs).

- **An LFP recorded at a particular position does not necessarily stem from neurons at the same position.** The amplitude of an extracellular spike (signature of an action potential) decays sharply with distance from the neuron it originates from (Figure 7.6). A spike recorded in, say, a particular cortical layer is therefore likely to stem from a neuron with its soma placed in the same layer. The same is not true for the LFP, which decays less steeply with distance from the sources. As in the example in Figure 8.18, the LFP recorded in the uppermost cortical layers can stem from neurons with their somas in the deeper cortical layers.

- **A negative current-source density (CSD) does not necessarily imply excitatory synaptic inputs onto neurons.** It follows from current conservation (Section 2.4) that the sum of the membrane currents across a neuron should always be zero. Inward currents (sinks) in one part of a neuron are therefore always accompanied by outward currents (sources) from other parts of the neuron. A negative CSD (Section 4.3) implies a local sink and may arise either due to excitatory synaptic currents nearby the electrode or to inhibitory synaptic currents in the same neuron further away. By the same arguments, a positive CSD (source) does not necessarily imply inhibitory synaptic inputs onto neurons. Example simulations that illustrate this point were shown in Figure 8.2.
- **Recorded LFPs do not necessarily stem from pyramidal cells.** A prerequisite for pyramidal cells to dominate the LFP (as they generally are expected to do when present) is that populations of such cells receive correlated inputs from synapses distributed non-uniformly over their dendrites (Section 8.2.2). With uniform inputs, their contribution to the LFP will be much smaller (Figure 8.5), and it is then unclear which cell type will dominate the LFP. Non-pyramidal cells, such as stellate cells, may also provide sizable LFP contributions provided that these cells receive non-uniformly placed synaptic inputs (Lindén et al. 2010, figure 1).

13.2   Applicability of the MC+VC Scheme

When working with mathematical models in the natural sciences, one typically needs to specify a number of parameters. In some branches of physics, the parameters are universal constants of nature that are known with a high degree of accuracy. The modeling challenge then lies not in determining parameter values, but rather in developing formalisms for solving complex equations that include these parameters. For example, solving the Schrödinger equation for multielectron atoms is a challenging (numerical) undertaking even if the involved parameters (masses and charges of electrons, protons, and neutrons; Planck's constant; and Coulomb's constant) are precisely known.

In biology, the situation is often reversed: the modeling formalism may be well-established while the determination of parameters is the limiting factor. This is arguably the case when it comes to the MC+VC scheme (Section 2.6.2) on which we have based most of the simulations and insights in this book.

Following the work of Hodgkin & Huxley (1952a) and the later work on dendritic modeling by Rall (1959, Segev et al. 1995), MC modeling based on cable theory has developed into the standard framework for mechanistic modeling of neurons. Its ability to predict membrane potentials and membrane currents has been validated against intracellular recordings in numerous experiments. The assumptions involved when doing MC modeling are nicely summarized in Koch (1999, chapter 2), and the framework is generally considered to be on a solid footing.

Likewise, VC theory is a well-established foundation for modeling many bioelectric phenomena (Plonsey & Barr 2007) and is also central in the field of bioimpedance

(Martinsen & Grimnes 2015). The theory is founded on Maxwell's equations, alongside the assumption that volume currents in tissue are ohmic, at least over the frequency range relevant for endogenous signals. The ohmic form $\mathbf{i}_t = -\sigma_t \nabla V_e$ has been (at least partially) validated in numerous experiments for length scales down to 0.1 mm or so (Ulbert et al. 2001, Logothetis et al. 2007, Miceli et al. 2017).

However, even if the MC+VC scheme provides a solid foundation for forward modeling, the parameters that go into MC+VC models are often poorly constrained by data.

13.2.1   Specification of Parameters

MC models require a parameterization of how conductances of synapses and ion channels vary along the neural morphology. Conductance parameters are difficult to determine experimentally with a high spatial resolution, especially along thin dendritic structures. As a further complication, conductance values are not constants of nature but vary, not only between different neuron types but also between individual neurons of the same type and even temporally within a given neuron. Most parameters in MC models are therefore normally not based on direct measurements, but rather adjusted by the modeler until the model reproduces some features of experimental data. The typical kind of data used for model tuning is patch clamp recordings from the soma as well as sometimes additional data sets that are included in order to better constrain the model (Hay et al. 2011, Almog & Korngreen 2014, Van Geit et al. 2016). Normally, the dendritic distributions of conductances are rather poorly constrained by the data sets used in such parameter optimization procedures, but these distributions can in principle always be better constrained by including more data sets in the optimization procedure.

The key parameter in VC models is the tissue conductivity σ_t. This parameter is, however, expected to vary between brains, brain regions, and even brain states. As we saw in Figure 5.3, there is quite some spread in the experimentally measured values of σ_t. Uncertainties regarding how σ_t varies with position can probably be mended through improved measurements. However, the common modeling assumption that brain tissue is a coarse-grained medium with a single-valued σ_t that is constant over the relevant region will probably break down at the micrometer length scale, especially very close to the nanometer-thick Debye layers surrounding cell membranes (Pods 2017). Effects of microscale variations in σ_t may be challenging to probe experimentally, but modeling such effects could be an attractive alternative. With a high-resolution anatomical model of a sufficiently large block of brain tissue – for example, reconstructed from electron-microscopic images – such a modeling study would in principle be straightforward, although computationally demanding. It would require finite element modeling (FEM) analogous to the study of liquid flow in a reconstructed brain-tissue block (Holter et al. 2017).

13.2.2   Experimental Comparison

Despite the challenges involved in accurately determining model parameters, models based on the MC+VC scheme have been shown to perform well when validated against experimental data. For example, the shapes of the computed spikes seen in Chapter 7 are clearly very similar to what is typically observed experimentally – for instance, those seen in Figure 7.1C. Likewise, the unitary LFPs computed in Hagen et al. (2017) agree excellently with what was measured in the sensory cortex in Swadlow et al. (2002) (see also Figure 8.12).

13.2.3   Ephaptic Interactions

Although parameter estimations may be the limiting factor in most MC+VC models, it should be noted that the modeling scheme also has shortcomings of a more principled nature. As we briefly discussed in Section 2.6.2, MC models are normally based on the assumption that intra- and extracellular ion concentrations and the extracellular potential V_e are constant. As such, the models do not account for *ephaptic* effects – that is, indirect non-synaptic effects that neurons can have on each other by causing changes in the extracellular environment. The main justification for neglecting ephaptic effects is that they are believed to be small. Ephaptic concentration effects are small because concentrations are believed to vary little under normal working conditions, as argued in Section 3.4. Ephaptic electric effects are small because V_e tends to be much smaller than the membrane potential V_m, as we have seen in multiple examples throughout this book.

Although electric ephaptic effects may be small, their role in neurodynamics cannot be categorically ruled out (Anastassiou & Koch 2015). Modeling studies addressing the population level have suggested that extracellular fields can synchronize spiking activity (Traub et al. 1985, Fröhlich & McCormick 2010). Modeling studies addressing a smaller spatial scale have shown that ephaptic coupling between axons packed in a fiber bundle can be quite prominent (Clark & Plonsey 1970, Bokil et al. 2001, Tveito et al. 2017, Shifman & Lewis 2019), that ephaptic interactions can affect coincidence detection of synaptic inputs (Goldwyn & Rinzel 2016), and that ephaptic interactions between the soma of one neuron and a nearby axon or dendrite can be quite strong (Holt & Koch 1999).

As stated above, ephaptic effects are not accounted for in the standard MC models. Note, however, that the effect on neural dynamics from an already-known (pre-computed or measured) extracellular potential V_e is in principle straightforward to include in the MC+VC framework, by simply imposing V_e as a boundary condition in the neuron modeling software. As a "hack," one may thus attempt to simulate ephaptic effects of V_e on V_m by using the two-step MC+VC scheme iteratively (Holt 1998, Gold et al. 2006). Gold et al. (2006) used this to explore *self-ephaptic* effects – that is, how a single neuron affects itself by changing its own extracellular environment. In the study, it was concluded that the inclusion of such self-ephaptic effects in a single-neuron model did not significantly affect the numerical results, as we have verified for an example simulation with the Hay model provided as an online resource (Code available via www .cambridge.org/electricbrainsignals).

Although concentration shifts are expectedly quite small during normal working conditions, dramatic changes in extracellular ion concentrations are a hallmark of many pathological conditions, including epilepsy, stroke, and spreading depression (Somjen

2004). Effects of intra- and extracellular ion-concentration changes on neural dynamics have also been examined in many modeling studies. These studies have demonstrated how changes in ionic concentrations lead to changes in ionic reversal potentials as well as how this in turn can have dramatic consequences for the firing properties of neurons (Kager et al. 2000, Ullah et al. 2009, Zandt et al. 2011, Hübel et al. 2014, Wei et al. 2014, Sætra et al. 2020, 2021). As we saw in Chapter 12, concentration changes can also give rise to extracellular diffusion potentials, which in principle could be detected in LFP measurements under some conditions.

In conclusion, the MC+VC scheme seems to be the best choice for modeling extracellular potentials for most conditions, but this scheme does not allow for self-consistent modeling of ephaptic effects or concentration effects. To model conditions where such effects are believed to be important, one should consider using one of the alternative schemes summarized in Section 2.6.3.

13.3   Outlook

We conclude the book by discussing the future outlook and anticipated challenges for the forward modeling of brain signals.

13.3.1   Areas of Application of Forward Modeling

We foresee four main areas of application of forward modeling of brain signals:

1. **Exploration of brain signals stemming from different kinds of neural activity to gain insights about what the signals can and cannot inform researchers about.**

    Over the last 20 years, our research group has spent a lot of time exploring links between neural activity and extracellular potentials. The biophysical connection is in principle straightforward, building on the well-established MC+VC scheme. Our experience has taught us that the detailed form of the brain signal predicted from theory does not always agree with our own prior expectations. We have thus learned that it is often better to do numerical simulations with tools such as LFPy than to trust our intuitions.

    In the present book, we have focused the LFP modeling examples on signals generated by geometrically aligned pyramidal neurons such as those found in the cortex and hippocampus. The cerebellum, with its characteristic geometric arranged Purkinje cells, is another area where sizable LFP signals are seen (Nicholson & Llinas 1971, Pellerin & Lamarre 1997). However, LFPs are also recorded in other areas lacking prominently aligned cell populations.

    In populations of cells with stellate structure, the LFP contributions from different neurons can be expected to largely cancel, unless the synaptic inputs are organized with a particular geometrical arrangement (Lindén et al. 2010). However, the spatial organization of the population of neurons may also affect the resulting population LFP. An example is the population of granule cells in the dentate gyrus, where modeling studies have suggested that the strong curvature of this brain area in rats may contribute to very large LFPs at particular electrode positions (Fernández-Ruiz et al. 2013).

The question of how various factors interact to shape the LFPs in differently structured brain areas is thus ripe for model-based explorations.

2. **Generation of benchmarking data for the development and testing of data analysis methods**.

A host of different methods have been developed for the analysis of electrophysiological data. Ideally, these should be validated against representative test data sets where the "ground truth" is known (Denker et al. 2012). Such benchmarking data is often difficult to obtain experimentally. For example, for the validation of spike-sorting algorithms, one needs to perform combined intracellular and extracellular spike recordings so that the same action potential is measured in both modalities. While such combined experimental data can be obtained (see e.g. Harris et al. (2000) or Yger et al. (2018)), they are, for technical reasons, limited to a small number of neurons recorded in particular situations. In contrast, model-based benchmarking data can be tailored to explore essentially any challenge met by a spike-sorting algorithm (Einevoll et al. 2012). An example of the generation of such benchmarking data was shown in Figure 7.14. For further examples, see Hagen et al. (2015) and Buccino & Einevoll (2021) and references therein.

The same approach can be used to explore methods for the analysis of other brain signals. Benchmarking data from the forward modeling of LFPs from cortical networks have, for example, been used to test methods for current-source density (CSD) analysis (Pettersen et al. 2008, Ness et al. 2015, Hagen et al. 2017), decomposition into independent components (Głąbska et al. 2014), and laminar population contributions (Głąbska et al. 2016) of LFPs.

A point to make here is that such benchmarking data do not necessarily have to be based on biophysical models that correspond in every detail to a particular biological system. A data analysis method claiming to be generally applicable should also work on model-based data even if the data does not perfectly match the experimental data of interest.

3. **Constraining neural network models.**

In physics, the traditional approach for building knowledge has been to suggest candidate models and use experiments to test and improve them. In neuroscience, this approach is already used to constrain single-neuron models based on intracellular recordings (Druckmann et al. 2007, Pozzorini et al. 2015, Teeter et al. 2018) as well as spike recordings (Gold et al. 2007).

Attempts have also been made to use spike recordings to constrain cortical network models, but this is difficult for several reasons. One reason is the relatively low number of spiking neurons ($\lesssim 100$) that one presently can record from simultaneously using a single linear multicontact electrode (Billeh et al. 2020, Siegle et al. 2021). Another reason is the variable nature of spiking from individual neurons, seen for example in the trial-to-trial variability in spike patterns in the sensory cortex under repeated presentations of the same stimulus (Rimehaug et al. 2023).

The LFP recorded in the sensory cortex from the same electrodes exhibits much less inter-trial variability than the spikes (Rimehaug et al. 2023). This reflects that cortical LFPs stem from synaptic inputs onto many cortical neurons due to afferent action potentials from large numbers of presynaptic neurons. While spikes mainly reflect outputs from neurons, the LFP mainly reflects synaptic inputs to neurons. The spike and LFP data therefore contain complementary information – for example, as taken advantage of in the so-called laminar population analysis (LPA) (Einevoll et al. 2007). The simultaneous use of both spike and LFP data in constraining network models might therefore improve model fitting and validation (Einevoll et al. 2019).

There are also other types of measurement that can be used to constrain models (see Figure 1.2). For network models, optical measures such as voltage-sensitive die imaging (VSDI) and two-photon calcium imaging are of particular interest since the former has a high temporal resolution while the latter has a high spatial resolution. Further, unlike for hemodynamics measures such as fMRI, the "measurement physics" – that is, a mathematical biophysics-based link between the activity of neurons and what is measured – is fairly well established for the optical techniques (Brette & Destexhe 2012).

4.   **Electric and magnetic stimulation.**

As briefly outlined in Section 4.6, the biophysical formalism used in the forward modeling of brain signals can, with minor additions, also be used to model the effects of electric and magnetic stimulation on neural dynamics. The modeling of such effects in neural network simulations could have important clinical applications given the growing use of such stimulation techniques to address mental health problems and in developing neural prostheses.

13.3.2   Future of Large-Scale Network Simulations

A major goal for the neuroscience community has been to make neural network simulations of the same size as whole brains in particular animals. A mouse, a favorite test animal, has on the order of 100 million neurons. Arguably, the "holy grail" is a simulation that mimics the neural activity in human brains containing on the order of 100 billion neurons, of which 15–20 billion are found in the cortex (Azevedo et al. 2009, Herculano-Houzel 2009). At the start of the 2020s, it is difficult to comfortably simulate networks with more than, say, 100,000 biophysically detailed multi-compartment (MC) neurons, roughly the number of neurons found for example in $1\,mm^3$ of cortical tissue. So, with this approach, we are still three orders of magnitude short of a mouse brain and six orders of magnitude short of a human brain. While the inclusion of 100,000 neurons representing a $1\,mm^3$ piece of brain tissue may be sufficient to model LFPs (Lindén et al. 2011, Reimann et al. 2013, Hagen et al. 2022, Rimehaug et al. 2023), it is clearly not sufficient to model human EEG signals in the general case.

To allow for studies of larger networks, we are therefore forced to make approximations. One option is to consider networks of point neurons such as integrate-and-fire neurons to model neural dynamics. This can reduce the computational demands by several orders of magnitude compared to what is required for MC neurons (Billeh et al. 2020). However, as outlined in this book, point neurons do not generate extracellular potentials (or extracellular currents), and "tricks" such as the hybrid and kernel methods described in Chapter 6 must be used to predict LFPs, EEG signals, and MEG signals. The applicability of such modeling tricks must be validated on ground-truth data from simulations with MC neurons.

While more research may be needed for the development and validation of suitable approximate methods, the program for developing such methods is straightforward in principle. Larger networks may also be simulated by replacing biophysically detailed MC neurons with reduced few-compartment models (see e.g. Amsalem et al. (2020)), but this does not necessarily preserve the correct spatiotemporal distribution of transmembrane currents in space. Approximate methods for extracellular signals may thus also be required for network models utilizing few-compartment models (Section 9.6.3).

In the foreseeable future, population firing-rate models seem to be the only viable option for whole-brain modeling of human brain dynamics (see e.g. The Virtual Brain (Sanz-Leon et al. 2013)). Approximate methods must also be developed to link the coarse-grained population variables in such models to contributions to LFP, EEG, and MEG signals. Again, this appears to be straightforward in principle as the approximate methods can be developed and validated based on ground-truth data from simulations of populations of MC neurons.

Note, however, that the challenge of computing electric and magnetic brain signals from network models – either based on MC neuron models, point-neuron models, or population firing-rate models – dwarfs compared to the challenge of making biophysics-based network models in the first place (Einevoll et al. 2019). The task of recording and compiling the experimental data needed to construct a "skeleton" network with plausible neuron models and synaptic connections is immense (Markram et al. 2015, Billeh et al. 2020); likewise is the task of recording comprehensive physiological data sets against which model predictions can be tested (Billeh et al. 2020, Siegle et al. 2021). The systematic tuning of the numerous model parameters to make the model predict a variety of physiological data is also very challenging (Billeh et al. 2020, Rimehaug et al. 2023).

While Albert Einstein could construct the theories of relativity more or less by himself, figuring out how the brain works will require a huge collaborative effort. As has been said: "It takes a world to understand the brain ..."

Appendix A
Frequency-Dependent Length Constant

For a DC current injection I_0 into the end of a semi-infinite cable, the steady-state membrane potential at a distance x along the cable will be proportional to

$$V_{\mathrm{m}}(x) \propto I_0 e^{-x/\lambda} , \tag{A.1}$$

where

$$\lambda = \sqrt{\frac{d r_{\mathrm{m}}}{4 r_{\mathrm{a}}}} \tag{A.2}$$

is the DC length constant (equation (3.32)), defined as the distance over which the steady-state potential decays by a factor $1/e$. Here, d, r_{a}, and r_{m} denote the diameter, axial resistivity, and specific membrane resistance of the cable, respectively.

In the case of an AC current injection $I_0 \sin(2\pi f t)$ with frequency f, the potential along the cable will be proportional to

$$V_{\mathrm{m}}(x,t) \propto I_0 \sin(2\pi f t + \theta) e^{-x/\lambda_{\mathrm{AC}}} , \tag{A.3}$$

where θ is a phase shift.

The amplitude decay factor $1/\lambda_{AC}(f)$ will in this case be frequency dependent. An expression for it can be derived by replacing r_{m} and r_{a} in equation (A.2) with their respective (generally complex) impedances, then taking the real part of the final expression (Eisenberg & Johnson 1970, Koch 1999). The frequency dependence of the cable comes from the capacitive membrane properties, which is accounted for in the membrane impedance $\hat{z}_{\mathrm{m}} = r_{\mathrm{m}}/(1 + j2\pi f \tau_{\mathrm{m}})$, where τ_{m} is the membrane time constant. Since the cytosol is purely resistive (with no frequency dependence), the axial impedance is simply r_{a}. With this, the complex decay factor becomes

$$\frac{1}{\lambda_{\mathrm{AC}}} = \mathrm{Re}\left\{ \sqrt{\frac{4 r_{\mathrm{a}}}{d r_{\mathrm{m}}}} \sqrt{1 + j2\pi f \tau_{\mathrm{m}}} \right\} = \mathrm{Re}\left\{ \frac{\sqrt{1 + j2\pi f \tau_{\mathrm{m}}}}{\lambda} \right\} . \tag{A.4}$$

The formula for the square root of a complex number $a + jb$ is

$$\sqrt{a + jb} = \pm \left(\sqrt{\frac{|a + jb| + a}{2}} + j \frac{b}{|b|} \sqrt{\frac{|a + jb| - a}{2}} \right) . \tag{A.5}$$

Using this formula, we take the positive root (we need a positive decay factor) of the complex factor $\sqrt{1 + j2\pi f \tau_m}$ in equation (A.4) and find that

$$\frac{1}{\lambda_{AC}} = \frac{1}{\lambda}\sqrt{\frac{\sqrt{1 + (2\pi f \tau_m)^2} + 1}{2}}.$$

(A.6)

The frequency-dependent length constant can then be defined as

$$\lambda_{AC} = \lambda \sqrt{\frac{2}{1 + \sqrt{1 + (2\pi f \tau_m)^2}}}.$$

(A.7)

An alternative derivation of λ_{AC} can be found in Pettersen & Einevoll (2008, appendix C).

Appendix B
Derivation of the Current-Dipole Approximation

A volume containing a discrete set of point-like current sinks and sources will set up an electric potential given by equation (4.8):

$$V_e(\mathbf{r}) = \frac{1}{4\pi\sigma_t} \sum_{n=1}^{N} \frac{I_n}{\|\mathbf{r} - \mathbf{r}_n\|} . \tag{B.1}$$

Here, I_n is a current source at location $\mathbf{r}_n$, and V_e is the electric potential at measurement point $\mathbf{r}$. In this appendix, we outline the multipole expansion, which is a reformulation of equation (B.1) that is exact as long as the distance from the center position used in the multipole expansion to the measurement point is larger than the distance between the center and the most distant current source (see Figure B.1).

B.1   Multipole Expansion

Analogous to how we can formulate the electric potential from a set of electric charges with the charge-multipole expansion (Reitz et al. 1993), we can derive the current-multipole expansion (in this book, simply referred to as "multipole expansion") for a set of current sources in a conductor. We let a number N current sources I_n be located at locations $\mathbf{r}_n$ in a volume and define a recording position $R = \|\mathbf{R}\| = \|\mathbf{r}_c - \mathbf{r}\|$ outside of the volume. For mathematical convenience, R is defined relatively to the center position,

$$\mathbf{r}_c = \sum_{n=1}^{N} \mathbf{r}_n / N \tag{B.2}$$

of the sources (see Figure B.1).

The first step is to write $1/\|\mathbf{r} - \mathbf{r}_n\| = 1/R_n$ from equation (B.1) as an infinite series, with R_n being the distance between current source I_n at $\mathbf{r}_n$ and the electrode position $\mathbf{r}$. By application of the cosine rule, we find that

$$R_n^2 = R^2 + r_{cn}^2 - 2Rr_{cn} \cos \theta_n , \tag{B.3}$$

where $r_{cn} = \|\mathbf{r}_{cn}\|$ is the distance between the volume midpoint $\mathbf{r}_c$ and current location $\mathbf{r}_n$, and θ_n is the angle between $\mathbf{r}_{cn}$ and $\mathbf{R}$ (Figure B.1). Equation (B.3) can be rewritten to the following expression for R_n:

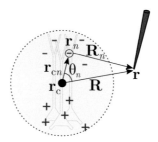

Figure B.1 Illustration of the definitions used for the multipole expansion. An electrode located in $\mathbf{r}$ measures the electric potential generated by a volume of point-current sinks and sources. The center of the volume $\mathbf{r}_c$ is a distance $R = \|\mathbf{R}\|$ away from the electrode. The location of current source n is $\mathbf{r}_n$, with a distance $\mathbf{r}_{cn}$ to the volume center. The distance vector from the current source to the electrode is denoted $\mathbf{R}_n$, and the angle between $\mathbf{R}$ and $\mathbf{r}_{cn}$ is denoted θ_n.

$$R_n^2 = R^2 \left[1 - \frac{r_{cn}}{R} 2 \cos \theta_n + \left(\frac{r_{cn}}{R} \right)^2 \right]$$

$$\implies R_n = R\sqrt{1 - 2h \cos \theta_n + h^2} \,,$$

where $h = \frac{r_{cn}}{R}$. Thus,

$$\frac{1}{\|\mathbf{r} - \mathbf{r}_n\|} = \frac{1}{R\sqrt{1 - 2h \cos \theta_n + h^2}} \,.$$

As $1/\sqrt{1 - 2h \cos \theta_n + h^2}$ is the generating function for the Legendre polynomials (Kreyszig 1993), we find that

$$\frac{1}{\sqrt{1 - 2h \cos \theta_n + h^2}} = \sum_{l=0}^{\infty} h^l P_l(\cos \theta_n), \quad \forall |h| = \left| \frac{r_{cn}}{R} \right| < 1$$

$$\implies \frac{1}{\|\mathbf{r} - \mathbf{r}_n\|} = \frac{1}{R} \sum_{l=0}^{\infty} \left(\frac{r_{cn}}{R} \right)^l P_l(\cos \theta_n), \quad \forall R > r_{cn} \,.$$

The symbol $\forall$ denotes *for all*, and P_l are the Legendre polynomials of degree l given by

$$P_0(\cos \theta_n) = 1$$
$$P_1(\cos \theta_n) = \cos \theta_n$$
$$P_2(\cos \theta_n) = \frac{3}{2} \cos^2 \theta_n - \frac{1}{2}$$
$$P_3(\cos \theta_n) = \frac{5}{2} \cos^3 \theta_n - \frac{3}{2} \cos \theta_n$$

$$\vdots$$

Insertion of the infinite series expression for $\frac{1}{\|\mathbf{r}-\mathbf{r}_n\|}$ into equation (B.1) yields the current-multipole expansion

$$V_e(R) = \frac{1}{4\pi\sigma_t} \frac{1}{R} \sum_{n=1}^{N} I_n \sum_{l=0}^{\infty} \left(\frac{r_{cn}}{R}\right)^l P_l(\cos\theta_n), \quad \forall R > r_{cn}. \tag{B.4}$$

For readability, we can rewrite this as

$$V_e(R) = \frac{1}{4\pi\sigma_t}\left[\frac{C_{\text{monopole}}}{R} + \frac{C_{\text{dipole}}}{R^2} + \frac{C_{\text{quadrupole}}}{R^3} + \frac{C_{\text{octopole}}}{R^4} + \ldots\right], \quad \forall R > r_{cn}, \tag{B.5}$$

where

$$C_{\text{monopole}} = \sum_{n=1}^{N} I_n \tag{B.6}$$

$$C_{\text{dipole}} = \sum_{n=1}^{N} I_n r_{cn} \cos\theta_n \tag{B.7}$$

$$C_{\text{quadrupole}} = \sum_{n=1}^{N} I_n r_{cn}^2 \left(\frac{3}{2}\cos^2\theta_n - \frac{1}{2}\right) \tag{B.8}$$

$$C_{\text{octopole}} = \sum_{n=1}^{N} I_n r_{cn}^3 \left(\frac{5}{2}\cos^3\theta_n - \frac{3}{2}\cos\theta_n\right). \tag{B.9}$$

Note that equation (B.5) is, in principle, exact and not an approximation for all $R > r_{cn}$, given that all terms are included in the calculation.

In the following sections, we examine how the first four terms of the current-multipole expansion contribute to the extracellular potential in neural tissue.

B.2   Monopole Contribution in Brain Tissue

No neuron (as a whole) can be a net current source or sink, meaning the transmembrane current sinks and sources will always sum to zero: $\sum_{n=1}^{N} I_n = 0$. The monopole contribution from neural sources will therefore always be zero:

$$V_e^{\text{monopole}}(R) = \frac{1}{4\pi\sigma_t}\frac{1}{R}\sum_{n=1}^{N} I_n = 0.$$

B.3   Dipole Contribution in Brain Tissue and the Current-Dipole Approximation

The dipole contribution is given by

$$V_e^{\text{dipole}}(R) = \frac{1}{4\pi\sigma_t}\frac{1}{R^2}\sum_{n=1}^{N} I_n r_{cn}\cos\theta_n.$$

Applying the definition of the scalar product $\mathbf{r}_{cn} \cdot \mathbf{R} = r_{cn} R \cos \theta_n$, we can rewrite this as

$$V_e^{\text{dipole}}(R) = \frac{1}{4\pi\sigma_t} \frac{1}{R^2} \sum_{n=1}^{N} I_n \mathbf{r}_{cn} \cdot \hat{\mathbf{R}}$$

$$= \frac{1}{4\pi\sigma_t} \frac{1}{R^2} \sum_{n=1}^{N} I_n (\mathbf{r}_n - \mathbf{r}_c) \cdot \hat{\mathbf{R}}$$

$$= \frac{1}{4\pi\sigma_t} \frac{1}{R^2} \left(\sum_{n=1}^{N} I_n \mathbf{r}_n \cdot \hat{\mathbf{R}} - \sum_{n=1}^{N} I_n \mathbf{r}_c \cdot \hat{\mathbf{R}} \right), \qquad (\text{B.10})$$

where $\hat{\mathbf{R}} = \mathbf{R}/R$. The second term in equation (B.10) vanishes because $\sum_{n=1}^{N} I_n = 0$, and if we recognize the current-dipole moment $\mathbf{P} = \sum_{n=1}^{N} I_n \mathbf{r}_n$ in the first term, we end up with

$$V_e^{\text{dipole}}(R) = \frac{1}{4\pi\sigma_t} \frac{\mathbf{P} \cdot \hat{\mathbf{R}}}{R^2} = \frac{1}{4\pi\sigma_t} \frac{p \cos \theta}{R^2}, \qquad (\text{B.11})$$

where θ is the angle between $\mathbf{P}$ and $\mathbf{R}$. This expression is the standard formulation of the current-dipole approximation.

When making the *current-dipole approximation*, one assumes that the extracellular potential is predicted by equation (B.11) alone. The rationale is that (i) the monopole contributions are zero, as we saw in Section B.2 and (ii) that the higher-order terms in equation (B.5) are much smaller than equation (B.11) for large R. To get an intuition about when the dipole approximation is justified, we will compute the quadrupole and octopole contributions ($V_e^{\text{quadrupole}}$, V_e^{octopole}) from a single current sink and source pair.

B.4   Quadrupole Contribution from a Sink-Source Pair

From equation (B.8), we see that the quadrupole contribution for the special case of a current sink I_1 at r_1 and a current source $I_2 = -I_1$ at r_2 is

$$V_e^{\text{quadrupole}}(R) = \frac{1}{4\pi\sigma_t} \frac{1}{R^3} \left[I_1 r_{c1}^2 \left(\frac{3}{2} \cos^2 \theta_1 - \frac{1}{2} \right) - I_1 r_{c2}^2 \left(\frac{3}{2} \cos^2 \theta_2 - \frac{1}{2} \right) \right].$$
$$(\text{B.12})$$

With the multipole center definition in equation (B.2), the center of the volume is halfway between the sink and the source, so that $r_{c1} = r_{c2}$ and

$$\theta_2 = \pi - \theta_1$$
$$\implies \cos \theta_2 = - \cos \theta_1$$
$$\implies \cos^2 \theta_2 = \cos^2 \theta_1 .$$

Inserting this into equation (B.12), we find that $V_e^{\text{quadrupole}} = 0$, meaning that there is no quadrupole contribution to the extracellular potential from a sink-source pair. Since all terms of Legendre polynomials for $l = 4, 6, 8, \ldots$ contain $\cos \theta_n$ raised to the power of an even number, multipole expansion term numbers 3, 5, 7, 9, ... are all equal to zero for a sink-source pair.

Note that these results for the multipole moments pertain to the situation where the origin is chosen to be at the midpoint between the sink and source in the sink-source pair. With other choices for the origin, the value for the quadrupole and higher-order multipole moments would be different (although the correct resulting potential will always be obtained if enough terms are included in the series).

B.5 Octopole Contribution from a Sink-Source Pair

For a sink-source current pair, I_1 at $\mathbf{r}_1$ and $I_2 = -I_1$ at $\mathbf{r}_2$, we have the following octopole contribution:

$$V_e^{\text{octopole}}(R) = \frac{1}{4\pi\sigma_t}\frac{1}{R^4}\left[I_1 r_{c1}^3\left(\frac{5}{2}\cos^3\theta_1 - \frac{3}{2}\cos\theta_1\right) - I_1 r_{c2}^3\left(\frac{5}{2}\cos^3\theta_2 - \frac{3}{2}\cos\theta_2\right)\right].$$
(B.13)

We again take $\mathbf{r}_c$ to be the midpoint between $\mathbf{r}_1$ and $\mathbf{r}_2$ so that $r_{c2} = r_{c1} = \frac{d}{2}$ and $\cos\theta_2 = -\cos\theta_1$. This implies that

$$V_e^{\text{octopole}}(R) = \frac{1}{4\pi\sigma_t R^4}\left[I_1\frac{d^3}{2^3}\left(\frac{5}{2}\cos^3\theta_1 - \frac{3}{2}\cos\theta_1\right) + I_1\frac{d^3}{2^3}\left(\frac{5}{2}\cos^3\theta_1 - \frac{3}{2}\cos\theta_1\right)\right]$$

$$= \frac{1}{4\pi\sigma_t}\frac{I_1 d^3}{R^4}\frac{5\cos^3\theta_1 - 3\cos\theta_1}{8}.$$
(B.14)

Further, we can rename θ_1 to θ and I_1 to I and then use that $p = Id$ to find the ratio between the octopole and dipole contributions to the extracellular potential from a sink-source pair:

$$\left|\frac{V_e^{\text{octopole}}}{V_e^{\text{dipole}}}\right|_{\text{max}} = \left|\frac{d^2}{R^2}\frac{5\cos^2\theta - 3}{8}\right|_{\text{max}} = \frac{3}{8}\frac{d^2}{R^2}.$$
(B.15)

In Nunez & Srinivasan (2006), it is suggested that the current-dipole approximation is applicable when $R > 3d$ or $R > 4d$. Inserting $R = 3d$ or $R = 4d$ into equation (B.15), we see that the octopole contribution from a sink-source pair is 1/24 or 3/128 of the dipole contribution, respectively.

Appendix C
Electric Stimulation

Electric stimulation of neurons works by perturbing the extracellular potentials V_e around the neurons, affecting the neural membrane potentials $V_m = V_i - V_e$. In general, V_e will respond to both an applied stimulus and to neural dynamics, and to fully account for stimulus effects, a self-consistent numerical scheme describing both the intra- and extracellular dynamics is required (Section 2.6.3). However, some key insights into how neurons are affected by electric stimuli can be obtained by assuming that V_e is determined exclusively by the stimulus. In the following sections, we consider two simple scenarios that allow us to investigate this analytically.

C.1   Electric Stimulation of a Dendritic Stick

The cable equation for a passive dendritic stick in the absence of any current inputs from synapses or electrodes is given by

$$c_m \frac{\partial V_m}{\partial t} = \frac{E_m - V_m}{r_m} + \frac{d}{4r_a} \frac{\partial^2 V_i}{\partial x^2}. \tag{C.1}$$

This is a slight modification of equation (3.29) from Chapter 3 in that the membrane potential V_m has been replaced by the intracellular potential V_i in the second term on the right. This term represents the effects from the axial currents inside the dendrite that depend on V_i, not V_m.[1] Since the membrane potential $V_m = V_i - V_e$, we can rewrite this equation as

$$c_m \frac{\partial V_m}{\partial t} = \frac{E_m - V_m}{r_m} + \frac{d}{4r_a} \frac{\partial^2 V_m}{\partial x^2} + \frac{d}{4r_a} \frac{\partial^2 V_e}{\partial x^2}. \tag{C.2}$$

The effect of the extracellular field set up by the stimulation electrode is represented by the last term on the right.

Since the cable equation is one-dimensional, it is only affected by the x-component of the field. We therefore consider a one-dimensional external field $\mathbf{E}_e = -dV_e/dx\, \mathbf{e_x}$, as can be set up between two parallel plate-like electrodes. A key insight from equation (C.2) is that a constant field $(d^2 V_e/dx^2 = 0)$ would have no effect on an infinitely long stick. Note, however, that real dendrites have finite lengths, and a straight dendritic

---

[1] In equation (3.29), V_e was assumed to be independent of x so that $\partial^2 V_i/\partial x^2 \approx \partial^2 V_m/\partial x^2$.

stick aligned with a constant electric field will still be affected by the field due to endpoint effects.

C.2   Two-Compartment Model

Our next toy example is a passive two-compartment model (Section 3.2.4) stimulated with an extracellular point electrode with distance r_1 from the midpoint of compartment 1 as well as distance r_2 from the midpoint of compartment 2. With the reference electrode placed infinitely far away, the imposed potential by a stimulation current I_{EC} will decay radially with distance r from the electrode (see equation (4.6)) so that the extracellular potential outside compartment $n = 1, 2$ equals

$$V_{en} = \frac{I_{EC}}{4\pi\sigma_t r_n} \, . \tag{C.3}$$

The membrane currents from the neural compartments can be written as (Section 3.1)

$$I_{m,n} = c_m \frac{dV_{mn}}{dt} + \frac{V_{mn} - E_m}{r_m} \, , \tag{C.4}$$

where the first and second terms on the right are the capacitive and leak currents, respectively. The axial current I_a between compartments 1 and 2 is given by Ohm's law

$$I_a = \frac{V_{i1} - V_{i2}}{4r_a L/(\pi d^2)} = \frac{-\Delta V_m - \Delta V_e}{4r_a L/(\pi d^2)} \, , \tag{C.5}$$

defined as positive when going from compartment 1 to compartment 2. The last equality is obtained by inserting $V_m = V_i - V_e$ and defining $\Delta V_m = V_{m2} - V_{m1}$ and $\Delta V_e = V_{e2} - V_{e1}$.

Current conservation requires that

$$I_{m1} = -I_a = \frac{\Delta V_m + \Delta V_e}{4r_a L/(\pi d^2)} \tag{C.6}$$

$$I_{m2} = +I_a = \frac{-\Delta V_m - \Delta V_e}{4r_a L/(\pi d^2)} \, . \tag{C.7}$$

Insertion of equation (C.4) into equations (C.6) and (C.7) gives

$$c_m \frac{dV_{m1}}{dt} + \frac{V_{m1} - E_m}{r_m} = \frac{\Delta V_m + \Delta V_e}{4r_a L/(\pi d^2)} \tag{C.8}$$

$$c_m \frac{dV_{m2}}{dt} + \frac{V_{m2} - E_m}{r_m} = \frac{-\Delta V_m - \Delta V_e}{4r_a L/(\pi d^2)} \, . \tag{C.9}$$

The response to an applied stimulus is obtained by inserting equation (C.3) into equations (C.8) and (C.9). To simplify the expression, we consider only the initial response to the stimulus and assume that the cell is near rest so that $V_{m1} \approx V_{m2} \approx E_m$, resulting in:

$$c_m \frac{dV_{m1}}{dt} = \frac{\Delta V_e}{4r_a L/(\pi d^2)} = \frac{d^2}{16r_a L}\frac{I_{EC}}{\sigma_t}\left(\frac{r_1 - r_2}{r_1 r_2}\right) \tag{C.10}$$

$$c_m \frac{dV_{m2}}{dt} = \frac{-\Delta V_e}{4r_a L/(\pi d^2)} = -\frac{d^2}{16r_a L}\frac{I_{EC}}{\sigma_t}\left(\frac{r_1 - r_2}{r_1 r_2}\right). \tag{C.11}$$

These equations tell us that the effect of the stimulus on the membrane potential in the two compartments will be of equal magnitude but of opposite sign, except in the special case when $r_1 = r_2$, in which case there will be no effect of the extracellular stimulation current. Further, only the difference in external potential along the cell matters, meaning that a uniform extracellular potential will have no effect, regardless of its magnitude.

Assuming that $r_2 > r_1$ in our example, we see that

$$\begin{matrix} r_2 > r_1 \\ I_{EC} > 0 \end{matrix} \quad \Rightarrow \quad \begin{matrix} \dfrac{dV_{m1}}{dt} < 0 \\[2mm] \dfrac{dV_{m2}}{dt} > 0 \end{matrix} \tag{C.12}$$

and

$$\begin{matrix} r_2 > r_1 \\ I_{EC} < 0 \end{matrix} \quad \Rightarrow \quad \begin{matrix} \dfrac{dV_{m1}}{dt} > 0 \\[2mm] \dfrac{dV_{m2}}{dt} < 0. \end{matrix} \tag{C.13}$$

A positive extracellular current ($I_{EC} > 0$) will thus lead to the closest compartment being hyperpolarized ($dV_{m1}/dt < 0$) and the more distant compartment being depolarized ($dV_{m2}/dt > 0$). A negative extracellular current ($I_{EC} < 0$) will have the opposite effect.

Going back to equation (C.10), we see that the change in membrane potential due to the extracellular current source is proportional to the stick cross-sectional area πd^2, implying that neurites of larger diameters are substantially more affected by extracellular current stimulations.

Appendix D

Derivation of the Point-Source Equation for Anisotropic Medium

We here present a derivation of the point-source equation for an anisotropic medium (equation (5.29)). We start with demanding current continuity for a point-current source I in position $\mathbf{r} = 0$:

$$\nabla \cdot \mathbf{i} = I\delta^3(\mathbf{r}) . \tag{D.1}$$

We integrate this over an arbitrary volume containing the source to get

$$\iiint \nabla \cdot \mathbf{i} \, dv = I . \tag{D.2}$$

We assume that the current density in any direction x depends on the field component in the same direction only, but the conductivity may differ between the directions. The current density is then given by

$$\mathbf{i} = -\left(\sigma_x \frac{\partial V_e}{\partial x}\mathbf{e}_x + \sigma_y \frac{\partial V_e}{\partial y}\mathbf{e}_y + \sigma_z \frac{\partial V_e}{\partial z}\mathbf{e}_z\right) \tag{D.3}$$

and its divergence by

$$\nabla \cdot \mathbf{i} = -\left(\sigma_x \frac{\partial^2 V_e}{\partial x^2} + \sigma_y \frac{\partial^2 V_e}{\partial y^2} + \sigma_z \frac{\partial^2 V_e}{\partial z^2}\right) . \tag{D.4}$$

To proceed, we make the coordinate substitutions (Parasnis 1986) $\xi = x/\sqrt{\sigma_x}$, $\eta = y/\sqrt{\sigma_y}$, and $\zeta = z/\sqrt{\sigma_z}$ so that

$$\nabla \cdot \mathbf{i} = -\tilde{\nabla}^2 V_e = -\left(\frac{\partial^2 V_e}{\partial \xi^2} + \frac{\partial^2 V_e}{\partial \eta^2} + \frac{\partial^2 V_e}{\partial \zeta^2}\right) , \tag{D.5}$$

where $\tilde{\nabla}$ is the divergence operator in the new coordinate system and

$$d\tilde{v} = d\xi d\eta d\zeta = \frac{dv}{\sqrt{\sigma_x \sigma_y \sigma_z}} \tag{D.6}$$

is the volume increment in the new coordinate system. With the new variables, equation (D.2) becomes

$$\iiint \tilde{\nabla}^2 V_e \, d\tilde{v} = -\frac{I}{\sqrt{\sigma_x \sigma_y \sigma_z}} . \tag{D.7}$$

We realize that equation (D.7) has the same form as equation (4.15) for an isotropic medium. Hence, in the new coordinate system, the tissue is isotropic so that the field will be spherically symmetric around the source, and we can follow the same recipe for solving it as we did when solving equation (4.15). That is, we use Gauss's theorem to convert the volume integral on the left-hand side to an integral over the surface $\tilde{s}$ enclosing the volume $\tilde{v}$ so that equation (D.7) becomes

$$\oiint \tilde{\nabla} V_e(\tilde{\mathbf{r}}) \cdot d\tilde{\mathbf{s}} = -\frac{I}{\sqrt{\sigma_x \sigma_y \sigma_z}} , \tag{D.8}$$

where

$$\tilde{\mathbf{r}} = \frac{x}{\sqrt{\sigma_x}} \mathbf{e}_\xi + \frac{y}{\sqrt{\sigma_y}} \mathbf{e}_\eta + \frac{z}{\sqrt{\sigma_z}} \mathbf{e}_\zeta \tag{D.9}$$

is the position vector in the $\xi \eta \zeta$-coordinate system. Due to spherical symmetry, this can be rewritten as

$$\oiint \frac{dV_e(\tilde{R})}{d\tilde{R}} d\tilde{\mathbf{s}} = -\frac{I}{\sqrt{\sigma_x \sigma_y \sigma_z}} , \tag{D.10}$$

where

$$\tilde{R} = |\tilde{\mathbf{r}}| = \sqrt{x^2/\sigma_x + y^2/\sigma_y + z^2/\sigma_z} \tag{D.11}$$

is the radius of the sphere (in $\xi \eta \zeta$), and $V_e(\tilde{R})$ is the same for all positions on its surface. Equation (D.10) has the solution

$$4\pi \tilde{R}^2 \frac{dV_e(\tilde{R})}{d\tilde{R}} = -\frac{I}{\sqrt{\sigma_x \sigma_y \sigma_z}} . \tag{D.12}$$

If we integrate this from ∞ to $\tilde{R}$, and we use $V_e(\infty) = 0$, we get

$$V_e(\tilde{R}) = \frac{I}{4\pi \tilde{R} \sqrt{\sigma_x \sigma_y \sigma_z}} . \tag{D.13}$$

Finally, we substitute back for $\tilde{R}$ (equation (D.11)) to obtain the final solution

$$V_e(x, y, z) = \frac{I}{4\pi \sqrt{\sigma_y \sigma_z x^2 + \sigma_x \sigma_z y^2 + \sigma_x \sigma_y z^2}} , \tag{D.14}$$

which is the same as equation (5.29) postulated in Section 5.6.

Appendix E
Statistical Measures

E.1   Mean

The *mean* value $\langle x \rangle_t$ of a time-dependent signal $x(t)$ of duration T is given by

$$\langle x \rangle_t = \frac{1}{T} \int_T x(t') \, dt' . \tag{E.1}$$

The notation $\langle \rangle_t$ denotes averaging over time.

E.2   Standard Deviation and Variance

The *variance* of a time-dependent signal $x(t)$ is given by

$$\mathrm{Var}(x) = \langle (x - \langle x \rangle_t)^2 \rangle_t = \frac{1}{T} \int_T (x(t') - \langle x \rangle_t)^2 \, dt' . \tag{E.2}$$

In the special case where the mean $\langle x \rangle_t = 0$, the variance reduces to

$$\langle x^2 \rangle_t = \frac{1}{T} \int_T x(t')^2 \, dt' . \tag{E.3}$$

The *standard deviation* (often denoted σ) is the square root of the variance (often denoted σ^2).

E.3   Covariance

The *covariance* of two time-dependent signals $x(t)$ and $y(t)$ is given by

$$\mathrm{Cov}(x, y) = \langle (x - \langle x \rangle_t)(y - \langle y \rangle_t) \rangle_t = \frac{1}{T} \int_T (x(t') - \langle x \rangle_t)(y(t') - \langle y \rangle_t) \, dt' . \tag{E.4}$$

In the special case where the mean values $\langle x \rangle_t$ and $\langle y \rangle_t$ are zero, the covariance reduces to

$$\langle xy \rangle_t = \frac{1}{T} \int_T x(t')y(t') \, dt' . \tag{E.5}$$

E.4   Correlation

Given two continuous signals $x(t)$ and $y(t)$, their *cross-correlation function* (CCF) can be defined as a function of time lag or displacement τ between the two signals as

$$R_{xy}(\tau) = \int_{-\infty}^{\infty} x(t)y(t + \tau)\, dt \ . \tag{E.6}$$

The CCF of a signal $x(t)$ by itself $R_{xx}(\tau)$ is often referred to as the *autocorrelation function*.

Sometimes, the cross-correlation function normalized by the standard deviations of the two signals will be encountered, defined as

$$\rho_{xy}(\tau) = \frac{R_{xy}(\tau)}{\sigma_x \sigma_y} \ , \tag{E.7}$$

where σ_x and σ_y denote the standard deviation of the two signals.

The normalized cross-correlation function is useful as it corresponds to a time-dependent, scale-free correlation (statistical dependence) measure that, for zero time lag ($\tau = 0$), corresponds to the scalar *Pearson correlation coefficient* (PCC), defined as

$$c_{xy} = \frac{\text{Cov}(x, y)}{\sigma_x \sigma_y} \ . \tag{E.8}$$

As such, we may treat $c_{xy}(\tau)$ as a time-dependent PCC, commonly used to identify which lag τ between signals maximizes their statistical dependence.

Appendix F
Fourier-Based Analyses

In this book, several relations and insights are obtained or analyzed in the frequency domain. In this appendix, we define some concepts, transformations, and relations used in this context. This appendix is not meant to provide a comprehensive review but only summarizes the most-used terms and relations encountered throughout the book.

F.1 Soft Introduction to Frequency Analysis

The extracellular potential $V_e(t)$ recorded over time is typically a wiggly and seemingly noisy curve. It can be difficult to interpret what this raw data tells us or to compare two different recordings of $V_e(t)$ and say something useful about how they differ from one another. A common way to analyze $V_e(t)$ (and time-dependent signals in general) is therefore to look at its frequency content, represented through a so-called *amplitude spectrum* or, alternatively, a *power spectrum*.

The basis for such an analysis is that any time-dependent signal can be expressed mathematically as a combination of independent and periodically oscillating functions, each with an individual frequency, amplitude, and phase.

To give an illustrative example, consider the signal in Figure F.1A. Although it might be hard to tell from looking at the graph alone, this signal is a sum of three sine functions:

$$x(t) = a_1 \sin(2\pi f_1 t) + a_2 \sin(2\pi f_2 t) + a_3 \sin(2\pi f_3 t), \qquad \text{(F.1)}$$

with amplitudes a_1, a_2, and a_3 of 8, 6, and 7 μV as well as frequencies f_1, f_2, and f_3 of 17, 149, and 500 Hz, respectively.

This information is easier to visualize via the signal's amplitude spectrum, plotted in Figure F.1B, which has three peaks that indicate the frequencies represented in the signal (horizontal axis) and their amplitudes (vertical axis).

The *power spectral density* (PSD), encountered many times in this book, contains the same information about the signal as the amplitude spectrum, only it is based on computing the squares of the amplitudes (as defined later).

The Fourier theory presented in Section F.2 describes how the frequency content $\hat{x}(f)$ can be computed from a signal $x(t)$ and also how to go in the other direction – that is, how one can compute the time-dependent signal $x(t)$ from its frequency content $\hat{x}(f)$.

Fourier theory can also be used in frequency filtering of signals, although we should note that other filtering methods exist. One can, for example, construct low-pass filtered

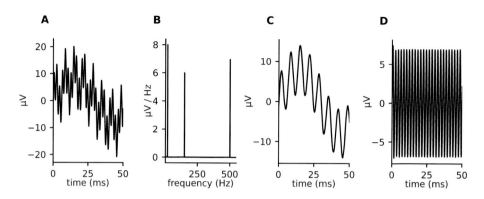

Figure F.1 Simple illustration of frequency analysis. A: Time-dependent signal defined in equation (F.1). **B**: Amplitude spectrum of signal. **C**: Low-pass filtered version ($f < 300\,\text{Hz}$) of signal. **D**: High-pass filtered version ($f > 300\,\text{Hz}$) of signal. Code available via www.cambridge .org/electricbrainsignals.

versions of a signal, containing mainly the frequency components below a certain cut-off frequency, or construct high-pass filtered versions, containing mainly the frequency components above a certain cut-off frequency. Low- and high-pass filtered versions of the example signal are shown in panels C and D, where in this case we have chosen a cut-off frequency of 300 Hz.

The low-pass filtered signal in this case effectively reduces to the sum of the two first sine functions (since f_1 and f_2 are below 300 Hz), while the high-pass filtered signal effectively reduces to only the last sine function (since f_3 is greater than 300 Hz). We chose 300 Hz in this example since this is an often-used cut-off frequency when analyzing extracellular signals recorded inside the brain. The low-frequency part is then what we call the local field potential (LFP). The high-frequency part is often called the multi-unit activity (MUA) and is typically used in the detection of spikes.

F.2   The Fourier Transform (FT)

A continuous function $x(t)$ may be represented in the frequency domain by the *Fourier transform* (FT) defined as

$$\hat{x}(f) = \int_{-\infty}^{\infty} x(t)e^{-2\pi jft}\, dt\,, \quad \forall f \in \mathbb{R}\,, \tag{F.2}$$

where $\hat{x}(f)$ is a complex function of the frequency f. The symbol $\forall$ denotes *for all*. The Fourier transforms of some commonly encountered functions are provided in Table F.1.

Conversely, we may transform a signal $\hat{x}(f)$ represented in the frequency domain back to the time domain using the *inverse Fourier transform* (IFT)

$$x(t) = \int_{-\infty}^{\infty} \hat{x}(f)e^{2\pi jft}\, df\,, \quad \forall t \in \mathbb{R}\,. \tag{F.3}$$

$x(t)$	$\hat{x}(f)$	$\lvert\hat{x}(f)\rvert^2$	Description
$\delta(t)$	1	1	delta function
1	$\delta(f)$	$\delta(f)^2$	constant
$e^{-t^2/2\sigma^2}$	$\sqrt{2\pi}\sigma e^{-2\pi^2\sigma^2 f^2}$	$2\pi\sigma^2 e^{-4\pi^2\sigma^2 f^2}$	Gaussian function
$e^{-t/\tau}\Theta(t)$	$\dfrac{j\tau}{j+2\pi f\tau}$	$\dfrac{\tau^2}{1+4\pi^2 f^2\tau^2}$	exponential function
$\dfrac{t}{\tau}e^{(1-t/\tau)}\Theta(t)$	$-\dfrac{e\tau}{(j+2\pi f\tau)^2}$	$\dfrac{e^2\tau^2}{(1+4\pi^2 f^2\tau^2)^2}$	α-function
$\mathrm{rect}(at)$	$\dfrac{\sin(\pi f/a)}{\pi f}$	$\dfrac{\sin(\pi f/a)^2}{\pi^2 f^2}$	rectangular function

Table F.1. Commonly encountered functions ($x(t)$), Fourier transforms ($\hat{x}(f)$), and their power spectral densities ($\lvert\hat{x}(f)\rvert^2$). $\Theta(t)$ is the Heaviside step function.

F.3   Power Spectral Density and Parseval's Theorem

The *power spectral density* (PSD) of a real function $x(t)$ is given by

$$\mathrm{PSD}(f) = \lvert\hat{x}(f)\rvert^2 . \tag{F.4}$$

Parseval's theorem,

$$\int_{-\infty}^{\infty} \lvert\hat{x}(f)\rvert^2\, df = \int_{-\infty}^{\infty} x(t)^2\, dt , \tag{F.5}$$

relates the integral of the PSD across all frequencies to the time integral of $x(t)^2$.

F.4   Convolution Theorem

Given two continuous signals $x(t)$ and $y(t)$, their convolution is defined as

$$r(t) = (x * y)(t) = \int_{-\infty}^{\infty} x(\tau)y(t-\tau)\, d\tau . \tag{F.6}$$

Here, the asterisk ($*$) denotes convolution. The symbol τ denotes a time shift (sometimes referred to as a time lag or displacement).

The convolution theorem states that in the frequency (Fourier) domain, the convolution of the signals is given as the product of their Fourier transforms:

$$\hat{r}(f) = \hat{x}(f)\hat{y}(f) . \tag{F.7}$$

F.5   Cross Power Spectral Density

Given two continuous signals $x(t)$ and $y(t)$ with cross-correlation $R_{xy}(\tau)$ (equation (E.6)), their *cross power spectral density* (CPSD) may be given as

$$S_{xy}(f) = \int_{-\infty}^{\infty} R_{xy}(\tau)e^{-2\pi jf\tau}\,d\tau\,, \qquad\qquad (F.8)$$

where its complex conjugate yields $S_{yx}(f) = S_{xy}^{*}(f)$.

The CPSD for a signal $x(t)$ by itself with autocorrelation $R_{xx}(\tau)$ gives the *autospectrum function*:

$$S_{xx}(f) = \int_{-\infty}^{\infty} R_{xx}(\tau)e^{-2\pi jf\tau}\,d\tau = \hat{R}_{xx}(f)\,. \qquad\qquad (F.9)$$

F.6   Coherence

The real-valued *coherence* between two signals $x(t)$ and $y(t)$ may be defined via their CPSD and autospectrum functions as

$$C_{xy}(f) = \frac{\left|S_{xy}(f)\right|^{2}}{S_{xx}(f)S_{yy}(f)}\,. \qquad\qquad (F.10)$$

Appendix G
Derivation of Formulas for Population Signals

G.1   General Expression for Population LFP

Following Lindén et al. (2011), we assume the single-neuron LFP contribution $V_{en}(t)$ from a neuron n to be a product of a temporal part $\xi_n(t)$ and a spatial part $F(r_n)$, where r_n is the distance between the neuron and the population center:

$$V_{en}(t) = \xi_n(t)F(r_n) . \tag{G.1}$$

$\xi_n(t)$ is a time-dependent function with a mean value of zero ($\langle\xi_n(t)^2\rangle_t{=}0$) and variance of unity ($\langle\xi_n(t)^2\rangle_t{=}1$). A mean value of zero for $\xi_n(t)$ implies a mean value of zero also for $V_{en}(t)$. A shape function $F(r_n)$ for computing LFPs at the depth of the neuron somas was given in equation (8.9), but other shape functions (tailored to the problem at hand) can also be used.

To study population signals, we introduce the (Pearson) correlation coefficient (equation (E.8)), quantifying the pair-wise temporal correlation

$$c_{Ve} = \frac{\langle\xi_n(t)\xi_m(t)\rangle_t}{\sqrt{\langle\xi_n(t)^2\rangle_t\langle\xi_m(t)^2\rangle_t}} = \langle\xi_n(t)\xi_m(t)\rangle_t , \quad n \neq m, \tag{G.2}$$

and assume that it has the same value for all pairs $n \neq m$. With this notation, *uncorrelated* single-neuron LFP sources correspond to $c_{Ve} = 0$, and completely *correlated* single-neuron LFP sources correspond to $c_{Ve} = 1$.

We now consider the compound LFP from all neurons positioned within a radius R from the population center:

$$V_e(t) = \sum_{r_n<R} V_{en}(t) = \sum_{r_n<R} \xi_n(t)F(r_n) . \tag{G.3}$$

The variance of the compound LFP can be found as

$$\begin{aligned}
V_{e\sigma}^2 &= \langle V_e(t)^2\rangle_t \\
&= \left\langle \sum_{r_n<R} V_{en}(t) \sum_{r_m<R} V_{em}(t)\right\rangle_t \\
&= \sum_{r_n<R}\sum_{r_m<R} \langle V_{en}(t)V_{em}(t)\rangle_t
\end{aligned}$$

$$= \sum_{r_n < R} \langle V_{en}(t)^2 \rangle_t + \sum_{r_n < R} \sum_{r_m \neq n < R} \langle V_{en}(t) V_{em}(t) \rangle_t$$

$$= \sum_{r_n < R} F(r_n)^2 \langle \xi_n(t)^2 \rangle_t + \sum_{r_n < R} \sum_{r_m \neq n < R} F(r_n) F(r_m) \langle \xi_n(t) \xi_m(t) \rangle_t$$

$$= \sum_{r_n < R} F(r_n)^2 + c_{Ve} \sum_{r_n < R} \sum_{r_m \neq n < R} F(r_n) F(r_m)$$

$$= (1 - c_{Ve}) \sum_{r_n < R} F(r_n)^2 + c_{Ve} \left(\sum_{r_n < R} F(r_n) \right)^2$$

$$= (1 - c_{Ve}) V_{eu\sigma}(R)^2 + c_{Ve} V_{ec\sigma}(R)^2 , \tag{G.4}$$

where we have introduced

$$V_{eu\sigma}(R) = \left(\sum_{r_n < R} F(r_n)^2 \right)^{1/2} \tag{G.5}$$

and

$$V_{ec\sigma}(R) = \sum_{r_n < R} F(r_n) . \tag{G.6}$$

We thus have

$$V_{e\sigma}(R, c_{Ve}) = \sqrt{(1 - c_{Ve}) V_{eu\sigma}(R)^2 + c_{Ve} V_{ec\sigma}(R)} , \tag{G.7}$$

in accordance with equation (8.24).

The sums in equations G.5 and G.6 can be approximated as integrals as follows:

$$V_{eu\sigma}(R) \approx \left(\int_0^R N(r) F(r)^2 \, dr \right)^{1/2} \tag{G.8}$$

(as in equation (8.19)) and

$$V_{ec\sigma}(R) \approx \int_0^R N(r) F(r) \, dr \tag{G.9}$$

(as in equation (8.25)).

G.2   General Expression for Population EEG

In analogy to equation (G.1) for LFPs, we assume the single-neuron EEG contribution $V_{en}(t)$ from a neuron n to be a product of a temporal part $\xi_n(t)$ and a spatial part F_n:

$$V_{en}(t) = \xi_n(t) F_n. \tag{G.10}$$

F_n, which is a shorthand notation for $F_n(\mathbf{r}, \mathbf{r}_n, \mathbf{P}_n)$ (equation (9.6)), is a function of both the position vector $\mathbf{r}_n$ and the current dipole moment $\mathbf{P}_n$ and can, for example, be given by equation (9.15), derived using the four-sphere head model. The compound EEG is thus given by

$$V_e(t) = \sum_n V_{en}(t) = \sum_n \xi_n(t) F_n. \tag{G.11}$$

In analogy with the derivation in Section G.1, the variance of the compound EEG is found to be

$$
\begin{aligned}
V_{e\sigma}^2 &= \langle (V_e(t)^2) \rangle_t \\
&= \sum_n \langle (V_{en}(t)^2) \rangle_t + \sum_n \sum_{m \neq n} \langle (V_{en}(t) V_{em}(t)) \rangle_t \\
&= \sum_n F_n^2 \langle (\xi_n(t)^2) \rangle_t + \sum_n \sum_{m \neq n} F_n F_{Pm} \langle (\xi_n(t) \xi_m(t)) \rangle_t \\
&= \sum_n F_n^2 + c_P \sum_n \sum_{m \neq n} F_n F_m \\
&= (1 - c_P) \sum_n F_n^2 + c_P \left(\sum_n F_n \right)^2 .
\end{aligned}
\tag{G.12}
$$

Here, c_P is the (Pearson) correlation coefficient, which is assumed to be the same for all pairs $n \neq m$:

$$c_P = \langle \xi_n(t) \xi_m(t) \rangle_t. \tag{G.13}$$

Appendix H
Equations for Computing Magnetic Fields

This appendix contains the derivation of two equations used to compute magnetic field contributions from tissue volume currents.

H.1

Derivation of Equation (11.5) for B_t

The contribution from the tissue-current density $\mathbf{i}_t(\mathbf{r})$ to the magnetic field is found by inserting $\mathbf{i}_t(\mathbf{r})$ into the Ampère-Laplace law (equation (11.1)):

$$\mathbf{B}_t(\mathbf{r}) = \frac{\mu_0}{4\pi} \iiint \frac{\mathbf{i}_t(\mathbf{r}') \times (\mathbf{r} - \mathbf{r}')}{\|\mathbf{r} - \mathbf{r}'\|^3} \, dv' \, . \tag{H.1}$$

The integral should be taken over the entire region of space where $\mathbf{i}_t(\mathbf{r})$ is nonzero.

We can write this equation in a different form by applying four calculus tricks (Hämäläinen et al. 1993, Ilmoniemi & Sarvas 2019). First, we make use of the identity

$$\nabla' \frac{1}{\|\mathbf{r} - \mathbf{r}'\|} = \frac{\mathbf{r} - \mathbf{r}'}{\|\mathbf{r} - \mathbf{r}'\|^3} \tag{H.2}$$

to write equation (H.1) as

$$\mathbf{B}_t(\mathbf{r}) = \frac{\mu_0}{4\pi} \iiint \mathbf{i}_t(\mathbf{r}') \times \nabla' \frac{1}{\|\mathbf{r} - \mathbf{r}'\|} \, dv' \, . \tag{H.3}$$

Next, we apply a rearranged product rule

$$\mathbf{i}_t(\mathbf{r}') \times \nabla' \frac{1}{\|\mathbf{r} - \mathbf{r}'\|} = \frac{\nabla' \times \mathbf{i}_t(\mathbf{r}')}{\|\mathbf{r} - \mathbf{r}'\|} - \nabla' \times \frac{\mathbf{i}_t(\mathbf{r}')}{\|\mathbf{r} - \mathbf{r}'\|} \, , \tag{H.4}$$

which allows us to write equation (H.3) in the form

$$\mathbf{B}_t(\mathbf{r}) = \frac{\mu_0}{4\pi} \iiint \frac{\nabla' \times \mathbf{i}_t(\mathbf{r}')}{\|\mathbf{r} - \mathbf{r}'\|} \, dv' - \frac{\mu_0}{4\pi} \iiint \nabla' \times \frac{\mathbf{i}_t(\mathbf{r}')}{\|\mathbf{r} - \mathbf{r}'\|} \, dv' \, . \tag{H.5}$$

The third calculus trick is to use the vector identity

$$\iiint \nabla' \times \mathbf{i}(\mathbf{r}') \, dv' = \oiint \mathbf{n} \times \mathbf{i}(\mathbf{r}') \, ds' = - \oiint \mathbf{i}(\mathbf{r}') \times d\mathbf{s}' \tag{H.6}$$

on the vector $\mathbf{i} = \mathbf{i}_t/\|\mathbf{r} - \mathbf{r}'\|$ to convert the second term in equation (H.5) to a surface integral:

$$\frac{\mu_0}{4\pi} \iiint \nabla' \times \frac{\mathbf{i}_t(\mathbf{r}')}{\|\mathbf{r} - \mathbf{r}'\|} \, dv' = -\frac{\mu_0}{4\pi} \oiint \frac{\mathbf{i}_t(\mathbf{r}')}{\|\mathbf{r} - \mathbf{r}'\|} \times d\mathbf{s}' . \tag{H.7}$$

Since the volume integral should be taken over the entire region where the $\mathbf{i}_t$ is nonzero, $\mathbf{i}_t$ must be zero on the surface enclosing the volume. Hence, the surface integral must be zero, and equation (H.5) reduces to

$$B_t(\mathbf{r}) = \frac{\mu_0}{4\pi} \iiint \frac{\nabla' \times \mathbf{i}_t(\mathbf{r}')}{\|\mathbf{r} - \mathbf{r}'\|} \, dv' . \tag{H.8}$$

We now insert the ohmic form $\mathbf{i}_t = -\sigma_t \nabla V_e$ (Chapter 4) for the tissue-current density. We then get

$$B_t(\mathbf{r}) = -\frac{\mu_0}{4\pi} \iiint \frac{\nabla' \times \sigma_t(\mathbf{r}')\nabla' V_e(\mathbf{r}')}{\|\mathbf{r} - \mathbf{r}'\|} \, dv' . \tag{H.9}$$

As a final calculus trick, we use the product rule

$$\nabla \times (f\mathbf{u}) = f\nabla \times \mathbf{u} + (\nabla f) \times \mathbf{u} \tag{H.10}$$

on the numerator in equation (H.9). We then get that

$$\nabla' \times \sigma_t(\mathbf{r}')\nabla' V_e(\mathbf{r}') = \nabla'\sigma_t(\mathbf{r}') \times \nabla' V_e(\mathbf{r}') , \tag{H.11}$$

since the second term $(\nabla' \times \nabla' V_e(\mathbf{r}'))$ is zero. With this, equation (H.9) takes the desired form

$$B_t(\mathbf{r}) = -\frac{\mu_0}{4\pi} \iiint \frac{\nabla'\sigma_t(\mathbf{r}') \times \nabla' V_e(\mathbf{r}')}{\|\mathbf{r} - \mathbf{r}'\|} \, dv' , \tag{H.12}$$

which is the same as equation (11.5).

H.2 Derivation of Equation (11.6) for B_t

A good starting point for deriving equation (11.6) is equation (H.9). By use of the vector identities

$$\nabla \times (\sigma_t \nabla V_e) = \nabla \sigma_t \times \nabla V_e = -\nabla \times (V_e \nabla \sigma_t) , \tag{H.13}$$

we can write this equation as

$$B_t(\mathbf{r}) = \frac{\mu_0}{4\pi} \iiint \frac{\nabla' \times \left(V_e(\mathbf{r}')\nabla'\sigma_t(\mathbf{r}')\right)}{\|\mathbf{r} - \mathbf{r}'\|} \, dv' . \tag{H.14}$$

Next, we use the vector identity

$$\frac{\nabla' \times \mathbf{i}}{\|\mathbf{r} - \mathbf{r}'\|} = \mathbf{i} \times \nabla' \frac{1}{\|\mathbf{r} - \mathbf{r}'\|} + \nabla' \times \frac{\mathbf{i}}{\|\mathbf{r} - \mathbf{r}'\|} \tag{H.15}$$

on the vector $V_e \nabla' \sigma_t$. The integral then splits into two parts:

$$\mathbf{B}_t(\mathbf{r}) = \frac{\mu_0}{4\pi} \iiint \left[V_e(\mathbf{r}') \nabla' \sigma_t(\mathbf{r}') \times \nabla' \frac{1}{\|\mathbf{r} - \mathbf{r}'\|} + \nabla' \times \left(\frac{V_e(\mathbf{r}') \nabla' \sigma_t(\mathbf{r}')}{\|\mathbf{r} - \mathbf{r}'\|} \right) \right] dv'.$$

$$(\text{H.16})$$

Like we did in equation (H.7), we can convert the last term to a surface integral. The surface can be placed at an arbitrary location provided that the surface encloses all the space where the currents are nonzero. Far away (outside the head or in an infinite homogeneous conductor), $\nabla' \sigma_t = 0$, and the surface-integral term vanishes. We are thus left with

$$\mathbf{B}_t(\mathbf{r}) = \frac{\mu_0}{4\pi} \iiint V_e(\mathbf{r}') \nabla' \sigma_t(\mathbf{r}') \times \nabla' \frac{1}{\|\mathbf{r} - \mathbf{r}'\|} dv', \qquad (\text{H.17})$$

which due to the identity in equation (H.2) can be written

$$\mathbf{B}_t(\mathbf{r}) = \frac{\mu_0}{4\pi} \iiint V_e(\mathbf{r}') \nabla' \sigma_t(\mathbf{r}') \times \frac{\mathbf{r} - \mathbf{r}'}{\|\mathbf{r} - \mathbf{r}'\|^3} dv', \qquad (\text{H.18})$$

which is the same as equation (11.6).

Appendix I
Derivation of the MC+ED Scheme

The MC+ED scheme (Figure 2.7B) adopts the two-step approach of the MC+VC scheme. The computation of neural current sources (step 1) using MC models is essentially the same in both schemes. The main difference is that the MC+VC allows all membrane currents in a compartment (leakage, capacitive, active ion specific) to be grouped into a total current source C_{all}, while the explicit modeling of extracellular ion concentrations (step 2) in the MC+ED scheme requires that all neural sources are made ion-specific. Hence, the MC+ED scheme requires that C_{all} is decomposed into one source f_k (units mol/m^3) per ion species k and an additional capacitive current-source density, C_{cap} (unit A/m^3), the only source term not accounted for in the set f_k.

The relationship between C_{all} and the sources required for MC+ED is

$$\frac{C_{all}}{\alpha} = C_{ion} + C_{cap}, \tag{I.1}$$

where

$$C_{ion} = F \sum_k z_k f_k \tag{I.2}$$

is the source term due to ionic membrane currents, and

$$C_{cap} = -\frac{\partial \rho_{em}}{\partial t} \tag{I.3}$$

is the source term due to capacitive membrane currents. The capacitive source term is proportional to the change in membrane charge per time unit, expressed in terms of a volume charge density ρ_{em} per extracellular tissue volume (exterior side of the membrane). The negative sign in equation (I.3) follows from the convention that an outwards capacitive current (i.e. a source) implies an accumulation of negative charge on the exterior side of the membrane. The reason for defining the capacitive source in this manner will become clear later. The introduction of the extracellular volume fraction (α) in equation (I.1) is merely a convention (see Appendix 12.4).

In step 2 of the MC+ED scheme, the dynamics of extracellular ion concentrations $[k]_e$ are computed from the Nernst-Planck continuity equation:

$$\frac{\partial [k]_e}{\partial t} = \nabla \cdot \left[D_k^* \nabla [k]_e + \frac{D_k^* z_k [k]_e}{\psi} \nabla V_e \right] + f_k, \tag{I.4}$$

with the ionic source term added on the right-hand side and where D_k^* is the effective diffusion constant in the extracellular medium (see Appendix 12.4).

Equation (I.4) gives us one equation for each ion species k and, to solve it for all ion species, we need an additional relation for the variable V_e. In the electroneutral electrodiffusive framework (Section 12.2.1.2), this constraint is based on the principle of bulk electroneutrality. To derive it, we first multiply both sides of equation (I.4) with $z_k F$ and take the sum over all ion species k. This gives us an equation for the charge density dynamics,

$$\frac{\partial \rho_e}{\partial t} = \nabla \cdot \left(F \sum_k z_k D_k^* \nabla [k]_e + \frac{\sigma_e}{\alpha} \nabla V_e \right) + F \sum_k z_k f_k , \tag{I.5}$$

where

$$\rho_e = F \sum_k z_k [k]_e + \rho_{e0} \tag{I.6}$$

is the extracellular charge density (per tissue unit volume), and

$$\sigma_e = \frac{\alpha F}{\psi} \sum_k D_k^* z_k^2 [k]_e \tag{I.7}$$

is the conductivity of the extracellular medium (as in equation (12.19)). In equation (I.6), ρ_{e0} represents a residual extracellular charge density (assumed to be constant in time and space), representing all ionic species that are not included in the set $[k]_e$.

Due to the assumption of bulk electroneutrality, there is a nonzero charge density only on the membrane, and we have $\rho_e = \rho_{em}$. With this, we can identify the left-hand side and the last term on the right-hand side of equation (I.5) as the capacitive (equation (I.3)) and ionic (equation (I.2)) current-source densities, respectively. Together they make up the total current-source density (equation (I.1)), which we can insert into equation (I.5) to obtain

$$\nabla (\sigma_e \nabla V_e) = -C_{all} - F \alpha \nabla \cdot \left(\sum_k z_k D_k^* \nabla [k]_e \right) . \tag{I.8}$$

This way of introducing the membrane charge into an otherwise electroneutral electrodiffusive framework has been previously referred to as the Kirchhoff-Nernst-Planck (KNP) framework (Solbrå et al. 2018). There exist other (more spatially resolved) methods for dealing with the membrane charge (see e.g. Mori (2006) or Mori & Peskin (2009)).

In summary, the MC+ED scheme (KNP version) involves the following two steps: (1) Use MC models to compute the ionic (f_k) and capacitive (C_{cap}) membrane sources. (2) Solve equation (I.4) for all extracellular ion concentrations, with equation (I.8) as a constraint determining the extracellular potential. Unlike the (standard) MC+VC scheme, the MC+ED scheme offers no analytical solutions to step 2, which must be solved numerically – for example, with finite element methods (Solbrå et al. 2018).

We note that equation (I.8) was derived in Appendix 12.4 following an easier path. The value of the more convoluted derivation given here is that it shows how the ionic concentrations always sum up to a net charge density that varies exclusively due to, and consistently with, the capacitive membrane current-source density (C_{cap}).

References

Aberra, A. S., Wang, B., Grill, W. M. & Peterchev, A. V. (2020), 'Simulation of transcranial magnetic stimulation in head model with morphologically-realistic cortical neurons', *Brain Stimulation* **13**(1), 175–189.

Adrian, E. (1928), *The Basis of Sensation*, W. W. Norton & Co, New York.

Adrian, E. D. & Moruzzi, G. (1939), 'Impulses in the pyramidal tract', *The Journal of Physiology* **97**(2), 153–199.

Agudelo-Toro, A. & Neef, A. (2013), 'Computationally efficient simulation of electrical activity at cell membranes interacting with self-generated and externally imposed electric fields', *Journal of Neural Engineering* **10**(2), 026019.

Aguilella, V., Mafé, S. & Pellicer, J. (1987), 'On the nature of the diffusion potential derived from Nernst-Planck flux equations by using the electroneutrality assumption', *Electrochimica Acta* **32**(3), 483–488.

Ahlfors, S. P. & Wreh II, C. (2015), 'Modeling the effect of dendritic input location on MEG and EEG source dipoles', *Medical & Biological Engineering & Computing* **53**(9), 879–887.

Akalin Acar, Z. & Makeig, S. (2013), 'Effects of forward model errors on EEG source localization', *Brain Topography* **26**(3), 378–396.

Akar, N. A., Cumming, B., Karakasis, V., Kusters, A., Klijn, W., Peyser, A. & Yates, S. (2019), 'Arbor – a morphologically-detailed neural network simulation library for contemporary high-performance computing architectures', in *2019 27th Euromicro International Conference on Parallel, Distributed and Network-Based Processing (PDP)*, IEEE.

Alcaraz, A., Nestorovich, E. M., López, M. L., García-Giménez, E., Bezrukov, S. M. & Aguilella, V. M. (2009), 'Diffusion, exclusion, and specific binding in a large channel: a study of ompf selectivity inversion', *Biophysical Journal* **96**(1), 56–66.

Allken, V., Chepkoech, J.-L., Einevoll, G. T. & Halnes, G. (2014), 'The subcellular distribution of T-type Ca2+ channels in interneurons of the lateral geniculate nucleus', *PLoS ONE* **9**(9), e107780.

Almog, M. & Korngreen, A. (2014), 'A quantitative description of dendritic conductances and its application to dendritic excitation in layer 5 pyramidal neurons', *Journal of Neuroscience* **34**(1), 182–196.

Almog, M. & Korngreen, A. (2016), 'Is realistic neuronal modeling realistic?', *Journal of Neurophysiology* **116**(5), 2180–2209.

Amari, S. (1977), 'Dynamics of pattern formation in lateral-inhibition type neural fields', *Biological Cybernetics* **27**, 77–87.

Amini, M., Hisdal, J. & Kalvøy, H. (2018), 'Applications of bioimpedance measurement techniques in tissue engineering', *Journal of Electrical Bioimpedance* **9**(1), 142–158.

Amit, D. J. & Brunel, N. (1997), 'Dynamics of a recurrent network of spiking neurons before and following learning', *Network: Computation in Neural Systems* **8**(4), 373–404.

Amsalem, O., Eyal, G., Rogozinski, N., Gevaert, M., Kumbhar, P., Schürmann, F. & Segev, I. (2020), 'An efficient analytical reduction of detailed nonlinear neuron models', *Nature Communications* **11**(1), 1–13.

Anastassiou, C. A., Buzsaki, C., Koch, C., Quiroga, R. & Panzeri, S. (2013), 'Biophysics of extracellular spikes', *Principles of Neural Coding* **15**, 146.

Anastassiou, C. A. & Koch, C. (2015), 'Ephaptic coupling to endogenous electric field activity: why bother?', *Current Opinion in Neurobiology* **31**, 95–103.

Anastassiou, C. A., Perin, R., Buzsáki, G., Markram, H. & Koch, C. (2015), 'Cell type- and activity-dependent extracellular correlates of intracellular spiking', *Journal of Neurophysiology* **114**(1), 608–623.

Andersen, R. A., Musallam, S. & Pesaran, B. (2004), 'Selecting the signals for a brain-machine interface', *Current Opinion in Neurobiology* **14**(6), 720–726.

Anderson, T. R., Huguenard, J. R. & Prince, D. A. (2010), 'Differential effects of Na+–K+ atpase blockade on cortical layer V neurons', *The Journal of Physiology* **588**(22), 4401–4414.

Antic, S. D., Zhou, W.-L., Moore, A. R., Short, S. M. & Ikonomu, K. D. (2010), 'The decade of the dendritic NMDA spike', *Journal of Neuroscience Research* **88**(14), 2991–3001.

Antonio, L. L., Anderson, M. L., Angamo, E. A., Gabriel, S., Klaft, Z.-J., Liotta, A., Salar, S., Sandow, N. & Heinemann, U. (2016), 'In vitro seizure like events and changes in ionic concentration', *Journal of Neuroscience Methods* **260**, 33–44.

Avery, J., Dowrick, T., Faulkner, M., Goren, N. & Holder, D. (2017), 'A versatile and reproducible multi-frequency electrical impedance tomography system', *Sensors* **17**(2), 280.

Ayata, C. & Lauritzen, M. (2015), 'Spreading depression, spreading depolarizations, and the cerebral vasculature', *Physiological Reviews* **95**(3), 953–993.

Azevedo, F. A., Carvalho, L. R., Grinberg, L. T., Farfel, J. M., Ferretti, R. E., Leite, R. E., Filho, W. J., Lent, R. & Herculano-Houzel, S. (2009), 'Equal numbers of neuronal and nonneuronal cells make the human brain an isometrically scaled-up primate brain', *The Journal of Comparative Neurology* **513**(5), 532–541.

Bakkum, D. J., Obien, M. E. J., Radivojevic, M., Jäckel, D., Frey, U., Takahashi, H. & Hierlemann, A. (2018), 'The axon initial segment is the dominant contributor to the neuron's extracellular electrical potential landscape', *Advanced Biosystems* **3**(2), 1800308.

Bangera, N. B., Schomer, D. L., Dehghani, N., Ulbert, I., Cash, S., Papavasiliou, S., Eisenberg, S. R., Dale, A. M. & Halgren, E. (2010), 'Experimental validation of the influence of white matter anisotropy on the intracranial EEG forward solution', *Journal of Computational Neuroscience* **29**(3), 371–387.

Baranauskas, G., Maggiolini, E., Vato, A., Angotzi, G., Bonfanti, A., Zambra, G., Spinelli, A. & Fadiga, L. (2012), 'Origins of 1/f2 scaling in the power spectrum of intracortical local field potential', *Journal of Neurophysiology* **107**(3), 984–994.

Baratham, V. L., Dougherty, M. E., Hermiz, J., Ledochowitsch, P., Maharbiz, M. M. & Bouchard, K. E. (2022), 'Columnar localization and laminar origin of cortical surface electrical potentials', *The Journal of Neuroscience* **42**(18), 3733–3748.

Barker, A. T., Jalinous, R. & Freeston, I. L. (1985), 'Non-invasive magnetic stimulation of human motor cortex', *Lancet* **1**(8437), 1106–1107.

Barry, P. H. & Lynch, J. W. (1991), 'Liquid junction potentials and small cell effects in patch-clamp analysis', *The Journal of Membrane Biology* **121**(2), 101–117.

Barthó, P., Hirase, H., Monconduit, L., Zugaro, M., Harris, K. D. & Buzsaki, G. (2004), 'Characterization of neocortical principal cells and interneurons by network interactions and extracellular features', *Journal of Neurophysiology* **92**(1), 600–608.

Baumann, S. B., Wozny, D. R., Kelly, S. K. & Meno, F. M. (1997), 'The electrical conductivity of human cerebrospinal fluid at body temperature', *IEEE Transactions on Biomedical Engineering* **44**(3), 220–223.

Bazelot, M., Dinocourt, C., Cohen, I. & Miles, R. (2010), 'Unitary inhibitory field potentials in the CA3 region of rat hippocampus', *Journal of Physiology* **588**(12), 2077–2090.

Beaulieu, C. (1993), 'Numerical data on neocortical neurons in adult rat, with special reference to the GABA population', *Brain Research* **609**(1–2), 284–292.

Bechhoefer, J. (2011), 'Kramers–Kronig, Bode, and the meaning of zero', *American Journal of Physics* **79**(10), 1053–1059.

Bédard, C. & Destexhe, A. (2008), 'A modified cable formalism for modeling neuronal membranes at high frequencies', *Biophysical Journal* **94**(4), 1133–1143.

Bédard, C. & Destexhe, A. (2009), 'Macroscopic models of local field potentials and the apparent 1/f noise in brain activity', *Biophysical Journal* **96**(7), 2589–2603.

Bédard, C. & Destexhe, A. (2012), Local field potentials, in R. Brette & A. Destexhe, eds., *Handbook of Neural Activity Measurement*, Cambridge University Press, Cambridge, pp. 136–191.

Bédard, C., Kröger, H. & Destexhe, A. (2006), 'Does the 1/f frequency scaling of brain signals reflect self-organized critical states?', *Physical Review Letters* **97**(11), 118102.

Beggs, J. M. & Plenz, D. (2003), 'Neuronal avalanches in neocortical circuits', *The Journal of Neuroscience* **23**(35), 11167–11177.

Belitski, A., Gretton, A., Magri, C., Murayama, Y., Montemurro, M. A., Logothetis, N. K. & Panzeri, S. (2008), 'Low-frequency local field potentials and spikes in primary visual cortex convey independent visual information', *Journal of Neuroscience* **28**(22), 5696–5709.

Beltrachini, L. (2019), 'A finite element solution of the forward problem in EEG for multipolar sources', *IEEE Transactions on Neural Systems and Rehabilitation Engineering* **27**(3), 368–377.

Benabid, A. L., Chabardes, S., Mitrofanis, J. & Pollak, P. (2009), 'Deep brain stimulation of the subthalamic nucleus for the treatment of Parkinson's disease', *The Lancet Neurology* **8**(1), 67–81.

Bender, K. J. & Trussell, L. O. (2012), 'The physiology of the axon initial segment', *Annual Review of Neuroscience* **35**, 249–265.

Berens, P., Keliris, G. A., Ecker, A. S., Logothetis, N. K. & Tolias, A. S. (2008), 'Comparing the feature selectivity of the gamma-band of the local field potential and the underlying spiking activity in primate visual cortex', *Frontiers in Systems Neuroscience* **2**, 2.

Bereshpolova, Y., Amitai, Y., Gusev, A. G., Stoelzel, C. R. & Swadlow, H. A. (2007), 'Dendritic backpropagation and the state of the awake neocortex', *Journal of Neuroscience* **27**(35), 9392–9399.

Berger, H. (1929), 'Über das Elektrenkephalogramm des Menschen', *Archiv für Psychiatrie und Nervenkrankheiten* **87**(35), 527–570.

Biasiucci, A., Franceschiello, B. & Murray, M. M. (2019), 'Electroencephalography', *Current Biology* **29**(3), R80–R85.

Billeh, Y. N., Cai, B., Gratiy, S. L., Dai, K., Iyer, R., Gouwens, N. W., Abbasi-Asl, R., Jia, X., Siegle, J. H., Olsen, S. R., Koch, C., Mihalas, S. & Arkhipov, A. (2020), 'Systematic integration of structural and functional data into multi-scale models of mouse primary visual cortex', *Neuron* **106**(3), 388–403.

Birbaumer, N., Elbert, T., Canavan, A. G. & Rockstroh, B. (1990), 'Slow potentials of the cerebral cortex and behavior', *Physiological Reviews* **70**(1), 1–41.

Bishop, C., Powell, S., Rutt, D. & Browse, N. (1986), 'Transcranial doppler measurement of middle cerebral artery blood flow velocity: a validation study', *Stroke* **17**(5), 913–915.

Blagoev, K., Mihaila, B., Travis, B., Alexandrov, L., Bishop, A., Ranken, D., Posse, S., Gasparovic, C., Mayer, A., Aine, C., Ulbert, I., Morita, M., Müller, W., Connor, J. & Halgren, E. (2007), 'Modelling the magnetic signature of neuronal tissue', *NeuroImage* **37**(1), 137–148.

Bliss, T. V. & Lømo, T. (1973), 'Long-lasting potentiation of synaptic transmission in the dentate area of the anaesthetized rabbit following stimulation of the perforant path', *The Journal of Physiology* **232**(2), 331–356.

Blomquist, P., Devor, A., Indahl, U. G., Ulbert, I., Einevoll, G. T. & Dale, A. M. (2009), 'Estimation of thalamocortical and intracortical network models from joint thalamic single-electrode and cortical laminar-electrode recordings in the rat barrel system', *PLoS Computational Biology* **5**(3), e1000328.

Bloomfield, S., Hamos, J. & Sherman, S. (1987), 'Passive cable properties and morphological correlates of neurones in the lateral geniculate nucleus of the cat', *The Journal of Physiology* **383**(1), 653–692.

Bojak, I., Oostendorp, T. F., Reid, A. T. & Kötter, R. (2010), 'Connecting mean field models of neural activity to EEG and fMRI data', *Brain Topography* **23**(2), 139–149.

Bokil, H., Laaris, N., Blinder, K., Ennis, M. & Keller, A. (2001), 'Ephaptic interactions in the mammalian olfactory system', *The Journal of Neuroscience* **21**(20), RC173.

Borges, F. S., Moreira, J. V. S., Takarabe, L. M., Lytton, W. W. & Dura-Bernal, S. (2022), 'Large-scale biophysically detailed model of somatosensory thalamocortical circuits in NetPyNE', *Frontiers in Neuroinformatics* **16**: 884245.

Bos, H., Diesmann, M. & Helias, M. (2016), 'Identifying anatomical origins of coexisting oscillations in the cortical microcircuit', *PLoS Computational Biology* **12**(10), e1005132.

Bossetti, C. A., Birdno, M. J. & Grill, W. M. (2008), 'Analysis of the quasi-static approximation for calculating potentials generated by neural stimulation', *Journal of Neural Engineering* **5**(1), 44–53.

Bowen, W. R. & Welfoot, J. S. (2002), 'Modelling the performance of membrane nanofiltration—critical assessment and model development', *Chemical Engineering Science* **57**(7), 1121–1137.

Bower, J. M. & Beeman, D. (1998), *The Book of GENESIS*, Springer, New York.

Brazier, M. A. B. (1949), 'The electrical fields at the surface of the head during sleep', *Electroencephalography and Clinical Neurophysiology* **1**(1–4), 195–204.

Brazier, M. A. B. (1963), *The Discoverers of the Steady Potentials of the Brain: Caton and Beck*, University of California. Brain Research Institute, Los Angeles.

Brazier, M. A. B. (1966), 'A study of the electrical fields at the surface of the head', *American Journal of EEG Technology* **6**(4), 114–128.

Bregman, H. (2021), *The Hidden Life of the Basal Ganglia: At the Base of Brain and Mind*, MIT Press, Cambridge, MA.

Brette, R. & Destexhe, A., eds. (2012), *Handbook of Neural Activity Measurement*, Cambridge University Press, Cambridge.

Brette, R., Rudolph, M., Carnevale, T., Hines, M., Beeman, D., Bower, J. M., Diesmann, M., Morrison, A., Goodman, P. H., Harris, F. C., Zirpe, M., Natschläger, T., Pecevski, D., Ermentrout, B., Djurfeldt, M., Lansner, A., Rochel, O., Vieville, T., Muller, E., Davison, A. P., El Boustani, S. & Destexhe, A. (2007), 'Simulation of networks of spiking neurons: a review of tools and strategies', *Journal of Computational Neuroscience* **23**(3), 349–398.

Brinchmann, C. (2021), Exploring the effect of ionic diffusion on extracellular potentials in the brain, Master's thesis, Norwegian University of Life Sciences, Ås.

Brunel, N. (2000), 'Dynamics of sparsely connected networks of excitatory and inhibitory spiking neurons', *Journal of Computational Neuroscience* **8**, 183–208.

Brunel, N. (2013), Modeling point neurons: From Hodgkin-Huxley to integrate-and-fire, *in* E. D. Schutter, ed., 'Computational Modeling Methods for Neuroscientists', MIT Press, Cambridge, MA, pp. 161–186.

Brunel, N. & van Rossum, M. C. W. (2007), 'Lapicque's 1907 paper: from frogs to integrate-and-fire', *Biological Cybernetics* **97**(5–6), 337–339.

Bruyns-Haylett, M., Luo, J., Kennerley, A. J., Harris, S., Boorman, L., Milne, E., Vautrelle, N., Hayashi, Y., Whalley, B. J., Jones, M., Berwick, J., Riera, J. & Zheng, Y. (2017), 'The neurogenesis of P1 and N1: a concurrent EEG/LFP study', *NeuroImage* **146**, 575–588.

Buccino, A. P., Damart, T., Bartram, J., Mandge, D., Xue, X., Zbili, M., Gänswein, T., Jaquier, A., Emmenegger, V., Markram, H., Hierlemann, A. & Geit, W. V. (2022), 'A multi-modal fitting approach to construct single-neuron models with patch clamp and high-density microelectrode arrays', *bioRxiv*.

Buccino, A. P. & Einevoll, G. T. (2021), 'Mearec: A fast and customizable testbench simulator for ground-truth extracellular spiking activity', *Neuroinformatics* **19**(1), 185–204.

Buccino, A. P., Garcia, S. & Yger, P. (2022), 'Spike sorting: new trends and challenges of the era of high-density probes', *Progress in Biomedical Engineering* **4**(2), 022005.

Buccino, A. P., Kordovan, M., Ness, T. V., Merkt, B., Häfliger, P. D., Fyhn, M., Cauwenberghs, G., Rotter, S. & Einevoll, G. T. (2018), 'Combining biophysical modeling and deep learning for multi-electrode array neuron localization and classification', *Journal of Neurophysiology* **120**(3), 1212–1232.

Buccino, A. P., Kuchta, M., Jæger, K. H., Ness, T. V., Berthet, P., Mardal, K.-A., Cauwenberghs, G. & Tveito, A. (2019), 'How does the presence of neural probes affect extracellular potentials?', *Journal of Neural Engineering* **16**(2), 026030.

Buccino, A. P., Yuan, X., Emmenegger, V., Xue, X., Gänswein, T. & Hierlemann, A. (2022), 'An automated method for precise axon reconstruction from recordings of high-density micro-electrode arrays', *Journal of Neural Engineering* **19**(2), 026026.

Buitenweg, J. R., Rutten, W. L. C. & Marani, E. (2003), 'Geometry-based finite-element modeling of the electrical contact between a cultured neuron and a microelectrode', *IEEE Transactions on Biomedical Engineering* **50**(4), 501–509.

Butson, C. R. & McIntyre, C. C. (2008), 'Current steering to control the volume of tissue activated during deep brain stimulation', *Brain Stimulation* **1**(1), 7–15.

Buzsáki, G., Anastassiou, C. A. & Koch, C. (2012), 'The origin of extracellular fields and currents–EEG, ECoG, LFP and spikes', *Nature Reviews. Neuroscience* **13**(6), 407–20.

Buzsaki, G., Bickford, R. G., Ponomareff, G., Thal, L. J., Mandel, R. & Gage, F. H. (1988), 'Nucleus basalis and thalamic control of neocortical activity in the freely moving rat', *The Journal of Neuroscience* **8**(11), 4007–4026.

Cagnan, H., Denison, T., McIntyre, C. & Brown, P. (2019), 'Emerging technologies for improved deep brain stimulation', *Nature Biotechnology* **37**(9), 1024–1033.

Cain, N., Iyer, R., Koch, C. & Mihalas, S. (2016), 'The computational properties of a simplified cortical column model', *PLoS Computational Biology* **12**(9), e1005045.

Campagnola, L., Seeman, S. C., Chartrand, T., Kim, L., Hoggarth, A., Gamlin, C., Ito, S., Trinh, J., Davoudian, P., Radaelli, C., Kim, M.-H., Hage, T., Braun, T., Alfiler, L., Andrade, J., Bohn, P., Dalley, R., Henry, A., Kebede, S., Mukora, A., Sandman, D., Williams, G., Larsen, R., Teeter, C., Daigle, T. L., Berry, K., Dotson, N., Enstrom, R., Gorham, M., Hupp, M., Lee, S. D., Ngo, K., Nicovich, R., Potekhina, L., Ransford, S., Gary, A., Goldy, J., McMillen, D., Pham, T., Tieu, M., Siverts, L. A., Walker, M., Farrell, C., Schroedter, M., Slaughterbeck, C., Cobb, C., Ellenbogen, R., Gwinn, R. P., Keene, C. D., Ko, A. L., Ojemann, J. G., Silbergeld, D. L., Carey, D., Casper, T., Crichton, K., Clark, M., Dee, N., Ellingwood, L., Gloe, J., Kroll, M., Sulc, J., Tung, H., Wadhwani, K., Brouner, K., Egdorf, T., Maxwell, M., McGraw, M., Pom, C. A., Ruiz, A., Bomben, J., Feng, D., Hejazinia, N., Shi, S., Szafer, A., Wakeman, W., Phillips, J., Bernard, A., Esposito, L., D'Orazi, F. D., Sunkin, S., Smith, K., Tasic, B., Arkhipov, A., Sorensen, S., Lein, E., Koch, C., Murphy, G., Zeng, H. & Jarsky, T. (2022), 'Local connectivity and synaptic dynamics in mouse and human neocortex', *Science* **375**, eabj5861.

Camuñas-Mesa, L. A. & Quiroga, R. Q. (2013), 'A detailed and fast model of extracellular recordings', *Neural Computation* **25**(5), 1191–1212.

Carnevale, T. & Hines, M. L. (2009), *The Neuron Book*, Cambridge University Press, Cambridge.

Caspers, H., Speckmann, E.-J. & Lehmenkühler, A. (1984), 'Electrogenesis of slow potentials of the brain', in T. Elbert, B. Rockstroh, W. Lutzenberger & N. Birbaumer, eds., *Self-Regulation of the Brain and Behavior*, Springer Berlin, Heidelberg, pp. 26–41.

Caton, R. (1875), 'The electric currents of the brain', *British Medical Journal* **2**, 278.

Catterall, W. A., Raman, I. M., Robinson, H. P., Sejnowski, T. J. & Paulsen, O. (2012), 'The Hodgkin-Huxley heritage: from channels to circuits', *Journal of Neuroscience* **32**(41), 14064–14073.

Cavallari, S., Panzeri, S. & Mazzoni, A. (2014), 'Comparison of the dynamics of neural interactions between current-based and conductance-based integrate-and-fire recurrent networks', *Frontiers in Neural Circuits* **8**, 12.

Chatzikalymniou, A. P. & Skinner, F. K. (2018), 'Deciphering the contribution of oriens-lacunosum/moleculare (OLM) cells to intrinsic theta rhythms using biophysical local field potential (LFP) models', *eNeuro* **5**(4), ENEURO.0146–18.2018.

Chen, D. & Eisenberg, R. (1993), 'Charges, currents, and potentials in ionic channels of one conformation', *Biophysical Journal* **64**(5), 1405–1421.

Chen, K. C. & Nicholson, C. (2000), 'Spatial buffering of potassium ions in brain extracellular space', *Biophysical Journal* **78**(6), 2776–2797.

Clark, G. M., Black, R., Dewhurst, D. J., Forster, I. C., Patrick, J. F. & Tong, Y. C. (1977), 'A multiple-electrode hearing prosthesis for cochlea implantation in deaf patients', *Medical Progress through Technology* **5**(3), 127–140.

Clark, J. W. & Plonsey, R. (1970), 'A mathematical study of nerve fiber interaction', *Biophysical Journal* **10**(10), 937–957.

Cohen, D. (1968), 'Magnetoencephalography: evidence of magnetic fields produced by alpha-rhythm currents', *Science* **161**(3843), 784–786.

Cohen, D. (1972), 'Magnetoencephalography: detection of the brain's electrical activity with a superconducting magnetometer', *Science* **175**(4022), 664–666.

Cohen, M. X. (2017), 'Where Does EEG Come From and What Does It Mean?', *Trends in Neurosciences* **40**(4), 208–218.

Coombes, S., beim Graben, P., Potthast, R. & Wright, J. (2014), *Neural Fields: Theory and Applications*, Springer Berlin, Heidelberg.

Cooper, R. (1946), 'The electrical properties of salt-water solutions over the frequency range 1–4000 mc/s', *Journal of the Institution of Electrical Engineers-Part III: Radio and Communication Engineering* **93**(22), 69–75.

Cordingley, G. & Somjen, G. (1978), 'The clearing of excess potassium from extracellular space in spinal cord and cerebral cortex', *Brain Research* **151**(2), 291–306.

Cressman, J. R., Ullah, G., Ziburkus, J., Schiff, S. J. & Barreto, E. (2009), 'The influence of sodium and potassium dynamics on excitability, seizures, and the stability of persistent states: I. Single neuron dynamics', *Journal of Computational Neuroscience* **26**(2), 159–70.

Cserpan, D., Meszéna, D., Wittner, L., Tóth, K., Ulbert, I., Somogyvári, Z. & Wójcik, D. K. (2017), 'Revealing the distribution of transmembrane currents along the dendritic tree of a neuron from extracellular recordings', *eLife* **6**, e29384.

Dai, K., Gratiy, S. L., Billeh, Y. N., Xu, R., Cai, B., Cain, N., Rimehaug, A. E., Stasik, A. J., Einevoll, G. T., Mihalas, S., Koch, C. & Arkhipov, A. (2020), 'Brain Modeling ToolKit: an open source software suite for multiscale modeling of brain circuits', *PLoS Computational Biology* **16**(11), e1008386.

Dale, A. M., Fischl, B. & Sereno, M. I. (1999), 'Cortical surface-based analysis: I. Segmentation and Surface Reconstruction', *NeuroImage* **9**(2), 179–194.

Darbas, M. & Lohrengel, S. (2019), 'Review on mathematical modelling of electroencephalography (EEG)', *Jahresbericht der Deutschen Mathematiker-Vereinigung* **121**(1), 3–39.

David, O. & Friston, K. J. (2003), 'A neural mass model for MEG/EEG: coupling and neuronal dynamics', *NeuroImage* **20**(3), 1743–1755.

Davison, A. P., Brüderle, D., Eppler, J. Kremkow, J., Muller, E., Pecevski, D., Perrinet, L. & Yger, P. (2008), 'PyNN: a common interface for neuronal network simulators', *Frontiers in Neuroinformatics* **2**, 388.

Dayan, P. & Abbott, L. F. (2001), *Theoretical Neuroscience*, MIT Press, Cambridge, MA.

de Curtis, M., Uva, L., Gnatkovsky, V. & Librizzi, L. (2018), 'Potassium dynamics and seizures: why is potassium ictogenic?', *Epilepsy Research* **143**, 50–59.

de Kamps, M. (2013), 'A generic approach to solving jump diffusion equations with applications to neural populations'. arXiv. https://doi.org/10.48550/arXiv.1309.1654.

de Munck, J. C. & Peters, M. J. (1993), 'A fast method to compute the potential in the multisphere model', *IEEE Transactions on Biomedical Engineering* **40**(11), 1166–1174.

De Schutter, E. (2009), *Computational Modeling Methods for Neuroscientists*, MIT Press, Cambridge, MA.

Deco, G., Jirsa, V. K., Robinson, P. A., Breakspear, M. & Friston, K. (2008), 'The dynamic brain: from spiking neurons to neural masses and cortical fields', *PLoS Computational Biology* **4**, e1000092.

Denker, M., Einevoll, G., Franke, F., Grün, S., Hagen, E., Kerr, J., Nawrot, M., Ness, T. B. & Wójcik, T. W. D. (2012), 'Report from 1st INCF workshop on validation of analysis methods', Technical report, International Neuroinformatics Coordinating Facility (INCF).

Derksen, H. E. & Verveen, A. A. (1966), 'Fluctuations of resting neural membrane potential', *Science* **151**(716), 1388–1389.

Destexhe, A., Contreras, D. & Steriade, M. (1999), 'Spatiotemporal analysis of local field potentials and unit discharges in cat cerebral cortex during natural wake and sleep states', *Journal of Neuroscience* **19**(11), 4595–4608.

Destexhe, A. & Huguenard, J. R. (2009), 'Modeling voltage-dependent channels', in E. D. Schutter, ed., *Computational Modeling Methods for Neuroscientists* MIT Press, Cambridge, MA, pp. 107–138.

Destexhe, A., Mainen, Z. F. & Sejnowski, T. J. (1994), 'Synthesis of models for excitable membranes, synaptic transmission and neuromodulation using a common kinetic formalism', *Journal of Computational Neuroscience* **1**(3), 195–230.

Deweese, M. R. & Zador, A. M. (2011), 'Whole cell recordings from neurons in the primary auditory cortex of rat in response to pure tones of different frequency and amplitude, along with recordings of nearby local field potential (lfp)' CRCNS. https://crcns.org/data-sets/ac/ac-2.

Diba, K., Lester, H. A. & Koch, C. (2004), 'Intrinsic noise in cultured hippocampal neurons: experiment and modeling.', *Journal of Neuroscience* **24**(43), 9723–9733.

Dietzel, I., Heinemann, U., Hofmeier, G. & Lux, H. (1982), 'Stimulus-induced changes in extracellular Na+ and Cl- concentration in relation to changes in the size of the extracellular space', *Experimental Brain Research* **46**(1), 73–84.

Dietzel, I., Heinemann, U. & Lux, H. (1989), 'Relations between slow extracellular potential changes, glial potassium buffering, and electrolyte and cellular volume changes during neuronal hyperactivity in cat', *Glia* **2**(1), 25–44.

Donahue, M. J., Kaszas, A., Turi, G. F., Rózsa, B., Slézia, A., Vanzetta, I., Katona, G., Bernard, C., Malliaras, G. G. & Williamson, A. (2018), 'Multimodal characterization of neural networks using highly transparent electrode arrays', *eNeuro* **5**(6), ENEURO.0187-18.2018.

Dowrick, T., Blochet, C. & Holder, D. (2015), 'In vivo bioimpedance measurement of healthy and ischaemic rat brain: implications for stroke imaging using electrical impedance tomography', *Physiological Measurement* **36**(6), 1273–1282.

Druckmann, S., Banitt, Y., Gidon, A., Schuermann, F., Markram, H. & Segev, I. (2007), 'A novel multiple objective optimization framework for constraining conductance-based neuron models by experimental data', *Frontiers in Neurosci* **1**(1), 7–18.

Drukarch, B., Holland, H. A., Velichkov, M., Geurts, J. J., Voorn, P., Glas, G. & de Regt, H. W. (2018), 'Thinking about the nerve impulse: a critical analysis of the electricity-centered conception of nerve excitability', *Progress in Neurobiology* **169**, 172–185.

Dubey, A. & Ray, S. (2016), 'Spatial spread of local field potential is band-pass in the primary visual cortex', *Journal of Neurophysiology* **116**(4), 1986–1999.

Dubey, A. & Ray, S. (2019), 'Cortical electrocorticogram (ECoG) is a local signal', *Journal of Neuroscience* **39**(22), 4299–4311.

Dubey, A. & Ray, S. (2020), 'Comparison of tuning properties of gamma and high-gamma power in local field potential (LFP) versus electrocorticogram (ECoG) in visual cortex', *Scientific Reports* **10**(1), 5422.

Dura-Bernal, S., Griffith, E. Y., Barczak, A., O'Connell, M. N., McGinnis, T., Schroeder, C. E., Lytton, W. W., Lakatos, P. & Neymotin, S. A. (2023), 'Data-driven multiscale model of macaque auditory thalamocortical circuits reproduces in vivo dynamics', *Cell Reports* **42**(11), 113378.

Dura-Bernal, S., Neymotin, S. A., Suter, B. A., Dacre, J., Moreira, J. V., Urdapilleta, E., Schiemann, J., Duguid, I., Shepherd, G. M. & Lytton, W. W. (2023), 'Multiscale model of primary motor cortex circuits predicts in vivo cell-type-specific, behavioral state-dependent dynamics', *Cell Reports* **42**(6), 112574.

Dura-Bernal, S., Suter, B. A., Gleeson, P., Cantarelli, M., Quintana, A., Rodriguez, F., Kedziora, D. J., Chadderdon, G. L., Kerr, C. C., Neymotin, S. A., McDougal, R. A., Hines, M., Shepherd, G. M. & Lytton, W. W. (2019), 'NetPyNE, a tool for data-driven multiscale modeling of brain circuits', *eLife* **8**: e44494.

Ecker, A. S., Berens, P., Keliris, G. A., Bethge, M., Logothetis, N. K. & Tolias, A. S. (2010), 'Decorrelated neuronal firing in cortical microcircuits', *Science* **327**(5965), 584–587.

Einevoll, G. T., Destexhe, A., Diesmann, M., Grün, S., Jirsa, V., de Kamps, M., Migliore, M., Ness, T. V., Plesser, H. E. & Schürmann, F. (2019), 'The scientific case for brain simulations', *Neuron* **102**(4), 735–744.

Einevoll, G. T., Franke, F., Hagen, E., Pouzat, C. & Harris, K. D. (2012), 'Towards reliable spike-train recordings from thousands of neurons with multielectrodes', *Current Opinion in Neurobiology* **22**(1), 11–17.

Einevoll, G. T., Kayser, C., Logothetis, N. K. & Panzeri, S. (2013), 'Modelling and analysis of local field potentials for studying the function of cortical circuits', *Nature Reviews Neuroscience* **14**(11), 770–785.

Einevoll, G. T., Lindén, H., Tetzlaff, T., Łęski, S. & Pettersen, K. H. (2013), 'Local field potential: biophysical origin and analysis', in R. Q. Quiroga & S. Panzeri, eds., *Principles of Neural Coding*, CRC Press, Taylor & Francis Group, Boca Raton, pp. 37–59.

Einevoll, G. T., Pettersen, K. H., Devor, A., Ulbert, I., Halgren, E. & Dale, A. M. (2007), 'Laminar population analysis: estimating firing rates and evoked synaptic activity from multielectrode recordings in rat barrel cortex', *Journal of Neurophysiology* **97**(3), 2174–2190.

Eisenberg, R. S. & Johnson, E. A. (1970), 'Three-dimensional electrical field problems in physiology', *Progress in Biophysics and Molecular Biology* **20**, 1–65.

Eisinger, R. S., Cernera, S., Gittis, A., Gunduz, A. & Okun, M. S. (2019), 'A review of basal ganglia circuits and physiology: application to deep brain stimulation', *Parkinsonism and Related Disorders* **59**, 9–20.

Elbohouty, M. (2013), Electrical conductivity of brain cortex slices in seizing and non-seizing states, PhD thesis, The University of Waikato.

Ellingsrud, A. J., Dukefoss, D. B., Enger, R., Halnes, G., Pettersen, K. & Rognes, M. E. (2022), 'Validating a computational framework for ionic electrodiffusion with cortical spreading depression as a case study', *eNeuro* **9**(2), ENEURO.0408-21.2022.

Ellingsrud, A. J., Solbrå, A., Einevoll, G. T., Halnes, G. & Rognes, M. E. (2020), 'Finite element simulation of ionic electrodiffusion in cellular geometries', *Frontiers in Neuroinformatics* **14**, 11.

Emmenegger, V., Obien, M. E. J., Franke, F. & Hierlemann, A. (2019), 'Technologies to study action potential propagation with a focus on hd-meas', *Frontiers in Cellular Neuroscience* **13**, 159.

Enger, R., Tang, W., Vindedal, G. F., Jensen, V., Helm, P. J., Sprengel, R., Looger, L. L. & Nagelhus, E. A. (2015), 'Dynamics of ionic shifts in cortical spreading depression', *Cerebral Cortex* **25**(11), 4469–4476.

Ermentrout, G. B. & Terman, D. H. (2010), *Mathematical Foundations of Neuroscience*, Springer-Verlag GmbH, New York.

Eyal, G., Verhoog, M. B., Testa-Silva, G., Deitcher, Y., Lodder, J. C., Benavides-Piccione, R., Morales, J., DeFelipe, J., de Kock, C. P., Mansvelder, H. D. & Segev, I. (2016), 'Unique membrane properties and enhanced signal processing in human neocortical neurons', *eLife* **5**: e16553.

Fee, M. S., Mitra, P. P. & Kleinfeld, D. (1996), 'Variability of extracellular spike waveforms of cortical neurons', *Journal of Neurophysiology* **76**(6), 3823–3833.

Feldberg, S. (2000), 'On the dilemma of the use of the electroneutrality constraint in electrochemical calculations', *Electrochemistry Communications* **2**(7), 453–456.

Ferguson, K. A., Chatzikalymniou, A. P. & Skinner, F. K. (2017), 'Combining theory, model, and experiment to explain how intrinsic theta rhythms are generated in an in vitro whole hippocampus preparation without oscillatory inputs', *eNeuro* **4**: ENEURO.0131–17.2017.

Ferguson, K. A., Huh, C. Y. L., Amilhon, B., Manseau, F., Williams, S. & Skinner, F. K. (2015), 'Network models provide insights into how oriens-lacunosum-moleculare and bistratified cell interactions influence the power of local hippocampal CA1 theta oscillations', *Frontiers in Systems Neuroscience* **9**, 110.

Fernández-Ruiz, A., Muñoz, S., Sancho, M., Makarova, J., Makarov, V. A. & Herreras, O. (2013), 'Cytoarchitectonic and dynamic origins of giant positive local field potentials in the dentate gyrus', *The Journal of Neuroscience* **33**(39), 15518–15532.

Fishman, H. M. (1973), 'Relaxation spectra of potassium channel noise from squid axon membranes', *Proceedings of the National Academy of Sciences of the United States of America* **70**(3), 876–879.

Foster, K. & Schwan, H. (1989), 'Dielectric properties of tissues and biological materials: a critical review', *Critical Reviews in Biomedical Engineering* **17**(1), 25–104.

Franks, W., Schenker, I., Schmutz, P. & Hierlemann, A. (2005), 'Impedance characterization and modeling of electrodes for biomedical applications', *IEEE Transactions on Biomedical Engineering* **52**(7), 1295–1302.

Fransson, P., Metsäranta, M., Blennow, M., Åden, U., Lagercrantz, H. & Vanhatalo, S. (2012), 'Early development of spatial patterns of power-law frequency scaling in fMRI resting-state and EEG data in the newborn brain', *Cerebral Cortex* **23**(3), 638–646.

Freeman, J. & Nicholson, C. (1975), 'Experimental optimization of current source-density technique for anuran cerebellum', *Journal of Neurophysiology* **38**(2), 369–382.

Freeman, W. J. (2009), 'Deep analysis of perception through dynamic structures that emerge in cortical activity from self-regulated noise', *Cognitive Neurodynamics* **3**(1), 105–116.

Freeman, W. J., Holmes, M. D., Burke, B. C. & Vanhatalo, S. (2003), 'Spatial spectra of scalp EEG and EMG from awake humans', *Clinical Neurophysiology* **114**(6), 1053–1068.

Freeman, W. J., Rogers, L. J., Holmes, M. D. & Silbergeld, D. L. (2000), 'Spatial spectral analysis of human electrocorticograms including the alpha and gamma bands', *Journal of Neuroscience Methods* **95**(2), 111–121.

Freestone, D. R., Karoly, P. J., Peterson, A. D., Kuhlmann, L., Lai, A., Goodarzy, F. & Cook, M. J. (2015), 'Seizure prediction: science fiction or soon to become reality?', *Current Neurology and Neuroscience Reports* **15**(11), 73.

Fröhlich, F. & McCormick, D. A. (2010), 'Endogenous electric fields may guide neocortical network activity', *Neuron* **67**(1), 129–43.

Fujisawa, S., Amarasingham, A., Harrison, M. T. & Buzsáki, G. (2008), 'Behavior-dependent short-term assembly dynamics in the medial prefrontal cortex', *Nature Neuroscience* **11**(7), 823–833.

Gabriel, C. (1996), 'Compilation of the dielectric properties of body tissues at rf and microwave frequencies', Technical report, King's College London (United Kingdom) Department of Physics.

Gabriel, S., Lau, R. W. & Gabriel, C. (1996), 'The dielectric properties of biological tissues: II. measurements in the frequency range 10 Hz to 20 GHz', *Physics in Medicine and Biology* **41**(11), 2251–2269.

Gardner-Medwin, A. (1980), 'Membrane transport and solute migration affecting the brain cell microenvironment', *Neurosciences Research Program Bulletin* **18**, 208–226.

Gardner-Medwin, A. (1983), 'Analysis of potassium dynamics in mammalian brain tissue', *The Journal of Physiology* **335**, 393–426.

Gardner-Medwin, A., Coles, J., Tsacopoulos, M. (1981), 'Clearance of extracellular potassium: evidence for spatial buffering by glial cells in the retina of the drone', *Brain Research* **209**(2), 452–457.

Geisler, C. D. & Gerstein, G. L. (1961), 'The surface EEG in relation to its sources', *Electroencephalography and Clinical Neurophysiology* **13**(6), 927–934.

Gentet, L. J., Avermann, M., Matyas, F., Staiger, J. F. & Petersen, C. C. (2010), 'Membrane potential dynamics of GABAergic neurons in the barrel cortex of behaving mice', *Neuron* **65**(3), 422–435.

Gentet, L. J., Stuart, G. J. & Clements, J. D. (2000), 'Direct measurement of specific membrane capacitance in neurons', *Biophysical Journal* **79**(1), 314–320.

Gerstein, G. L. (1960), 'Analysis of firing patterns in single neurons', *Science* **131**(3416), 1811–1812.

Gerstner, W., Kistler, W. M., Naud, R. & Paninsky, L. (2014), *Neuronal Dynamics*, Cambridge University Press, Cambridge.

Geselowitz, D. B. (1970), 'On the magnetic field generated outside an inhomogeneous volume conductor by internal current sources', *IEEE Transactions on Magnetics* **6**(2), 346–347.

Geselowitz, D. B. (1967), 'On bioelectric potentials in an inhomogeneous volume conductor', *Biophysical Journal* **7**(1), 1–11.

Głąbska, H., Chintaluri, C. & Wójcik, D. K. (2017), 'Collection of simulated data from a thalamocortical network model', *Neuroinformatics* **15**(1), 87–99.

Głąbska, H., Potworowski, J., Łęski, S. & Wójcik, D. K. (2014), 'Independent components of neural activity carry information on individual populations', *PLoS ONE* **9**(8), e105071.

Głąbska, H. T., Norheim, E., Devor, A., Dale, A. M., Einevoll, G. T. & Wójcik, D. K. (2016), 'Generalized laminar population analysis (gLPA) for interpretation of multielectrode data from cortex', *Frontiers in Neuroinformatics* **10**, 1.

Glomb, K., Cabral, J., Cattani, A., Mazzoni, A., Raj, A. & Franceschiello, B. (2022), 'Computational models in electroencephalography', *Brain Topography* **35**(1), 142–161.

Goethals, S. & Brette, R. (2020), 'Theoretical relation between axon initial segment geometry and excitability', *eLife* **9**: e53432.

Gold, C., Girardin, C. C., Martin, K. A. C. & Koch, C. (2009), 'High-amplitude positive spikes recorded extracellularly in cat visual cortex', *Journal of Neurophysiology* **102**(6), 3340–3351.

Gold, C., Henze, D. A. & Koch, C. (2007), 'Using extracellular action potential recordings to constrain compartmental models', *Journal of Computational Neuroscience* **23**(1), 39–58.

Gold, C., Henze, D. A., Koch, C. & Buzsáki, G. (2006), 'On the origin of the extracellular action potential waveform: a modeling study', *Journal of Neurophysiology* **95**(5), 3113–3128.

Goldman, D. E. (1943), 'Potential, impedance, and rectification in membranes', *The Journal of General Physiology* **27**(1), 37–60.

Goldwyn, J. H. & Rinzel, J. (2016), 'Neuronal coupling by endogenous electric fields: cable theory and applications to coincidence detector neurons in the auditory brain stem', *Journal of Neurophysiology* **115**(4), 2033–2051.

Goto, T., Hatanaka, R., Ogawa, T., Sumiyoshi, A., Riera, J. & Kawashima, R. (2010), 'An evaluation of the conductivity profile in the somatosensory barrel cortex of Wistar rats', *Journal of Neurophysiology* **104**(6), 3388–3412.

Gouwens, N. W., Berg, J., Feng, D., Sorensen, S. A., Zeng, H., Hawrylycz, M. J., Koch, C. & Arkhipov, A. (2018), 'Systematic generation of biophysically detailed models for diverse cortical neuron types', *Nature Communications* **9**(1), 710.

Gramfort, A., Papadopoulo, T., Olivi, E. & Clerc, M. (2010), 'OpenMEEG: opensource software for quasistatic bioelectromagnetics', *BioMedical Engineering OnLine* **9**: 45.

Gratiy, S. L., Halnes, G., Denman, D., Hawrylycz, M. J., Koch, C., Einevoll, G. T. & Anastassiou, C. A. (2017), 'From Maxwell's equations to the theory of current-source density analysis', *European Journal of Neuroscience* **45**(8), 1013–1023.

Grodzinsky, F. (2011), *Fields, Forces, and Flows in Biological Systems.*, Garland Science, Taylor & Francis Group, London & New York.

Grynszpan, F. & Geselowitz, D. B. (1973), 'Model studies of the magnetocardiogram', *Biophysical Journal* **13**(9), 911–925.

Hagen, E., Dahmen, D., Stavrinou, M. L., Lindén, H., Tetzlaff, T., Van Albada, S. J., Grün, S., Diesmann, M. & Einevoll, G. T. (2016), 'Hybrid scheme for modeling local field potentials from point-neuron networks', *Cerebral Cortex* **26**(12), 4461–4496.

Hagen, E., Fossum, J. C., Pettersen, K. H., Alonso, J. M., Swadlow, H. A. & Einevoll, G. T. (2017), 'Focal local field potential signature of the single-axon monosynaptic thalamocortical connection', *Journal of Neuroscience* **37**(20), 5123–5143.

Hagen, E., Magnusson, S. H., Ness, T. V., Halnes, G., Babu, P. N., Linssen, C., Morrison, A. & Einevoll, G. T. (2022), 'Brain signal predictions from multi-scale networks using a linearized framework', *PLOS Computational Biology* **18**(8), e1010353.

Hagen, E., Næss, S., Ness, T. V. & Einevoll, G. T. (2018), 'Multimodal modeling of neural network activity: computing LFP, ECoG, EEG, and MEG signals with LFPy 2.0', *Frontiers in Neuroinformatics* **12**, 92.

Hagen, E. & Ness, T. V. (2023), 'LFPy/ElectricBrainSignals: ElectricBrainSignals-1.0.0rc1'. Zenodo. https://doi.org/10.5281/zenodo.8255422.

Hagen, E., Ness, T. V., Khosrowshahi, A., Sørensen, C., Fyhn, M., Hafting, T., Franke, F. & Einevoll, G. T. (2015), 'ViSAPy: A Python tool for biophysics-based generation of virtual spiking activity for evaluation of spike-sorting algorithms', *Journal of Neuroscience Methods* **245**, 182–204.

Haj-Yasein, N. N., Jensen, V., Østby, I., Omholt, S. W., Voipio, J., Kaila, K., Ottersen, O. P., Hvalby, Ø. & Nagelhus, E. A. (2012), 'Aquaporin-4 regulates extracellular space volume dynamics during high-frequency synaptic stimulation: a gene deletion study in mouse hippocampus', *Glia* **60**(6), 867–874.

Hallermann, S., de Kock, C. P. J., Stuart, G. J. & Kole, M. H. P. (2012), 'State and location dependence of action potential metabolic cost in cortical pyramidal neurons', *Nature Neuroscience* **15**(7), 1007–1014.

Hallett, M. (2007), 'Transcranial magnetic stimulation: a primer', *Neuron* **55**(2), 187–199.

Halnes, G., Augustinaite, S., Heggelund, P., Einevoll, G. T. & Migliore, M. (2011), 'A multi-compartment model for interneurons in the dorsal lateral geniculate nucleus', *PLoS Computational Biology* **7**(9), e1002160.

Halnes, G., Mäki-Marttunen, T., Keller, D., Pettersen, K. H., Andreassen, O. A. & Einevoll, G. T. (2016), 'Effect of ionic diffusion on extracellular potentials in neural tissue', *PLoS Computational Biology* **12**(11), e1005193.

Halnes, G., Mäki-Marttunen, T., Pettersen, K. H., Andreassen, O. A. & Einevoll, G. T. (2017), 'Ion diffusion may introduce spurious current sources in current-source density (CSD) analysis', *Journal of Neurophysiology* **118**(1), 114–120.

Halnes, G., Østby, I., Pettersen, K. H., Omholt, S. W. & Einevoll, G. T. (2013), 'Electrodiffusive model for astrocytic and neuronal ion concentration dynamics', *PLoS Computational Biology* **9**(12), e1003386.

Hämäläinen, M., Hari, R., Ilmoniemi, R. J., Knuutila, J. & Lounasmaa, O. V. (1993), 'Magnetoencephalography – theory, instrumentation, and applications to noninvasive studies of the working human brain', *Reviews of Modern Physics* **65**(413), 414–460.

Hansen, A. J. & Zeuthen, T. (1981), 'Extracellular ion concentrations during spreading depression and ischemia in the rat brain cortex', *Acta Physiologica Scandinavica* **113**(4), 437–445.

Hari, R. & Ilmoniemi, R. J. (1986), 'Cerebral magnetic fields', *Critical Reviews in Biomedical Engineering* **14**(2), 93–126.

Harnett, M. T., Magee, J. C. & Williams, S. R. (2015), 'Distribution and function of HCN channels in the apical dendritic tuft of neocortical pyramidal neurons', *Journal of Neuroscience* **35**(3), 1024–1037.

Harris, K. D., Henze, D. A., Csicsvari, J., Hirase, H. & Buzsaki, G. (2000), 'Accuracy of tetrode spike separation as determined by simultaneous intracellular and extracellular measurements', *Journal of Neurophysiology* **84**(1), 401–414.

Harris, K. D. & Shepherd, G. M. G. (2015), 'The neocortical circuit: themes and variations', *Nature Neuroscience* **18**(2), 170–181.

Hasted, J., Ritson, D. & Collie, C. (1948), 'Dielectric properties of aqueous ionic solutions. parts I and II', *The Journal of Chemical Physics* **16**(1), 1–21.

Havstad, J. W. (1976), *Electrical Impedance of Cerebral Cortex: An Experimental and Theoretical Investigation*, Stanford University.

Hay, E., Hill, S., Schürmann, F., Markram, H. & Segev, I. (2011), 'Models of neocortical layer 5b pyramidal cells capturing a wide range of dendritic and perisomatic active properties', *PLoS Computational Biology* **7**(7), e1002107.

Heinemann, U., Schaible, H. G. & Schmidt, R. F. (1990), 'Changes in extracellular potassium concentration in cat spinal cord in response to innocuous and noxious stimulation of legs with healthy and inflamed knee joints', *Experimental Brain Research* **79**(2), 283–292.

Helias, M., Kunkel, S., Masumoto, G., Igarashi, J., Eppler, J. M., Ishii, S., Fukai, T., Morrison, A. & Diesmann, M. (2012), 'Supercomputers ready for use as discovery machines for neuroscience', *Frontiers in Neuroinformatics* **6**, 26.

Helias, M., Tetzlaff, T. & Diesmann, M. (2013), 'Echoes in correlated neural systems', *New Journal of Physics* **15**(2), 023002.

Helmholtz, H. (1853), 'Ueber einige gesetze der vertheilung elektrischer ströme in körperlichen leitern, mit anwendung auf die thierisch-elektrischen versuche (schluss.)', *Annalen der Physik und Chemie* **165**(7), 353–377.

Henderson, P. (1907), 'An equation for the calculation of potential difference at any liquid junction boundary', *Zeitschrift für Physikalische Chemie* **59**, 118–127.

Henriquez, C. S. (1993), 'Simulating the electrical behavior of cardiac tissue using the bidomain model', *Critical Reviews in Biomedical Engineering* **21**(1), 1–77.

Henze, D. A., Borhegy, Z., Csicsvari, J., Mamiya, A., Harris, K. D. & Buzsaki, G. (2000), 'Intracellular features predicted by extracellular recordings in the hippocampus in vivo', *Journal of Neuroscience* **84**(1), 390–400.

Herculano-Houzel, S. (2009), 'The human brain in numbers: a linearly scaled-up primate brain', *Frontiers in Human Neuroscience* **3**, 31.

Herrera, B., Sajad, A., Woodman, G. F., Schall, J. D. & Riera, J. J. (2020), 'A minimal biophysical model of neocortical pyramidal cells: implications for frontal cortex microcircuitry and field potential generation', *Journal of Neuroscience* **40**(44), 8513–8529.

Herrera, B., Westerberg, J. A., Schall, M. S., Maier, A., Woodman, G. F., Schall, J. D. & Riera, J. J. (2022), 'Resolving the mesoscopic missing link: biophysical modeling of eeg from cortical columns in primates', *NeuroImage* **263**, 119593.

Herreras, O. & Makarova, J. (2020), 'Mechanisms of the negative potential associated with Leão's spreading depolarization: a history of brain electrogenesis', *Journal of Cerebral Blood Flow & Metabolism* **40**(10), 1934–1952.

Herreras, O. & Somjen, G. (1993), 'Analysis of potential shifts associated with recurrent spreading depression and prolonged unstable spreading depression induced by microdialysis of elevated K+ in hippocampus of anesthetized rats', *Brain Research* **610**(2), 283–294.

Hill, M., Rios, E., Sudhakar, S. K., Roossien, D. H., Caldwell, C., Cai, D., Ahmed, O. J., Lempka, S. F. & Chestek, C. A. (2018), 'Quantitative simulation of extracellular single unit recording from the surface of cortex', *Journal of Neural Engineering* **15**(5), 056007.

Hille, B. (2001), *Ion Channels of Excitable Membranes*, 3rd ed., Sinauer Associates: Sunderland, MA.

Hines, M. L. & Carnevale, N. T. (1997), 'The NEURON simulation environment', *Neural Computation* **9**(6), 1179–1209.

Hines, M. L. & Carnevale, N. T. (2001), 'Neuron: a tool for neuroscientists', *The Neuroscientist* **7**(2), 123–135.

Hines, M. L., Davison, A. P. & Muller, E. (2009), 'NEURON and Python', *Frontiers in Neuroinformatics* **3**, 1.

Hines, M. L., Morse, T., Migliore, M., Carnevale, N. T. & Shepherd, G. M. (2004), 'ModelDB: a database to support computational neuroscience', *Journal of Computational Neuroscience* **17**(1), 7–11.

Hjorth, J., Hellgren Kotaleski, J. & Kozlov, A. (2021), 'Predicting synaptic connectivity for large-scale microcircuit simulations using Snudda', *Neuroinformatics* **19**(4), 685–701.

Hladky, S. B. & Barrand, M. A. (2014), 'Mechanisms of fluid movement into, through and out of the brain: evaluation of the evidence', *Fluids and Barriers of the CNS* **11**(1), 1–32.

Hodgkin, A. L. & Huxley, A. F. (1952*a*), 'A quantitative description of membrane current and its application to conduction and excitation in nerve', *The Journal of Physiology* **117**(4), 500–544.

Hodgkin, A. L. & Huxley, A. F. (1952*b*), 'The components of membrane conductance in the giant axon of loligo', *The Journal of Physiology* **116**(4), 473–496.

Hodgkin, A. L. & Katz, B. (1949), 'The effect of sodium ions on the electrical activity of the giant axon of the squid', *The Journal of Physiology* **108**(1), 37.

Hoeltzell, P. B. & Dykes, R. W. (1979), 'Conductivity in the somatosensory cortex of the cat–evidence for cortical anisotropy', *Brain Research* **177**(1), 61–82.

Holt, G. & Koch, C. (1999), 'Electrical interactions via the extracellular potential near cell bodies', *Journal of Computational Neuroscience* **6**, 169–184.

Holt, G. R. (1998), A critical reexamination of some assumptions and implications of cable theory in neurobiology, PhD thesis, California Institute of Technology.

Holter, K. E., Kehlet, B., Devor, A., Sejnowski, T. J., Dale, A. M., Omholt, S. W., Ottersen, O. P., Nagelhus, E. A., Mardal, K.-A. & Pettersen, K. H. (2017), 'Interstitial solute transport in 3d reconstructed neuropil occurs by diffusion rather than bulk flow', *Proceedings of the National Academy of Sciences* **114**(37), 9894–9899.

Hu, H., Vervaeke, K., Graham, L. J. & Storm, J. F. (2009), 'Complementary theta resonance filtering by two spatially segregated mechanisms in CA1 hippocampal pyramidal neurons', *The Journal of Neuroscience* **29**(46), 14472–14483.

Hu, H., Vervaeke, K. & Storm, J. F. (2007), 'M-channels (kv7/kcnq channels) that regulate synaptic integration, excitability, and spike pattern of CA1 pyramidal cells are located in the perisomatic region', *The Journal of Neuroscience* **27**(8), 1853–1867.

Huang, Y., Dmochowski, J. P., Su, Y., Datta, A., Rorden, C. & Parra, L. C. (2013), 'Automated MRI segmentation for individualized modeling of current flow in the human head', *Journal of Neural Engineering* **10**(6), 997–1003.

Huang, Y. & Parra, L. C. (2015), 'Fully automated whole-head segmentation with improved smoothness and continuity, with theory reviewed', *PLoS ONE* **10**(5), 1–34.

Huang, Y., Parra, L. C. & Haufe, S. (2016), 'The New York Head – A precise standardized volume conductor model for EEG source localization and tES targeting', *NeuroImage* **140**, 150–162.

Hubel, D. H. & Wiesel, T. N. (1959), 'Receptive fields of single neurones in the cat's striate cortex', *The Journal of Physiology* **148**(3), 574–591.

Hübel, N., Schöll, E. & Dahlem, M. A. (2014), 'Bistable dynamics underlying excitability of ion homeostasis in neuron models', *PLoS Computational Biology* **10**(5), e1003551.

Hunt, M. J., Falinska, M., Łęski, S., Wójcik, D. K. & Kasicki, S. (2010), 'Differential effects produced by ketamine on oscillatory activity recorded in the rat hippocampus, dorsal striatum and nucleus accumbens', *Journal of Psychopharmacology* **25**(6), 808–821.

Ilmoniemi, R., Hämäläinen, M. & Knuutila, J. (1985), 'The forward and inverse problems in the spherical model', in H. Weinberg, G. Stroink & T. W. Katila, eds., *Biomagnetism: Applications & Theory*, Pergamon Press, New York, pp. 278–282.

Ilmoniemi, R. J., Mäki, H., Saari, J., Salvador, R. & Miranda, P. C. (2016), 'The frequency-dependent neuronal length constant in transcranial magnetic stimulation', *Frontiers in Cellular Neuroscience* **10**, 194.

Ilmoniemi, R. J. & Sarvas, J. (2019), *Brain Signals: Physics and Mathematics of MEG and EEG*, MIT Press, Cambridge, MA.

Ishai, P. B., Talary, M. S., Caduff, A., Levy, E. & Feldman, Y. (2013), 'Electrode polarization in dielectric measurements: a review', *Measurement Science and Technology* **24**(10), 102001.

Izhikevich, E. M. (2007), *Dynamical Systems in Neuroscience*, MIT Press, Cambridge, MA.

Jackson, J. D. (1998), *Classical Electrodynamics*, 3rd ed., John Wiley & Sons, Inc., New York.

Jacobson, G. A., Diba, K., Yaron-Jakoubovitch, A., Oz, Y., Koch, C., Segev, I. & Yarom, Y. (2005), 'Subthreshold voltage noise of rat neocortical pyramidal neurones', *Journal of Physiology* **564**(Pt 1), 145–160.

Jankowski, M. M., Islam, M. N. & O'Mara, S. M. (2017), 'Dynamics of spontaneous local field potentials in the anterior claustrum of freely moving rats', *Brain Research* **1677**, 101–117.

Jansen, B. H. & Rit, V. G. (1995), 'Electroencephalogram and visual evoked potential generation in a mathematical model of coupled cortical columns', *Biological Cybernetics* **73**(4), 357–366.

Jasielec, J. J. (2021), 'Electrodiffusion phenomena in neuroscience and the Nernst–Planck–Poisson equations', *Electrochem* **2**(2), 197–215.

Jatoi, M. A. & Kamel, N. (2017), *Brain Source Localization Using EEG Signal Analysis*, CRC Press, Taylor & Francis Group, Boca Raton.

Jirsa, V. K., Jantzen, K. J., Fuchs, A. & Kelso, J. A. (2002), 'Spatiotemporal forward solution of the EEG and MEG using network modeling', *IEEE Transactions on Medical Imaging* **21**(5), 493–504.

Johnston, D. & Wu, S. M.-S. (1994), *Foundations of cellular neurophysiology*, MIT Press, Cambridge, MA.

Jolivet, R., Lewis, T. J. & Gerstner, W. (2004), 'Generalized integrate-and-fire models of neuronal activity approximate spike trains of a detailed model to a high degree of accuracy', *Journal of Neurophysiology* **92**(2), 959–976.

Jones, S. R., Pritchett, D. L., Sikora, M. A., Stufflebeam, S. M., Hämäläinen, M. & Moore, C. I. (2009), 'Quantitative analysis and biophysically realistic neural modeling of the MEG mu rhythm: rhythmogenesis and modulation of sensory-evoked responses', *Journal of Neurophysiology* **102**(6), 3554–72.

Jones, S. R., Pritchett, D. L., Stufflebeam, S. M., Hämäläinen, M. & Moore, C. I. (2007), 'Neural correlates of tactile detection: a combined magnetoencephalography and biophysically based computational modeling study', *Journal of Neuroscience* **27**(40), 10751–10764.

Jordan, J., Ippen, T., Helias, M., Kitayama, I., Sato, M., Igarashi, J., Diesmann, M. & Kunkel, S. (2018), 'Extremely scalable spiking neuronal network simulation code: from laptops to exascale computers', *Frontiers in Neuroinformatics* **12**, 2.

Joucla, S. & Yvert, B. (2012), 'Modeling extracellular electrical neural stimulation: from basic understanding to MEA-based applications', *Journal of Physiology* **106**(3–4), 146–158.

Joucla, S., Yvert, B., Glière, A. & Yvert, B. (2014), 'Current approaches to model extracellular electrical neural microstimulation', *Frontiers in Neuroscience* **8**, 1–12.

Jun, J. J., Steinmetz, N. A., Siegle, J. H., Denman, D. J., Bauza, M., Barbarits, B., Lee, A. K., Anastassiou, C. A., Andrei, A., Aydın, Ç., Barbic, M., Blanche, T. J., Bonin, V., Couto, J., Dutta, B., Gratiy, S. L., Gutnisky, D. A., Häusser, M., Karsh, B., Ledochowitsch, P., Lopez, C. M., Mitelut, C., Musa, S., Okun, M., Pachitariu, M., Putzeys, J., Rich, P. D., Rossant, C., Sun, W.-L., Svoboda, K., Carandini, M., Harris, K. D., Koch, C., O'Keefe, J. & Harris, T. D. (2017), 'Fully integrated silicon probes for high-density recording of neural activity', *Nature* **551**(7679), 232–236.

Kager, H., Wadman, W. J. & Somjen, G. G. (2000), 'Simulated seizures and spreading depression in a neuron model incorporating interstitial space and ion concentrations', *Journal of Neurophysiology* **84**(1), 495–512.

Kajikawa, Y. & Schroeder, C. E. (2011), 'How local is the local field potential?', *Neuron* **72**(5), 847–858.

Kalmbach, B. E., Buchin, A., Long, B., Close, J., Nandi, A., Miller, J. A., Bakken, T. E., Hodge, R. D., Chong, P., de Frates, R., Dai, K., Maltzer, Z., Nicovich, P. R., Keene, C. D., Silbergeld, D. L., Gwinn, R. P., Cobbs, C. Ko, A. L., Ojemann, J. G., Koch, C., Anastassiou, C. A., Lein, E. S. & Ting, J. T. (2018), 'h-channels contribute to divergent intrinsic membrane properties of supragranular pyramidal neurons in human versus mouse cerebral cortex', *Neuron* **100**(5), 1194–1208.

Katzner, S., Nauhaus, I., Benucci, A., Bonin, V., Ringach, D. L. & Carandini, M. (2009), 'Local origin of field potentials in visual cortex', *Neuron* **61**(1), 35–41.

Khodagholy, D., Gelinas, J. N., Thesen, T., Doyle, W., Devinsky, O., Malliaras, G. G. & Buzsáki, G. (2015), 'Neurogrid: recording action potentials from the surface of the brain', *Nature Neuroscience* **18**(2), 310–315.

Kiebel, S. J., Garrido, M. I., Moran, R. J. & Friston, K. J. (2008), 'Dynamic causal modelling for EEG and MEG', *Cognitive Neurodynamics* **2**(2), 121–136.

Kinney, J. P., Spacek, J., Bartol, T. M., Bajaj, C. L., Harris, K. M. & Sejnowski, T. J. (2013), 'Extracellular sheets and tunnels modulate glutamate diffusion in hippocampal neuropil', *Journal of Comparative Neurology* **521**(2), 448–464.

Klimesch, W., Doppelmayr, M., Russegger, H., Pachinger, T. & Schwaiger, J. (1998), 'Induced alpha band power changes in the human EEG and attention', *Neuroscience Letters* **244**(2), 73–76.

Koch, C. (1984), 'Cable theory in neurons with active, linearized membranes', *Biological Cybernetics* **50**(1), 15–33.

Koch, C. (1999), *Biophysics of Computation: Information Processing in Single Neurons.*, 1st ed., Oxford University Press: New York.

Koch, C. & Hepp, K. (2006), 'Quantum mechanics in the brain', *Nature* **440**(7084), 611–611.

Koch, C., Rapp, M. & Segev, I. (1996), 'A brief history of time (constants)', *Cerebral Cortex* **6**(2), 93–101.

Koch, C. & Segev, I. (1998), *Methods in Neuronal Modeling: From Ions to Networks*, 2nd ed., MIT Press, Cambridge, MA.

Koessler, L., Colnat-Coulbois, S., Cecchin, T., Hofmanis, J., Dmochowski, J. P., Norcia, A. M. & Maillard, L. G. (2017), 'In-vivo measurements of human brain tissue conductivity using focal electrical current injection through intracerebral multicontact electrodes', *Human Brain Mapping* **38**(2), 974–986.

Kohl, C., Parviainen, T. & Jones, S. R. (2022), 'Neural mechanisms underlying human auditory evoked responses revealed by human neocortical neurosolver', *Brain Topography* **35**(1), 19–35.

Kole, M. H. P., Hallermann, S. & Stuart, G. J. (2006), 'Single Ih channels in pyramidal neuron dendrites: properties, distribution, and impact on action potential output', *Journal of Neuroscience* **26**(6), 1677–1687.

Kovac, S., Speckmann, E.-J. & Gorji, A. (2018), 'Uncensored EEG: the role of DC potentials in neurobiology of the brain', *Progress in Neurobiology* **165–167**, 51–65.

Kraig, R. & Nicholson, C. (1978), 'Extracellular ionic variations during spreading depression', *Neuroscience* **3**(11), 1045–1059.

Krassowska, W. & Neu, J. C. (1994), 'Response of a single cell to an external electric field', *Biophysical Journal* **66**(6), 1768–1776.

Kreiman, G., Hung, C. P., Kraskov, A., Quiroga, R. Q., Poggio, T. & DiCarlo, J. J. (2006), 'Object selectivity of local field potentials and spikes in the macaque inferior temporal cortex', *Neuron* **49**(3), 433–445.

Kreyszig, E. (1993), *Advanced Engineering Mathematics*, 7th ed., John Wiley & Sons, New York.

Kriener, B., Enger, H., Tetzlaff, T., Plesser, H. E., Gewaltig, M.-O. & Einevoll, G. T. (2014), 'Dynamics of self-sustained asynchronous-irregular activity in random networks of spiking neurons with strong synapses', *Frontiers in Computational Neuroscience* **8**, 136.

Krishnaswamy, P., Obregon-Henao, G., Ahveninen, J., Khan, S., Babadi, B., Iglesias, J. E., Hämäläinen, M. S. & Purdon, P. L. (2017), 'Sparsity enables estimation of both subcortical and cortical activity from MEG and EEG', *Proceedings of the National Academy of Sciences of the United States of America* **114**, E10465–E10474.

Krív, N., Syková, E. & Vyklický, L. (1975), 'Extracellular potassium changes in the spinal cord of the cat and their relation to slow potentials, active transport and impulse transmission', *The Journal of Physiology* **249**(1), 167–182.

Kubota, Y., Sohn, J. & Kawaguchi, Y. (2018), 'Large volume electron microscopy and neural microcircuit analysis', *Frontiers in Neural Circuits* **12**, 98.

Kumar, S. S., Gänswein, T., Buccino, A. P., Xue, X., Bartram, J., Emmenegger, V. & Hierlemann, A. (2022), 'Tracking axon initial segment plasticity using high-density microelectrode arrays: a computational study', *Frontiers in Neuroinformatics* **16**: 957255.

Kunkel, S., Schmidt, M., Eppler, J. M., Plesser, H. E., Masumoto, G., Igarashi, J., Ishii, S., Fukai, T., Morrison, A., Diesmann, M. & Helias, M. (2014), 'Spiking network simulation code for petascale computers', *Frontiers in Neuroinformatics* **8**, 78.

Kuokkanen, P. T., Ashida, G., Kraemer, A., McColgan, T., Funabiki, K., Wagner, H., Köppl, C., Carr, C. E. & Kempter, R. (2018), 'Contribution of action potentials to the extracellular field potential in the nucleus laminaris of barn owl', *Journal of Neurophysiology* **119**(4), 1422–1436.

Lapicque, L. (1907), 'Recherches quantitatives sur l'excitation électrique des nerfs traitée comme une polarisation', *Journal of Physiol Pathol Générale* **9**, 620–635.

Larkum, M. (2013), 'A cellular mechanism for cortical associations: an organizing principle for the cerebral cortex', *Trends in Neurosciences* **36**, 141–151.

Larkum, M. E., Kaiser, K. M. M. & Sakmann, B. (1999), 'Calcium electrogenesis in distal apical dendrites of layer 5 pyramidal cells at a critical frequency of back-propagating action potentials', *Proceedings of the National Academy of Sciences* **96**(25), 14600–14604.

Larkum, M. E., Nevian, T., Sandler, M., Polsky, A. & Schiller, J. (2009), 'Synaptic integration in tuft dendrites of layer 5 pyramidal neurons: a new unifying principle', *Science* **325**(5941), 756–760.

Larsen, B. R., Stoica, A. & MacAulay, N. (2016), 'Managing brain extracellular K+ during neuronal activity: the physiological role of the Na+/K+-ATPase subunit isoforms', *Frontiers in Physiology* **7**, 141.

Lauritzen, M. & Hansen, A. J. (1992), 'The effect of glutamate receptor blockade on anoxic depolarization and cortical spreading depression', *Journal of Cerebral Blood Flow and Metabolism* **12**(2), 223–229.

Lemon, R. N., Baker, S. N. & Kraskov, A. (2021), 'Classification of cortical neurons by spike shape and the identification of pyramidal neurons', *Cerebral Cortex* **31**(11): 5131–5138.

Lempka, S. F., Johnson, M. D., Moffitt, M. A., Otto, K. J., Kipke, D. R. & McIntyre, C. C. (2011), 'Theoretical analysis of intracortical microelectrode recordings', *Journal of Neural Engineering* **8**(4), 045006.

Lempka, S. F. & McIntyre, C. C. (2013), 'Theoretical analysis of the local field potential in deep brain stimulation applications', *PLoS ONE* **8**(3), e59839.

Léonetti, M. & Dubois-Violette, E. (1998), 'Theory of electrodynamic instabilities in biological cells', *Physical Review Letters* **81**(9), 1977–1980.

Łęski, S., Lindén, H., Tetzlaff, T., Pettersen, K. H. & Einevoll, G. T. (2013), 'Frequency dependence of signal power and spatial reach of the local field potential', *PLoS Computational Biology* **9**(7), e1003137.

Light, G. A. & Näätänen, R. (2013), 'Mismatch negativity is a breakthrough biomarker for understanding and treating psychotic disorders', *Proceedings of the National Academy of Sciences* **110**(38), 15175–15176.

Lindén, H., Hagen, E., Łęski, S., Norheim, E. S., Pettersen, K. H. & Einevoll, G. T. (2014), 'LFPy: a tool for biophysical simulation of extracellular potentials generated by detailed model neurons', *Frontiers in Neuroinformatics* **7**(41), 1–15.

Lindén, H., Pettersen, K. H. & Einevoll, G. T. (2010), 'Intrinsic dendritic filtering gives low-pass power spectra of local field potentials', *Journal of Computational Neuroscience* **29**(3), 423–44.

Lindén, H., Tetzlaff, T., Potjans, T. C., Pettersen, K. H., Grün, S., Diesmann, M. & Einevoll, G. T. (2011), 'Modeling the spatial reach of the LFP', *Neuron* **72**(5), 859–872.

Lindsay, K. A., Rosenberg, J. R. & Tucker, G. (2004), 'From Maxwell's equations to the cable equation and beyond', *Progress in Biophysics and Molecular Biology* **85**(1), 71–116.

Liu, J. & Newsome, W. T. (2006), 'Local field potential in cortical area MT: stimulus tuning and behavioral correlations', *Journal of Neuroscience* **26**(30), 7779–7790.

Logg, A., Mardal, K.-A. & Wells, G. N. (2012), *Automated Solution of Differential Equations by the Finite Element Method*, Vol. 84 of *Lecture Notes in Computational Science and Engineering*, Springer Berlin Heidelberg, Berlin, Heidelberg.

Logothetis, N. K., Kayser, C. & Oeltermann, A. (2007), 'In vivo measurement of cortical impedance spectrum in monkeys: implications for signal propagation', *Neuron* **55**(5), 809–823.

Lopes da Silva, F. (2013), 'EEG and MEG: Relevance to neuroscience', *Neuron* **80**(5), 1112–1128.

Lopreore, C. L., Bartol, T. M., Coggan, J. S., Keller, D. X., Sosinsky, G. E., Ellisman, M. H. & Sejnowski, T. J. (2008), 'Computational modeling of three-dimensional electrodiffusion in biological systems: application to the node of Ranvier', *Biophysical Journal* **95**(6), 2624–2635.

Luo, J., Macias, S., Ness, T. V., Einevoll, G. T., Zhang, K. & Moss, C. F. (2018), 'Neural timing of stimulus events with microsecond precision', *PLoS Biology* **16**(10), 1–22.

Lyshevski, S. (2007), *Nano and Molecular Electronics Handbook*, CRC Press, Taylor and Francis Group, Boca Raton.

Lytton, W. H. (2002), *From Computer to Brain – Foundations of Computational Neuroscience*, Springer, New York.

Magee, J. C. (1998), 'Dendritic hyperpolarization-activated currents modify the integrative properties of hippocampal CA1 pyramidal neurons', *The Journal of Neuroscience* **18**(19), 7613–7624.

Magee, J. C. & Grienberger, C. (2020), 'Synaptic plasticity forms and functions', *Annual Review of Neuroscience* **43**, 95–117.

Mainen, Z. F. & Sejnowski, T. J. (1995), 'Reliability of spike timing in neocortical neurons', *Science* **268**(5216), 1503–1506.

Mäki-Marttunen, T., Kaufmann, T., Elvsåshagen, T., Devor, A., Djurovic, S., Westlye, L. T., Linne, M.-L., Rietschel, M., Schubert, D., Borgwardt, S., Efrim-Budisteanu, M., Bettella, F., Halnes, G. & Hagen, E. (2019), 'Biophysical psychiatry – how computational neuroscience can help to understand the complex mechanisms of mental disorders', *Frontiers in Psychiatry* **10**(534), 1–14.

Mäki-Marttunen, T., Krull, F., Bettella, F., Hagen, E., Næss, S., Ness, T. V., Moberget, T., Elvsåshagen, T., Metzner, C., Devor, A., Edwards, A. G., Fyhn, M., Djurovic, S., Anders, M. D., Andreassen, O. A. & Einevoll, G. T. (2019), 'Alterations in schizophrenia-associated genes can lead to increased power in delta oscillations', *Cerebral Cortex* **29**(2), 875–891.

Manita, S., Suzuki, T., Homma, C., Matsumoto, T., Odagawa, M., Yamada, K., Ota, K., Matsubara, C., Inutsuka, A., Sato, M., Ohkura, M., Yamanaka, A., Yanagawa, Y., Nakai, J., Hayashi, Y., Larkum, M. E. & Murayama, M. (2015), 'A top-down cortical circuit for accurate sensory perception', *Neuron* **86**(5), 1304–1316.

Markram, H., Muller, E., Ramaswamy, S., Reimann, M. W., Abdellah, M., Sanchez, C. A., Ailamaki, A., Alonso-Nanclares, L., Antille, N., Arsever, S., Kahou, G. A. A., Berger, T. K., Bilgili, A., Buncic, N., Chalimourda, A., Chindemi, G., Courcol, J.-D., Delalondre, F., Delattre, V., Druckmann, S., Dumusc, R., Dynes, J., Eilemann, S., Gal, E., Gevaert, M. E., Ghobril, J.-P., Gidon, A., Graham, J. W., Gupta, A., Haenel, V., Hay, E., Heinis, T., Hernando, J. B., Hines, M., Kanari, L., Keller, D., Kenyon, J., Khazen, G., Kim, Y., King, J. G., Kisvarday, Z., Kumbhar, P., Lasserre, S., Le Bé, J.-V., Magalhães, B. R. C., Merchán-Pérez, A., Meystre, J., Morrice , B. R., Muller, J., Muñoz-Céspedes, A., Muralidhar, S., Muthurasa, K., Nachbaur, D., Newton, T. H., Nolte, M., Ovcharenko, A., Palacios, J., Pastor, L., Perin, R., Ranjan, R., Riachi, I., Rodríguez, J.-R., Riquelme, J. L., Rössert, C., Sfyrakis, K., Shi, Y., Shillcock, J. C., Silberg, G., Silva, R., Tauheed, F., Telefont, M., Toledo-Rodriguez, M., Tränkler, T., Van Geit, W., Díaz, J. V., Walker, R., Wang, Y., Zaninetta, S. M., DeFelipe, J., Hill, S. L., Segev, I. & Schürmann, F. (2015), 'Reconstruction and simulation of neocortical microcircuitry', *Cell* **163**(2), 456–492.

Markram, H., Toledo-Rodriguez, M., Wang, Y., Gupta, A., Silberg, G. & Wu, C. (2004), 'Interneurons of the neocortical inhibitory system', *Nature Reviews Neuroscience* **5**(10), 793–807.

Martínez-Cañada, P., Ness, T. V., Einevoll, G. T., Fellin, T. & Panzeri, S. (2021), 'Computation of the electroencephalogram (EEG) from network models of point neurons', *PLoS Computational Biology* **17**(4), e1008893.

Martinez, J., Pedreira, C., Ison, M. J. & Quiroga, R. Q. (2009), 'Realistic simulation of extracellular recordings', *Journal of Neuroscience Methods* **184**(2), 285–293.

Martinsen, Ø. G. & Grimnes, S. (2015), *Bioimpedance and Bioelectricity Basics*, Academic Press (Elsevier), Cambridge, MA.

Mauro, A., Conti, F., Dodge, F. & Schor, R. (1970), 'Subthreshold behavior and phenomenological impedance of the squid giant axon', *The Journal of General Physiology* **55**(4), 497–523.

Mazzoni, A., Brunel, N., Cavallari, S., Logothetis, N. K. & Panzeri, S. (2011), 'Cortical dynamics during naturalistic sensory stimulations: experiments and models', *Journal of Physiology* **105**(1–3), 2–15.

Mazzoni, A., Lindén, H., Cuntz, H., Lansner, A., Panzeri, S. & Einevoll, G. T. (2015), 'Computing the local field potential (LFP) from integrate-and-fire network models', *PLoS Computational Biology* **11**(12), e1004584.

McCann, H., Pisano, G. & Beltrachini, L. (2019), 'Variation in reported human head tissue electrical conductivity values', *Brain Topography* **32**(5), 825–858.

McColgan, T., Liu, J., Kuokkanen, P. T., Carr, C. E., Wagner, H. & Kempter, R. (2017), 'Dipolar extracellular potentials generated by axonal projections', *eLife* **6**: e26106.

McCreery, D. B. & Agnew, W. F. (1983), 'Changes in extracellular potassium and calcium concentration and neural activity during prolonged electrical stimulation of the cat cerebral cortex at defined charge densities', *Experimental Neurology* **79**(2), 371–396.

McCulloch, W. S. & Pitts, W. (1943), 'A logical calculus of the ideas immanent in nervous activity', *The Bulletin of Mathematical Biophysics* **5**(4), 115–133.

McIntyre, C. C. & Grill, W. M. (2001), 'Finite element analysis of the current-density and electric field generated by metal microelectrodes', *Annals of Biomedical Engineering* **29**(3), 227–235.

Mechler, F. & Victor, J. D. (2012), 'Dipole characterization of single neurons from their extracellular action potentials', *Journal of Computational Neuroscience* **32**(1), 73–100.

Mechler, F., Victor, J. D., Ohiorhenuan, I., Schmid, A. M. & Hu, Q. (2011), 'Three-dimensional localization of neurons in cortical tetrode recordings', *Journal of Neurophysiology* **106**(2), 828–848.

Meffin, H., Tahayori, B., Grayden, D. B. & Burkitt, A. N. (2012), 'Modeling extracellular electrical stimulation: I. derivation and interpretation of neurite equations', *Journal of Neural Engineering* **9**(6), 065005.

Meffin, H., Tahayori, B., Sergeev, E. N., Mareels, I. M. Y., Grayden, D. B. & Burkitt, A. N. (2014), 'Modelling extracellular electrical stimulation: III. derivation and interpretation of neural tissue equations', *Journal of Neural Engineering* **11**(6), 065004.

Meunier, C. & Segev, I. (2002), 'Playing the devil's advocate: is the Hodgkin–Huxley model useful?', *Trends in Neurosciences* **25**(11), 558–563.

Miceli, S., Ness, T. V., Einevoll, G. T. & Schubert, D. (2017), 'Impedance spectrum in cortical tissue: implications for propagation of LFP signals on the microscopic level', *eNeuro* **4**(1), ENEURO.0291-16.2016.

Migliore, M., Cook, E., Jaffe, D., Turner, D. & Johnston, D. (1995), 'Computer simulations of morphologically reconstructed CA3 hippocampal neurons', *Journal of Neurophysiology* **73**(3), 1157–1168.

Migliore, M. & Shepherd, G. M. (2002), 'Emerging rules for the distributions of active dendritic conductances', *Nature Reviews Neuroscience* **3**(5), 362–370.

Miller, K. J., Sorensen, L. B., Ojemann, J. G. & den Nijs, M. (2009), 'Power-law scaling in the brain surface electric potential', *PLoS Computational Biology* **5**(12), e1000609.

Miller, P. (2018), *An Introductory Course in Computational Neuroscience*, MIT Press, Cambridge, MA.

Milstein, J., Mormann, F., Fried, I. & Koch, C. (2009), 'Neuronal shot noise and brownian 1/f 2 behavior in the local field potential', *PLoS ONE* **4**(2), e4338.

Milton, J. G. (2012), 'Neuronal avalanches, epileptic quakes and other transient forms of neurodynamics', *European Journal of Neuroscience* **36**(2), 2156–2163.

Miranda, P. C., Callejón-Leblic, M. A., Salvador, R. & Ruffini, G. (2018), 'Realistic modeling of transcranial current stimulation: the electric field in the brain', *Current Opinion in Biomedical Engineering* **8**, 20–27.

Mitzdorf, U. (1985), 'Current source-density method and application in cat cerebral cortex: investigation of evoked potentials and EEG phenomena', *Physiological Reviews* **65**(1), 37–100.

Mizuseki, K., Sirota, A., Pastalkova, E. & Buzsáki, G. (2009), 'Theta oscillations provide temporal windows for local circuit computation in the entorhinal-hippocampal loop', *Neuron* **64**(2), 267–280.

Moffitt, M. A. & McIntyre, C. C. (2005), 'Model-based analysis of cortical recording with silicon microelectrodes', *Clinical Neurophysiology* **116**(9), 2240–2250.

Molina-Martínez, B., Jentsch, L.-V., Ersoy, F., van der Moolen, M., Donato, S., Ness, T. V., Heutink, P., Jones, P. D. & Cesare, P. (2022), 'A multimodal 3D neuro-microphysiological system with neurite-trapping microelectrodes', *Biofabrication* **14**(2), 025004.

Monai, H., Inoue, M., Miyakawa, H. & Aonishi, T. (2012), 'Low-frequency dielectric dispersion of brain tissue due to electrically long neurites', *Physical Review E* **86**(6), 061911.

Mondragón-González, S. L. & Burguière, E. (2017), 'Bio-inspired benchmark generator for extracellular multi-unit recordings', *Scientific Reports* **7**, 43253.

Monfared, O., Tahayori, B., Freestone, D., Nešić, D., Grayden, D. B. & Meffin, H. (2020), 'Determination of the electrical impedance of neural tissue from its microscopic cellular constituents', *Journal of Neural Engineering* **17**(1), 016037.

Mori, Y. (2006), A three-dimensional model of cellular electrical activity, PhD thesis, New York University.

Mori, Y. (2015), 'A multidomain model for ionic electrodiffusion and osmosis with an application to cortical spreading depression', *Physica D: Nonlinear Phenomena* **308**, 94–108.

Mori, Y., Fishman, G. I. & Peskin, C. S. (2008), 'Ephaptic conduction in a cardiac strand model with 3D electrodiffusion', *Proceedings of the National Academy of Sciences of the United States of America* **105**(17), 6463–6468.

Mori, Y. & Peskin, C. (2009), 'A numerical method for cellular electrophysiology based on the electrodiffusion equations with internal boundary conditions at membranes', *Communications in Applied Mathematics and Computational Science* **4**(1), 85–134.

Moulin, C., Glière, A., Barbier, D., Joucla, S., Yvert, B., Mailley, P. & Guillemaud, R. (2008), 'A new 3-D finite-element model based on thin-film approximation for microelectrode array recording of extracellular action potential', *IEEE Transactions on Biomedical Engineering* **55**(2 Pt 1), 683–692.

Mukamel, R. & Fried, I. (2012), 'Human intracranial recordings and cognitive neuroscience', *Annual Review of Psychology* **63**, 511–537.

Murakami, S., Hirose, A. & Okada, Y. C. (2003), 'Contribution of ionic currents to magnetoencephalography (MEG) and electroencephalography (EEG) signals generated by guinea-pig CA3 slices', *The Journal of Physiology* **553**(Pt 3), 975–985.

Murakami, S. & Okada, Y. (2006), 'Contributions of principal neocortical neurons to magnetoencephalography and electroencephalography signals', *The Journal of Physiology* **575**(Pt 3), 925–936.

Murakami, S., Zhang, T., Hirose, A. & Okada, Y. C. (2002), 'Physiological origins of evoked magnetic fields and extracellular field potentials produced by guinea-pig CA3 hippocampal slices', *The Journal of Physiology* **544**(1), 237–251.

Næss, S., Chintaluri, C., Ness, T. V., Dale, A. M., Einevoll, G. T. & Wójcik, D. K. (2017), 'Corrected four-sphere head model for EEG signals', *Frontiers in Human Neuroscience* **11**, 1–7.

Næss, S., Halnes, G., Hagen, E., Hagler, D. J., Dale, A. M., Einevoll, G. T. & Ness, T. V. (2021), 'Biophysically detailed forward modeling of the neural origin of EEG and MEG signals', *NeuroImage* **225**(117467), 2020.07.01.181875.

Nagarajan, S. S. & Durand, D. M. (1996), 'A generalized cable equation for magnetic stimulation of axons', *IEEE Transactions on Biomedical Engineering* **43**(3), 304–312.

Nair, J., Klaassen, A.-L., Arato, J., Vyssotski, A. L., Harvey, M. & Rainer, G. (2018), 'Basal forebrain contributes to default mode network regulation', *Proceedings of the National Academy of Sciences* **115**(6), 1352–1357.

Neher, E. (1992), 'Correction for liquid junction potentials in patch clamp experiments', *Methods in Enzymology* **207**, 123–131.

Nelson, M. J. & Pouget, P. (2010), 'Do electrode properties create a problem in interpreting local field potential recordings?', *Journal of Neurophysiology* **103**(5), 2315–2317.

Nelson, M. J., Pouget, P., Nilsen, E. A., Patten, C. D. & Schall, J. D. (2008), 'Review of signal distortion through metal microelectrode recording circuits and filters', *Journal of Neuroscience Methods* **169**(1), 141–157.

Ness, T. V., Chintaluri, C., Potworowski, J., Łęski, S., Głąbska, H., Wójcik, D. K. & Einevoll, G. T. (2015), 'Modelling and analysis of electrical potentials recorded in microelectrode arrays (MEAs)', *Neuroinformatics* **13**(4), 403–426.

Ness, T. V., Remme, M. W. H. & Einevoll, G. T. (2016), 'Active subthreshold dendritic conductances shape the local field potential', *Journal of Physiology* **594**(13), 3809–3825.

Ness, T. V., Remme, M. W. H. & Einevoll, G. T. (2018), 'h-type membrane current shapes the local field potential from populations of pyramidal neurons', *Journal of Neuroscience* **38**(26), 6011–6024.

Neymotin, S. A., Daniels, D. S., Caldwell, B., McDougal, R. A., Carnevale, N. T., Jas, M., Moore, C. I., Hines, M. L., Hämäläinen, M. & Jones, S. R. (2020), 'Human Neocortical Neurosolver (HNN), a new software tool for interpreting the cellular and network origin of human MEG/EEG data', *eLife* **9**, e51214.

Nicholson, C. (1973), 'Theoretical analysis of field potentials in anisotropic ensembles of neuronal elements', *IEEE Transactions on Biomedical Engineering* **20**(4), 278–288.

Nicholson, C. (2001), 'Diffusion and related transport mechanisms in brain tissue', *Reports on Progress in Physics* **64**(7), 815.

Nicholson, C. & Freeman, J. A. (1975), 'Theory of current source-density analysis and determination of conductivity tensor for anuran cerebellum', *Journal of Neurophysiology* **38**(2), 356–368.

Nicholson, C. & Llinas, R. (1971), 'Field potentials in the alligator cerebellum and theory of their relationship to Purkinje cell dendritic spikes', *Journal of Neurophysiology* **34**(4), 509–531.

Nicholson, C. & Phillips, J. (1981), 'Ion diffusion modified by tortuosity and volume fraction in the extracellular microenvironment of the rat cerebellum', *The Journal of Physiology* **321**(1), 225–257.

Nicholson, C. & Syková, E. (1998), 'Extracellular space structure revealed by diffusion analysis', *Trends in Neurosciences* **21**(5), 207–215.

Nicholson, C., ten Bruggencate, G., Stockle, H. & Steinberg, R. (1978), 'Calcium and potassium changes in extracellular microenvironment of cat cerebellar cortex', *Journal of Neurophysiology* **41**(4), 1026–1039.

Nicholson, P. W. (1965), 'Specific impedance of cerebral white matter', *Experimental Neurology* **13**(4), 386–401.

Niedermeyer, E. (2003), 'The clinical relevance of EEG interpretation', *Clinical Electroencephalography* **34**(3), 93–98.

Normann, R. A., Maynard, E. M., Rousche, P. J. & Warren, D. J. (1999), 'A neural interface for a cortical vision prosthesis', *Vision Research* **39**(15), 2577–2587.

Nunez, P. L. (1974), 'The brain wave equation: a model for the EEG', *Mathematical Biosciences* **21**(3–4), 279–297.

Nunez, P. L., Nunez, M. D. & Srinivasan, R. (2019), 'Multi-scale neural sources of EEG: genuine, equivalent, and representative. A tutorial review', *Brain Topography* **32**(2), 193–214.

Nunez, P. L. & Srinivasan, R. (2006), *Electric Fields of the Brain: The Neurophysics of EEG*, Oxford University Press, USA, New York.

Obien, M. E. J., Hierlemann, A. & Frey, U. (2019), 'Accurate signal-source localization in brain slices by means of high-density microelectrode arrays', *Scientific Reports* **9**(1), 1–19.

O'Connell, R. & Mori, Y. (2016), 'Effects of glia in a triphasic continuum model of cortical spreading depression', *Bulletin of Mathematical Biology* **78**(10), 1943–1967.

Offner, F. F. (1991), 'Ion flow through membranes and the resting potential of cells', *The Journal of Membrane Biology* **123**(2), 171–182.

Okada, Y. C., Huang, J.-C., Rice, M. E., Tranchina, D. & Nicholson, C. (1994), 'Origin of the apparent tissue conductivity in the molecular and granular layers of the in vitro turtle cerebellum and the interpretation of current source-density analysis', *Journal of Neurophysiology* **72**(2), 742–753.

Okun, M. & Lampl, I. (2008), 'Instantaneous correlation of excitation and inhibition during ongoing and sensory-evoked activities', *Nature Neuroscience* **11**(5), 535–537.

Olsson, T., Broberg, M., Pope, K., Wallace, A., Mackenzie, L., Blomstrand, F., Nilsson, M. & Willoughby, J. (2006), 'Cell swelling, seizures and spreading depression: an impedance study', *Neuroscience* **140**(2), 505–515.

Orkand, R. K., Nicholls, J. G. & Kuffler, S. W. (1966), 'Effect of nerve impulses on the membrane potential of glial cells in the central nervous system of amphibia', *Journal of Neurophysiology* **29**(4), 788–806.

Ostojic, S. (2014), 'Two types of asynchronous activity in networks of excitatory and inhibitory spiking neurons', *Nature Neuroscience* **17**(4), 594–600.

Øyehaug, L., Østby, I., Lloyd, C. M., Omholt, S. W. & Einevoll, G. T. (2012), 'Dependence of spontaneous neuronal firing and depolarisation block on astroglial membrane transport mechanisms', *Journal of Computational Neuroscience* **32**(1), 147–165.

Pakkenberg, B., Pelvig, D., Marner, L., Bundgaard, M. J., Gundersen, H. J. G., Nyengaard, J. R. & Regeur, L. (2003), 'Aging and the human neocortex', *Experimental Gerontology* **38**(1–2), 95–99.

Palva, S. & Palva, J. M. (2011), 'Functional roles of alpha-band phase synchronization in local and large-scale cortical networks', *Frontiers in Psychology* **2**(SEP), 1–15.

Parasnis, D. S. (1986), *Principles of Applied Geophysics*, 4th ed., Chapman and Hall, New York.

Pashut, T., Wolfus, S., Friedman, A., Lavidor, M., Bar-Gad, I., Yeshurun, Y. & Korngreen, A. (2011), 'Mechanisms of magnetic stimulation of central nervous system neurons', *PLoS Computational Biology* **7**(3), e1002022.

Pellerin, J. P. & Lamarre, Y. (1997), 'Local field potential oscillations in primate cerebellar cortex during voluntary movement', *Journal of Neurophysiology* **78**(6), 3502–3507.

Perelman, Y. & Ginosar, R. (2006), 'Analog frontend for multichannel neuronal recording system with spike and LFP separation', *Journal of Neuroscience Methods* **153**(1), 21–26.

Perram, J. W. & Stiles, P. J. (2006), 'On the nature of liquid junction and membrane potentials', *Physical Chemistry Chemical Physics* **8**(36), 4200–4213.

Pesaran, B., Vinck, M., Einevoll, G. T., Sirota, A., Fries, P., Siegel, M., Truccolo, W., Schroeder, C. E. & Srinivasan, R. (2018), 'Investigating large-scale brain dynamics using field potential recordings: analysis and interpretation', *Nature Neuroscience* **21**(7), 903–919.

Petersen, C. C. (2017), 'Whole-cell recording of neuronal membrane potential during behavior', *Neuron* **95**(6), 1266–1281.

Pethig, R. (1987), 'Dielectric properties of body tissues', *Clinical Physics and Physiological Measurement* **8**(4A), 5.

Pettersen, K. H., Devor, A., Ulbert, I., Dale, A. M. & Einevoll, G. T. (2006), 'Current-source density estimation based on inversion of electrostatic forward solution: effects of finite extent of neuronal activity and conductivity discontinuities', *Journal of Neuroscience Methods* **154**(1–2), 116–133.

Pettersen, K. H. & Einevoll, G. T. (2008), 'Amplitude variability and extracellular low-pass filtering of neuronal spikes', *Biophysical Journal* **94**(3), 784–802.

Pettersen, K. H., Hagen, E. & Einevoll, G. T. (2008), 'Estimation of population firing rates and current source densities from laminar electrode recordings', *Journal of Computational Neuroscience* **24**(3), 291–313.

Pettersen, K. H., Lindén, H., Dale, A. M. & Einevoll, G. T. (2012), 'Extracellular spikes and CSD', in R. Brette & A. Destexhe, eds., *Handbook of Neural Activity Measurements*, Cambridge University Press, Cambridge, pp. 92–135.

Pettersen, K. H., Lindén, H., Tetzlaff, T. & Einevoll, G. T. (2014), 'Power laws from linear neuronal cable theory: power spectral densities of the soma potential, soma membrane current and single-neuron contribution to the EEG', *PLoS Computational Biology* **10**(11), e1003928.

Pfurtscheller, G. & Cooper, R. (1975), 'Frequency dependence of the transmission of the EEG from cortex to scalp', *Electroencephalography and Clinical Neurophysiology* **38**(1), 93–96.

Piastra, M. C., Nüßing, A., Vorwerk, J., Clerc, M., Engwer, C. & Wolters, C. H. (2021), 'A comprehensive study on electroencephalography and magnetoencephalography sensitivity to cortical and subcortical sources', *Human Brain Mapping* **42**(4), 978–992.

Pickard, W. F. (1976), 'Generalizations of the Goldman-Hodgkin-Katz Equation', *Mathematical Biosciences* **30**(1–2), 99–111.

Pietrobon, D. & Moskowitz, M. A. (2014), 'Chaos and commotion in the wake of cortical spreading depression and spreading depolarizations', *Nature Reviews Neuroscience* **15**(6), 379–393.

Pitts, W. (1952), 'Investigation on synaptic transmission', in von Foerster H, ed., *Cybernetics: Circular Causal and Feedback Mechanisms in Biological and Social Systems* (Trans. 9th Conf.), Josiah Macy Jr. Foundation, New York, pp. 159–166.

Planck, M. (1890), 'Ueber die potentialdifferenz zwischen zwei verdünnten lösungen binärer electrolyte', *Annalen der Physik* **276**(8), 561–576.

Plesser, H. E., Eppler, J. M., Morrison, A., Diesmann, M. & Gewaltig, M.-O. (2007), 'Efficient parallel simulation of large-scale neuronal networks on clusters of multiprocessor computers', in A.-M. Kermarrec, L. Bougé & T. Priol, eds., *Euro-Par 2007 Parallel Processing*, Springer Berlin Heidelberg, Berlin, Heidelberg, pp. 672–681.

Plonsey, R. & Barr, R. C. (2007), *Bioelectricity: a Quantitative Approach*, Springer Science & Business Media, New York.

Plonsey, R. & Heppner, D. B. (1967), 'Considerations of quasi-stationarity in electrophysiological systems', *The Bulletin of Mathematical Biophysics* **29**(4), 657–664.

Pods, J. (2017), 'A comparison of computational models for the extracellular potential of neurons', *Journal of Integrative Neuroscience* **16**(1), 19–32.

Pospischil, M., Toledo-Rodriguez, M., Monier, C., Piwkowska, Z., Bal, T., Frégnac, Y., Markram, H. & Destexhe, A. (2008), 'Minimal Hodgkin-Huxley type models for different classes of cortical and thalamic neurons', *Biological Cybernetics* **99**(4–5), 427–441.

Potjans, T. C. & Diesmann, M. (2014), 'The cell-type specific cortical microcircuit: relating structure and activity in a full-scale spiking network model', *Cerebral Cortex* **24**(3), 785–806.

Potworowski, J., Jakuczun, W., Łęski, S. & Wójcik, D. (2012), 'Kernel current source density method', *Neural Computation* **24**(2), 541–575.

Pozzorini, C., Mensi, S., Hagens, O., Naud, R., Koch, C. & Gerstner, W. (2015), 'Automated high-throughput characterization of single neurons by means of simplified spiking models', *PLoS Computational Biology* **11**(6), e1004275.

Qian, N. & Sejnowski, T. (1989), 'An electro-diffusion model for computing membrane potentials and ionic concentrations in branching dendrites, spines and axons', *Biological Cybernetics* **62**(1), 1–15.

Quiroga, R. Q. (2007), 'Spike sorting', *Scholarpedia* **2**(12), 3583.

Raimondo, J. V., Burman, R. J., Katz, A. A. & Akerman, C. J. (2015), 'Ion dynamics during seizures', *Frontiers in Cellular Neuroscience* **9**, 419.

Rall, W. (1959), 'Branching dendritic trees and motoneuron membrane resistivity', *Experimental Neurology* **1**(5), 491–527.

Rall, W. (1962), 'Electrophysiology of a dendritic neuron model', *Biophysical Journal* **2**(2 Pt 2), 145–167.

Rall, W. (1977), 'Core conductor theory and cable properties of neurons', in J. M. Brookhart & V. B. Mountcastle, eds., *Handbook of Physiology*, American Physiological Society, Bethesda, pp. 39–97.

Rall, W. (1989), 'Cable theory for dendritic neurons', in C. Koch & I. Segev, eds., *Methods in Neuronal Modeling*, MIT Press, Cambridge, MA, pp. 9–92.

Rall, W. & Shepherd, G. M. (1968), 'Theoretical reconstruction of field potentials and dendrodendritic synaptic interactions in olfactory bulb', *Journal of Neurophysiology* **31**(6), 884–915.

Ramaswamy, S., Courcol, J.-D., Abdellah, M., Adaszewski, S. R., Antille, N., Arsever, S., Atenekeng, G., Bilgili, A., Brukau, Y., Chalimourda, A., Chindemi, G., Delalondre, F., Dumusc, R., Eilemann, S., Gevaert, M. E., Gleeson, P., Graham, J. W., Hernando, J. B., Kanari, L., Katkov, Y., Keller, D., King, J. G., Ranjan, R., Reimann, M. W., Rössert, C., Shi, Y., Shillcock, J. C., Telefont, M., Geit, W. V., Diaz, J. V., Walker, R., Wang, Y., Zaninetta, S. M., DeFelipe, J., Hill, S. L., Muller, J., Segev, I., Schürmann, F., Muller, E. B. & Markram, H. (2015), 'The neocortical microcircuit collaboration portal: a resource for rat somatosensory cortex', *Frontiers in Neural Circuits* **9**, 44.

Ranck, J. B. (1963), 'Specific impedance of rabbit cerebral cortex', *Experimental Neurology* **7**(2), 144–152.

Ranta, R., Le Cam, S., Tyvaert, L. & Louis-Dorr, V. (2017), 'Assessing human brain impedance using simultaneous surface and intracerebral recordings', *Neuroscience* **343**, 411–422.

Rasmussen, R., Nicholas, E., Petersen, N. C., Dietz, A. G., Xu, Q., Sun, Q. & Nedergaard, M. (2019), 'Cortex-wide changes in extracellular potassium ions parallel brain state transitions in awake behaving mice', *Cell Reports* **28**(5), 1182–1194.e4.

Rasmussen, R., O'Donnell, J., Ding, F. & Nedergaard, M. (2020), 'Interstitial ions: A key regulator of state-dependent neural activity?', *Progress in Neurobiology*, **193**, 101802.

Rattay, F. (1999), 'The basic mechanism for the electrical stimulation of the nervous system', *Neuroscience* **89**, 335–346.

Ray, S. & Bhalla, U. S. (2008), 'PyMOOSE: Interoperable scripting in Python for MOOSE', *Frontiers in Neuroinformatics* **2**, 6.

Ray, S. & Maunsell, J. (2011), 'Different origins of gamma rhythm and high-gamma activity in macaque visual cortex', *PLoS Biology* **9**(4), e10001610.

Reimann, M., King, J., Muller, E., Ramaswamy, S. & Markram, H. (2015), 'An algorithm to predict the connectome of neural microcircuits', *Frontiers in Computational Neuroscience* **9**, 120.

Reimann, M. W., Anastassiou, C. A., Perin, R., Hill, S. L., Markram, H. & Koch, C. (2013), 'A biophysically detailed model of neocortical local field potentials predicts the critical role of active membrane currents', *Neuron* **79**(2), 375–390.

Reitz, J. R., Milford, F. J. & Christy, R. W. (1993), *Foundations of Electromagnetic Theory*, 4th ed., Addison-Wesley Publishing Company, Reading, MA.

Remme, M. & Rinzel, J. (2011), 'Role of active conductances in subthreshold input integration', *Journal of Compuational Neuroscience* **31**(1), 13–30.

Renart, A., de la Rocha, J., Bartho, P., Hollender, L., Parga, N., Reyes, A. & Harris, K. D. (2010), 'The asynchronous state in cortical circuits', *Science* **327**(5965), 587–590.

Richardson, M. J. (2004), 'Effects of synaptic conductance on the voltage distribution and firing rate of spiking neurons', *Physical Review E* **69**(5), 051918.

Richter, F. & Lehmenkühler, A. (1993), 'Spreading depression can be restricted to distinct depths of the rat cerebral cortex', *Neuroscience Letters* **152**(1–2), 65–68.

Rimehaug, A. E., Stasik, A. J., Hagen, E., Billeh, Y. N., Siegle, J. H., Dai, K., Olsen, S. R., Koch, C., Einevoll, G. T. & Arkhipov, A. (2023), 'Uncovering circuit mechanisms of current sinks and sources with biophysical simulations of primary visual cortex', *eLife* **12**, e87169.

Rinzel, J. (1990), 'Discussion: electrical excitability of cells, theory and experiment: review of the Hodgkin-Huxley foundation and an update', *Bulletin of Mathematical Biology* **52**(1–2), 3–23.

Ritter, P., Schirner, M., Mcintosh, A. R. & Jirsa, V. K. (2013), 'The Virtual Brain Integrates computational modeling and multimodal neuroimaging', *Brain Connectivity* **3**(2), 121–145.

Robinson, D. A. (1968), 'The electrical properties of metal microelectrodes', *Proceedings of the IEEE* **56**(6), 1065–1071.

Robinson, R. A. & Stokes, R. H. (2002), *Electrolyte Solutions*, Courier Corporation, North Chelmsford, MA.

Rogers, N., Thunemann, M., Devor, A. & Gilja, V. (2020), 'Impact of brain surface boundary conditions on electrophysiology and implications for electrocorticography', *Frontiers in Neuroscience* **14**(763).

Romeni, S., Valle, G., Mazzoni, A. & Micera, S. (2020), 'Tutorial: a computational framework for the design and optimization of peripheral neural interfaces', *Nature Protocols* **15**(10), 3129–3153.

Rosanally, S., Mazza, F. & Hay, E. (2023), 'Implications of reduced inhibition in schizophrenia on simulated human prefrontal microcircuit activity and EEG'. bioRxiv. https://doi.org/10.1101/2023.08.11.553052.

Rosenberg, S. A. (1969), 'A computer evaluation of equations for predicting the potential across biological membranes', *Biophysical Journal* **9**(4), 500.

Rössert, C., Pozzorini, C., Chindemi, G., Davison, A. P., Eroe, C., King, J., Newton, T. H., Nolte, M., Ramaswamy, S., Reimann, M. W., Wybo, W., Gewaltig, M.-O., Gerstner, W., Markram, H., Segev, I. & Muller, E. (2016), 'Automated point-neuron simplification of data-driven microcircuit models'. arXiv. https://doi.org/10.48550/arXiv.1604.00087.

Roth, A. & van Rossum, M. C. W. (2009), 'Modeling synapses', in E. De Schutter, ed., *Computational Modeling Methods for Neuroscientists*, The MIT Press, Cambridge, MA, pp. 139–160.

Roth, B. & Basser, P. (1990), 'A model of the stimulation of a nerve fiber by electromagnetic induction', *IEEE Transactions on Biomedical Engineering* **37**(6), 588–597.

Rotter, S. & Diesmann, M. (1999), 'Exact digital simulation of time-invariant linear systems with applications to neuronal modeling', *Biological Cybernetics* **81**(5–6), 381–402.

Rudolph, M., Pelletier, J. G., Paré, D. & Destexhe, A. (2005), 'Characterization of synaptic conductances and integrative properties during electrically induced EEG-activated states in neocortical neurons in vivo', *Journal of Neurophysiology* **94**(4), 2805–2821.

Rush, S. & Driscoll, D. A. (1969), 'EEG electrode sensitivity-an application of reciprocity', *IEEE Transactions on Biomedical Engineering* **16**(1), 15–22.

Sætra, M. J., Einevoll, G. T. & Halnes, G. (2020), 'An electrodiffusive, ion conserving Pinsky-Rinzel model with homeostatic mechanisms', *PLoS Computational Biology* **16**(4), e1007661.

Saha, S., Mamun, K. A., Ahmed, K., Mostafa, R., Naik, G. R., Darvishi, S., Khandoker, A. H. & Baumert, M. (2021), 'Progress in brain computer interface: challenges and opportunities', *Frontiers in Systems Neuroscience* **15**, 578875.

Sala, F. & Hernández-Cruz, A. (1990), 'Calcium diffusion modeling in a spherical neuron. Relevance of buffering properties', *Biophysical Journal* **57**(2), 313–324.

Sanei, S. & Chambers, J. A. (2021), *EEG Signal Processing and Machine Learning*, John Wiley and Sons, Hoboken, NJ.

Sanz-Leon, P., Knock, S. A., Spiegler, A. & Jirsa, V. K. (2015), 'Mathematical framework for large-scale brain network modeling in The Virtual Brain', *NeuroImage* **111**, 385–430.

Sanz-Leon, P., Knock, S. A., Woodman, M. M., Domide, L., Mersmann, J., Mcintosh, A. R. & Jirsa, V. (2013), 'The Virtual Brain: a simulator of primate brain network dynamics', *Frontiers in Neuroinformatics* **7**, 10.

Sarvas, J. (1987), 'Basic mathematical and electromagnetic concepts of the biomagnetic inverse problem', *Physics in Medicine & Biology* **32**(1), 11.

Savtchenko, L. P., Poo, M. M. & Rusakov, D. A. (2017), 'Electrodiffusion phenomena in neuroscience: a neglected companion', *Nature Reviews Neuroscience* **18**(10), 598.

Scheffer-Teixeira, R., Belchior, H., Leao, R., Ribeiro, S. & Tort, A. (2013), 'On high-frequency field oscillations (>100 Hz) and the spectral leakage of spiking activity', *Journal of Neuroscience* **33**(4), 1535–1539.

Schiller, J., Major, G., Koester, H. J. & Schiller, Y. (2000), 'NMDA spikes in basal dendrites of cortical pyramidal neurons', *Nature* **404**(6775), 285–289.

Schomburg, E. W., Anastassiou, C. A., Buzsaki, G. & Koch, C. (2012), 'The spiking component of oscillatory extracellular potentials in the rat hippocampus', *Journal of Neuroscience* **32**(34), 11798–11811.

Schroeder, C. E., Lindsley, R. W., Specht, C., Marcovici, A., Smiley, J. F. & Javitt, D. C. (2001), 'Somatosensory input to auditory association cortex in the macaque monkey', *Journal of Neurophysiology* **85**, 1322–1327.

Schroeder, C. E., Mehta, A. D. & Givre, S. J. (1998), 'A spatiotemporal profile of visual system activation revealed by current source density analysis in the awake macaque', *Cerebral Cortex* **8**(7), 575–592.

Schuecker, J., Diesmann, M. & Helias, M. (2015), 'Modulated escape from a metastable state driven by colored noise', *Physical Review. E, Statistical, Nonlinear, and Soft Matter Physics* **92**, 052119.

Schwalger, T., Deger, M. & Gerstner, W. (2017), 'Towards a theory of cortical columns: from spiking neurons to interacting neural populations of finite size', *PLoS Computational Biology* **13**(4), e1005507.

Schwan, H. P. (1957), 'Electrical properties of tissue and cell suspensions', in J. Lawrence & C. A. Tobias, eds., *Advances in Biological and Medical Physics*, Vol. 5, Academic Press (Elsevier), Cambridge, MA, pp. 147–209.

Schwan, H. P. (1992), 'Linear and nonlinear electrode polarization and biological materials', *Annals of Biomedical Engineering* **20**(3), 269–288.

Seeber, M., Cantonas, L. M., Hoevels, M., Sesia, T., Visser-Vandewalle, V. & Michel, C. M. (2019), 'Subcortical electrophysiological activity is detectable with high-density EEG source imaging', *Nature Communications* **10**(1), 753.

Segev, I., Rinzel, J. & Shepherd, G. M. E. (1995), *The Theoretical Foundation of Dendritic Function - Selected Papers of Wilfrid Rall with Commentaries*, MIT Press, Cambridge, MA.

Senk, J., Hagen, E., van Albada, S. J. & Diesmann, M. (2018), 'Reconciliation of weak pairwise spike-train correlations and highly coherent local field potentials across space'. arXiv. https://doi.org/10.48550/arXiv.1805.10235.

Senk, J., Korvasová, K., Schuecker, J., Hagen, E., Tetzlaff, T., Diesmann, M. & Helias, M. (2020), 'Conditions for wave trains in spiking neural networks', *Physical Review Research* **2**(2), 023174.

Shadlen, M. N. & Newsome, W. T. (1998), 'The variable discharge of cortical neurons: implications for connectivity, computation, and information coding', *The Journal of Neuroscience* **18**(10), 3870–3896.

Sharott, A. (2014), 'Local field potential, methods of recording', in D. Jaeger & R. Jung, eds., *Encyclopedia of Computational Neuroscience*, Springer New York, New York, NY, pp. 1–3.

Sherman, M. A., Lee, S., Law, R., Haegens, S., Thorn, C. A., Hämäläinen, M. S., Moore, C. I. & Jones, S. R. (2016), 'Neural mechanisms of transient neocortical beta rhythms: converging evidence from humans, computational modeling, monkeys, and mice', *Proceedings of the National Academy of Sciences* **113**(33): E4885–4894.

Shifman, A. R. & Lewis, J. E. (2019), 'Elfenn: a generalized platform for modeling ephaptic coupling in spiking neuron models', *Frontiers in Neuroinformatics* **13**, 35.

Siegel, M., Donner, T. H. & Engel, A. K. (2012), 'Spectral fingerprints of large-scale neuronal interactions', *Nature Reviews Neuroscience* **13**(2), 121–134.

Siegle, J. H., Jia, X., Durand, S., Gale, S., Bennett, C., Graddis, N., Heller, G., Ramirez, T. K., Choi, H., Luviano, J. A., Groblewski, P. A., Ahmed, R., Arkhipov, A., Bernard, A., Billeh, Y. N., Brown, D., Buice, M. A., Cain, N., Caldejon, S., Casal, L., Cho, A., Chvilicek, M., Cox, T. C., Dai, K., Denman, D. J., de Vries, S. E. J., Dietzman, R., Esposito, L., Farrell, C., Feng, D., Galbraith, J., Garrett, M., Gelfand, E. C., Hancock, N., Harris, J. A., Howard, R., Hu, B., Hytnen, R., Iyer, R., Jessett, E., Johnson, K., Kato, I., Kiggins, J., Lambert, S., Lecoq, J., Ledochowitsch, P., Lee, J. H., Leon, A., Li, Y., Liang, E., Long, F., Mace, K., Melchior, J., Millman, D., Mollenkopf, T., Nayan, C., Ng, L., Ngo, K., Nguyen, T., Nicovich, P. R., North, K., Ocker, G. K., Ollerenshaw, D., Oliver, M., Pachitariu, M., Perkins, J., Reding, M., Reid, D., Robertson, M., Ronellenfitch, K., Seid, S., Slaughterbeck, C., Stoecklin, M., Sullivan, D., Sutton, B., Swapp, J., Thompson, C., Turner, K., Wakeman, W., Whitesell, J. D., Williams, D., Williford, A., Young, R., Zeng, H., Naylor, S., Phillips, J. W., Reid, R. C., Mihalas, S., Olsen, S. R. & Koch, C. (2021), 'Survey of spiking in the mouse visual system reveals functional hierarchy', *Nature* **592**(7852), 86–92.

Sinha, M. & Narayanan, R. (2015), 'HCN channels enhance spike phase coherence and regulate the phase of spikes and lfps in the theta-frequency range', *Proceedings of the National Academy of Sciences of the United States of America* **112**(17), E2207–E2216.

Sinha, M. & Narayanan, R. (2021), 'Active dendrites and local field potentials: biophysical mechanisms and computational explorations', *Neuroscience* **489**, 111–142.

Siwy, Z. & Fuliński, A. (2002), 'Origin of $1/f^\alpha$ noise in membrane channel currents', *Physical Review Letters* **89**(15), 158101.

Skaar, J.-E. W., Stasik, A. J., Hagen, E., Ness, T. V. & Einevoll, G. T. (2020), 'Estimation of neural network model parameters from local field potentials (LFPs)', *PLoS Computational Biology* **16**(3), e1007725.

Skinner, F. K., Rich, S., Lunyov, A. R., Lefebvre, J. & Chatzikalymniou, A. P. (2021), 'A hypothesis for theta rhythm frequency control in CA1 microcircuits', *Frontiers in Neural Circuits* **15**, 643360.

Skou, J. C. (1957), 'The influence of some cations on an adenosine triphosphatase from peripheral nerves', *Biochimica et Biophysica Acta* **23**(2), 394–401.

Softky, W. & Koch, C. (1993), 'The highly irregular firing of cortical cells is inconsistent with temporal integration of random EPSPs', *The Journal of Neuroscience* **13**(1), 334–350.

Sokalski, T. & Lewenstam, A. (2001), 'Application of Nernst–Planck and Poisson equations for interpretation of liquid-junction and membrane potentials in real-time and space domains', *Electrochemistry Communications* **3**(3), 107–112.

Solbrå, A., Bergersen, A. W., van den Brink, J., Malthe-Sørenssen, A., Einevoll, G. T. & Halnes, G. (2018), 'A Kirchhoff-Nernst-Planck framework for modeling large scale extracellular electrodiffusion surrounding morphologically detailed neurons', *PLoS Computational Biology* **14**(10), 1–26.

Somjen, G., Aitken, P., Giacchino, J. & McNamara, J. (1986), 'Interstitial ion concentrations and paroxysmal discharges in hippocampal formation and spinal cord', *Advances in Neurology* **44**, 663–680.

Somjen, G. G. (2004), *Ions in the Brain: Normal Function, Seizures, and Stroke*, Oxford University Press, Oxford.

Somogyvari, Z., Cserpán, D., Ulbert, I. & Erdi, P. (2012), 'Localization of single-cell current sources based on extracellular potential patterns: the spike csd method', *The European Journal of Neuroscience* **36**(10), 3299–3313.

Song, Z., Cao, X. & Huang, H. (2018), 'Electroneutral models for dynamic Poisson-Nernst-Planck systems', *Physical Review E* **97**(1), 012411.

Speckmann, E.-J., Altrup, U., Lücke, A. & Köhling, R. (1994), 'Principles of electrogenesis of slow field potentials in the brain', in. H.-J. Heinze, T. F. Münte, G. R. Mangun, eds., *Cognitive Electrophysiology*, Birkhäuser Boston, Boston, MA, pp. 288–299.

Spruston, N. (2008), 'Pyramidal neurons: dendritic structure and synaptic integration', *Nature Reviews Neuroscience* **9**(3), 206–221.

Srinivasan, R., Nunez, P. L. & Silberstein, R. B. (1998), 'Spatial filtering and neocortical dynamics: estimates of EEG coherence', *IEEE Transactions on Biomedical Engineering* **45**(7), 814–826.

Stanton, M. (1983), 'Origin and magnitude of transmembrane resting potential in living cells', *Philosophical Transactions of the Royal Society of London. B, Biological Sciences* **301**(1104), 85–141.

Sterratt, D., Graham, B., Gillies, A., Einevoll, G. T. & Willshaw, D. (2023), *Principles of Computational Modelling in Neuroscience*, 2nd ed., Cambridge University Press, Cambridge.

Stimberg, M., Brette, R. & Goodman, D. F. (2019), 'Brian 2, an intuitive and efficient neural simulator', *eLife* **8**, e47314.

Stuart, G., Spruston, N. & Häusser, M. (2007), *Dendrites*, Oxford University Press, Oxford.

Stumpf, M. P. H. & Porter, M. A. (2012), 'Critical truths about power laws', *Science* **335**(6069), 665–666.

Sundnes, J., Nielsen, B. F., Mardal, K. A., Cai, X., Lines, G. T. & Tveito, A. (2006), 'On the computational complexity of the bidomain and the monodomain models of electrophysiology', *Annals of Biomedical Engineering* **34**(7), 1088–1097.

Suzuki, M. & Larkum, M. E. (2017), 'Dendritic calcium spikes are clearly detectable at the cortical surface', *Nature Communications* **8**(276), 1–10.

Swadlow, H. A., Gusev, A. G. & Bezdudnaya, T. (2002), 'Activation of a cortical column by a thalamocortical impulse', *Journal of Neuroscience* **22**(17), 7766–7773.

Syková, E. & Nicholson, C. (2008), 'Diffusion in brain extracellular space', *Physiological Reviews* **88**(4), 1277–1340.

Sypert, G. & Ward, A. (1974), 'Changes in extracellular potassium activity during neocortical propagated seizures', *Experimental Neurology* **45**(1), 19–41.

Sætra, M. J., Einevoll, G. T. & Halnes, G. (2021), 'An electrodiffusive neuron-extracellular-glia model for exploring the genesis of slow potentials in the brain', *PLoS Computational Biology* **17**(7), e1008143.

Tahayori, B., Meffin, H., Dokos, S., Burkitt, A. N. & Grayden, D. B. (2012), 'Modeling extracellular electrical stimulation: II. computational validation and numerical results', *Journal of Neural Engineering* **9**(6), 065006.

Tahayori, B., Meffin, H., Sergeev, E. N., Mareels, I. M. Y., Burkitt, A. N. & Grayden, D. B. (2014), 'Modelling extracellular electrical stimulation: IV. effect of the cellular composition of neural tissue on its spatio-temporal filtering properties', *Journal of Neural Engineering* **11**(6), 065005.

Tao, A., Tao, L. & Nicholson, C. (2005), 'Cell cavities increase tortuosity in brain extracellular space', *Journal of Theoretical Biology* **234**(4), 525–536.

Tao, L. & Nicholson, C. (2004), 'Maximum geometrical hindrance to diffusion in brain extracellular space surrounding uniformly spaced convex cells', *Journal of Theoretical Biology* **229**(1), 59–68.

Taxidis, J., Anastassiou, C. A., Diba, K. & Koch, C. (2015), 'Local field potentials encode place cell ensemble activation during hippocampal sharp wave ripples', *Neuron* **87**(3), 590–604.

Taylor, A. L., Goaillard, J.-M. & Marder, E. (2009), 'How multiple conductances determine electrophysiological properties in a multicompartment model', *Journal of Neuroscience* **29**(17), 5573–5586.

Teeter, C., Iyer, R., Menon, V., Gouwens, N., Feng, D., Berg, J., Szafer, A., Cain, N., Zeng, H., Hawrylycz, M., Koch, C. & Mihalas, S. (2018), 'Generalized leaky integrate-and-fire models classify multiple neuron types', *Nature Communications* **9**(1), 709.

Teleńczuk, B., Dehghani, N., Quyen, M. L. V., Cash, S. S., Halgren, E., Hatsopoulos, N. G. & Destexhe, A. (2017), 'Local field potentials primarily reflect inhibitory neuron activity in human and monkey cortex', *Scientific Reports* **7**(1), 1–16.

Teleńczuk, B., Teleńczuk, M. & Destexhe, A. (2020*a*), 'A kernel-based method to calculate local field potentials from networks of spiking neurons', *Journal of Neuroscience Methods* **344**, 108871.

Telenczuk, M., Brette, R., Destexhe, A. & Telenczuk, B. (2018), 'Contribution of the axon initial segment to action potentials recorded extracellularly', *eNeuro* **5**: ENEURO.0068–18.2018.

Teleńczuk, M., Teleńczuk, B. & Destexhe, A. (2020*b*), 'Modelling unitary fields and the single-neuron contribution to local field potentials in the hippocampus', *Journal of Physiology* **598**(18), 3957–3972.

Telkes, I., Jimenez-Shahed, J., Viswanathan, A., Abosch, A. & Ince, N. F. (2016), 'Prediction of STN-DBS electrode implantation track in Parkinson's disease by using local field potentials', *Frontiers in Neuroscience* **10**, 1–16.

Telkes, I., Viswanathan, A., Jimenez-Shahed, J., Abosch, A., Ozturk, M., Gupte, A., Jankovic, J. & Ince, N. (2018), 'Local field potentials of subthalamic nucleus contain electrophysiological footprints of motor subtypes of Parkinson's disease', *Proceedings of the National Academy of Sciences* **115**(36), E8567–E8576.

Tetzlaff, T., Helias, M., Einevoll, G. T. & Diesmann, M. (2012), 'Decorrelation of neural-network activity by inhibitory feedback', *PLoS Computational Biology* **8**(8), e1002596.

Thio, B. J., Aberra, A. S., Dessert, G. E. & Grill, W. M. (2022), 'Ideal current dipoles are appropriate source representations for simulating neurons for intracranial recordings', *Clinical Neurophysiology* **145**, 26–35.

Thorbergsson, P. T., Garwicz, M., Schouenborg, J. & Johansson, A. J. (2012), 'Computationally efficient simulation of extracellular recordings with multielectrode arrays', *Journal of Neuroscience Methods* **211**(1), 133–144.

Thunemann, M., Hossain, L., Ness, T. V., Rogers, N., Lee, K., Lee, S. H., Kılıç, K., Oh, H., Economo, M. N., Gilja, V., Einevoll, G. T., Dayeh, S. A. & Devor, A. (2022), 'Imaging through windansee electrode arrays reveals a small fraction of local neurons following surface MUA'. bioRxiv. https://doi.org/10.1101/2022.09.01.506113.

Thunemann, M., Ness, T. V., Kilic, K., Ferri, C. G., Sakadzic, S., Dale, A. M., Fainman, Y., Boas, D. A., Einevoll, G. T. & Devor, A. (2018), 'Does light propagate better along pyramidal apical dendrites in cerebral cortex?'. Poster presentation. https://doi.org/10.1364/TRANSLATIONAL.2018.JW3A.56.

Tivadar, R. I. & Murray, M. M. (2019), 'A primer on electroencephalography and event-related potentials for organizational neuroscience', *Organizational Research Methods* **22**(1), 69–94.

Toll, J. S. (1956), 'Causality and the dispersion relation: logical foundations', *Physical Review* **104**(6), 1760–1770.

Tomsett, R. J., Ainsworth, M., Thiele, A., Sanayei, M., Chen, X., Gieselmann, M. A., Whittington, M. A., Cunningham, M. O. & Kaiser, M. (2015), 'Virtual electrode recording tool for extracellular potentials (vertex): comparing multi-electrode recordings from simulated and biological mammalian cortical tissue', *Brain Structure and Function* **220**(4), 2333–2353.

Torres, D., Makarova, J., Ortuño, T., Benito, N., Makarov, V. A. & Herreras, O. (2019), 'Local and volume-conducted contributions to cortical field potentials', *Cerebral Cortex* **29**(12), 5234–5254.

Touboul, J. & Destexhe, A. (2017), 'Power-law statistics and universal scaling in the absence of criticality', *Physical Review E* **95**(1), 012413.

Tracey, B. & Williams, M. (2011), 'Computationally efficient bioelectric field modeling and effects of frequency-dependent tissue capacitance', *Journal of Neural Engineering* **8**(3), 036017.

Trainito, C., von Nicolai, C., Miller, E. K. & Siegel, M. (2019), 'Extracellular spike waveform dissociates four functionally distinct cell classes in primate cortex', *Current Biology* **29**(18), 2973–2982.e5.

Trappenberg, T. P. (2002), *Fundamentals of Computational Neuroscience*, Oxford University Press, Oxford.

Traub, R. D., Contreras, D., Cunningham, M. O., Murray, H., LeBeau, F. E. N., Roopun, A., Bibbig, A., Wilent, W. B., Higley, M. J. & Whittington, M. A. (2005), 'Single-column thalamocortical network model exhibiting gamma oscillations, sleep spindles, and epileptogenic bursts', *Journal of Neurophysiology* **93**(4), 2194–2232.

Traub, R., Dudek, F., Taylor, C. P. & Knowles, W. D. (1985), 'Simulation of hippocampal afterdischarges synchronized by electrical interactions', *Neuroscience* **14**(4), 1033–1038.

Tripp, J. (1983), 'Physical concepts and mathematical models', in S. J. Williamson, G.-L. Romani, L. Kaufman & I. Modena, eds., *Biomagnetism*, Springer, New York, pp. 101–139.

Tuttle, A., Diaz, J. R. & Mori, Y. (2019), 'A computational study on the role of glutamate and NMDA receptors on cortical spreading depression using a multidomain electrodiffusion model', *PLoS Computational Biology* **15**(12), e1007455.

Tveito, A., Jæger, K. H., Lines, G. T., Paszkowski, Ł., Sundnes, J., Edwards, A. G., Māki-Marttunen, T., Halnes, G. & Einevoll, G. T. (2017), 'An evaluation of the accuracy of classical models for computing the membrane potential and extracellular potential for neurons', *Frontiers in Computational Neuroscience* **11**, 27.

Tveito, A., Mardal, K.-A. & Rognes, M. E. (2021), *Modeling Excitable Tissue: The EMI Framework*, Springer Nature, Cham, Switzerland.

Uhlirova, H., Kılıç, K., Tian, P., Sakadz, S., Saisan, P. A., Gagnon, L., Thunemann, M., Nizar, K., Yasseen, M. A., Jr, D. J. H., Vandenberghe, M., Djurovic, S., Andreassen, O. A., Silva, G. A., Masliah, E., Kleinfeld, D., Vinogradov, S., Buxton, R. B., Einevoll, G. T., Boas, D. A., Dale, A. M. & Devor, A. (2016), 'The roadmap for estimation of cell-type-specific neuronal activity from non-invasive measurements', *Proceedings of the Royal Society of London. Series B Biological Sciences* **371**(1705), 20150356.

Ulbert, I., Halgren, E., Heit, G. & Karmos, G. (2001), 'Multiple microelectrode-recording system for human intracortical applications', *Journal of Neuroscience Methods* **106**(1), 69–79.

Ullah, G., Cressman, J. R., Barreto, E. & Schiff, S. J. (2009), 'The influence of sodium and potassium dynamics on excitability, seizures, and the stability of persistent states. II. Network and glial dynamics', *Journal of Computational Neuroscience* **26**(2), 171–183.

Van Geit, W., Gevaert, M., Chindemi, G., Rössert, C., Courcol, J.-D., Muller, E., Schürmann, F., Segev, I. & Markram, H. (2016), 'BluePyOpt: leveraging open source software and cloud infrastructure to optimise model parameters in neuroscience', *Frontiers in Neuroinformatics* **10**, 1–18.

Van Uitert, R., Weinstein, D. & Johnson, C. (2003), 'Volume currents in forward and inverse magnetoencephalographic simulations using realistic head models', *Annals of Biomedical Engineering* **31**(1), 21–31.

van Vreeswijk, C. & Sompolinsky, H. (1997), 'Irregular firing in cortical circuits with inhibition/excitation balance', in J. M. Bower, ed., *Computational Neuroscience*, Springer, US, New York, pp. 209–213.

Vermaas, M., Piastra, M. C., Oostendorp, T. F., Ramsey, N. F. & Tiesinga, P. H. E. (2020*a*), 'FEMfuns: a volume conduction modeling pipeline that includes resistive, capacitive or dispersive tissue and electrodes', *Neuroinformatics* **18**(4), 569–580.

Vermaas, M., Piastra, M. C., Oostendorp, T., Ramsey, N. & Tiesinga, P. H. (2020*b*), 'When to include ecog electrode properties in volume conduction models', *Journal of Neural Engineering* **17**(5), 056031.

Vissani, M., Palmisano, C., Volkmann, J., Pezzoli, G., Micera, S., Isaias, I. U. & Mazzoni, A. (2021), 'Impaired reach-to-grasp kinematics in Parkinsonian patients relates to dopamine-dependent, subthalamic beta bursts', *npj Parkinson's Disease* **7**(1), 53.

Viswam, V., Obien, M. E. J., Franke, F., Frey, U. & Hierlemann, A. (2019), 'Optimal electrode size for multi-scale extracellular-potential recording from neuronal assemblies', *Frontiers in Neuroscience* **13**, 1–23.

Vorwerk, J., Cho, J.-H., Rampp, S., Hamer, H., Knosche, T. R. & Wolters, C. H. (2014), 'A guideline for head volume conductor modeling in EEG and MEG', *NeuroImage* **100**, 590–607.

Wagner, T., Eden, U., Rushmore, J., Russo, C. J., Dipietro, L., Fregni, F., Simon, S., Rotman, S., Pitskel, N. B., Ramos-Estebanez, C., Pascual-Leone, A., Grodzinsky, A. J., Zahn, M. & Valero-Cabré, A. (2014), 'Impact of brain tissue filtering on neurostimulation fields: a modeling study', *NeuroImage* **85**(3), 1048–1057.

Waters, J., Schaefer, A. & Sakmann, B. (2005), 'Backpropagating action potentials in neurones: measurement, mechanisms and potential functions', *Progress in Biophysics and Molecular Biology* **87**(1), 145–170.

Wei, Y., Ullah, G. & Schiff, S. J. (2014), 'Unification of neuronal spikes, seizures, and spreading depression', *Journal of Neuroscience* **34**(35), 11733–11743.

Whittingstall, K. & Logothetis, N. K. (2013), 'Physiological foundations of neural signals', *Principles of Neural Coding* **15**, 146.

Wilson, C. J. (1984), 'Passive cable properties of dendritic spines and spiny neurons', *Journal of Neuroscience* **4**(1), 281–297.

Wilting, J. & Priesemann, V. (2019), '25 years of criticality in neuroscience - established results, open controversies, novel concepts', *Current Opinion in Neurobiology* **58**, 105–111.

Wolters, C. H., Anwander, A., Tricoche, X., Weinstein, D., Koch, M. A. & MacLeod, R. S. (2006), 'Influence of tissue conductivity anisotropy on EEG/MEG field and return current computation in a realistic head model: a simulation and visualization study using high-resolution finite element modeling', *NeuroImage* **30**(3), 813–826.

Wybo, W. A., Jordan, J., Ellenberger, B., Mengual, U. M., Nevian, T. & Senn, W. (2021), 'Data-driven reduction of dendritic morphologies with preserved dendro-somatic responses', *eLife* **10**, 1–26.

Wybo, W. A. M., Stiefel, K. M. & Torben-Nielsen, B. (2013), 'The green's function formalism as a bridge between single- and multi-compartment modeling', *Biological Cybernetics* **107**(6), 685–694.

Xing, D., Yeh, C.-I. & Shapley, R. M. (2009), 'Spatial spread of the local field potential and its laminar variation in visual cortex', *Journal of Neuroscience* **29**(37), 11540–11549.

Xu, N. L., Harnett, M. T., Williams, S. R., Huber, D., O'Connor, D. H., Svoboda, K. & Magee, J. C. (2012), 'Nonlinear dendritic integration of sensory and motor input during an active sensing task', *Nature* **492**, 247–251.

Yaron-Jakoubovitch, A., Jacobson, G. A., Koch, C., Segev, I. & Yarom, Y. (2008), 'A paradoxical isopotentiality: a spatially uniform noise spectrum in neocortical pyramidal cells', *Frontiers in Cellular Neuroscience* **2**, 3.

Yger, P., Spampinato, G. L., Esposito, E., Lefebvre, B., Deny, S., Gardella, C., Stimberg, M., Jetter, F., Zeck, G., Picaud, S., Duebel, J. & Marre, O. (2018), 'A spike sorting toolbox for up to thousands of electrodes validated with ground truth recordings in vitro and in vivo', *eLife* **7**, e34518.

Zandt, B.-J., ten Haken, B., van Dijk, J. G. & van Putten, M. J. (2011), 'Neural dynamics during anoxia and the "wave of death"', *PLoS ONE* **6**(7), e22127.

Zempel, J. M., Politte, D. G., Kelsey, M., Verner, R., Nolan, T. S., Babajani-Feremi, A., Prior, F. & Larson-Prior, L. J. (2012), 'Characterization of scale-free properties of human electrocorticography in awake and slow wave sleep states', *Frontiers in Neurology* **3**, 76.

Zeng, F.-G., Rebscher, S., Harrison, W., Sun, X. & Feng, H. (2008), 'Cochlear implants: system design, integration, and evaluation', *IEEE Reviews in Biomedical Engineering* **1**, 115–142.

Zeng, H. & Sanes, J. R. (2017), 'Neuronal cell-type classification: challenges, opportunities and the path forward', *Nature Reviews Neuroscience* **18**(9), 530–546.

Zhuchkova, E., Remme, M. W. H. & Schreiber, S. (2013), 'Somatic versus dendritic resonance: differential filtering of inputs through non-uniform distributions of active conductances', *PLoS ONE* **8**(11), e78908.

Ziegler, E., Chellappa, S. L., Gaggioni, G., Ly, J. Q., Vandewalle, G., André, E., Geuzaine, C. & Phillips, C. (2014), 'A finite-element reciprocity solution for EEG forward modeling with realistic individual head models', *NeuroImage* **103**, 542–551.

Zimmermann, J. & van Rienen, U. (2021), 'Ambiguity in the interpretation of the low-frequency dielectric properties of biological tissues', *Bioelectrochemistry* **140**, 107773.

Index

Printed in the United States
by Baker & Taylor Publisher Services